Metaphysic in Three Books Ontology, Cosmology, and Psychology

Rudolf Hermann Lotze

BIBLIOBAZAAR

Clarendon Press Series

METAPHYSIC

IN THREE BOOKS

ONTOLOGY, COSMOLOGY, AND PSYCHOLOGY

BY

HERMANN LOTZE

ENGLISH TRANSLATION

EDITED BY

BERNARD BOSANQUET, M.A.

FELLOW OF UNIVERSITY COLLEGE, OXFORD

Oxford

AT THE CLARENDON PRESS

1884

AUTHOR'S PREFACE.

THE publication of this second volume has been delayed by a variety of hindrances, which caused a lengthened interruption of its passage through the press. In the meantime several works have appeared which I should have been glad to notice; but it was impossible, for the above reason, to comment upon them in the appropriate parts of my book; and I therefore reserve what I have to say about them.

I can promise nothing in respect of the third volume but that, should I have strength to finish it, it will be confined to a discussion of the main problems of Practical Philosophy, Aesthetic, and the Philosophy of Religion. I shall treat each of these separately, and without the lengthiness which was unavoidable in the present volume owing to a divergence from prevalent views.

THE AUTHOR.

GÖTTINGEN: *December* 23, 1878.

EDITOR'S PREFACE.

THE Translation of the Metaphysic has been executed, like that of the Logic, by several hands. The whole of Book I (Ontology) and the chapter 'Of Time' (Book II, ch. iii) were translated by the late Mr. T. H. Green, Whyte's Professor of Moral Philosophy at Oxford; chapters i, ii, and iv, of Book II by Mr. B. Bosanquet, Fellow of University College, Oxford; chapters v–viii (inclusive) of Book II by the Rev. C. A. Whittuck, Fellow of Brasenose College, Oxford; and the whole of Book III by Mr. A. C. Bradley, Fellow of Balliol College, Oxford. The Index and Table of Contents were added by the Editor.

The entire translation has been revised by the Editor, who is responsible in every case for the rendering finally adopted. The Editor has to thank Mr. J. C. Wilson, of Oriel College, Oxford, for ample and ready assistance when consulted on passages involving the technical language of Mathematics or Physics; if the Author's meaning in such places has been intelligibly conveyed, this result is wholly due to Mr. Wilson's help.

In conveying his assent to the proposal of an English translation, the Author expressed a wish to work out Book III of the Metaphysic (the Psychology) more fully, but had not time to carry out his intention. For the third volume of the Author's 'System of Philosophy,' alluded to in the Preface, no materials were found after his death sufficiently advanced for publication, excepting a paper subsequently published in 'Nord und Süd' (June 1882), under the title 'Die Principien der Ethik.' The Author's views on the subjects reserved for the volume in question may be gathered in part from his earlier work 'Mikrokosmus,' which will soon, it may be hoped, be made accessible to English readers, and more fully from his lectures recently published under the titles 'Grundzüge der Aesthetik,' 'der Praktischen Philosophie,' and 'der Religionsphilosophie.'

TABLE OF CONTENTS.

BOOK I.

On the Connexion of Things.

INTRODUCTION.

CHAPTER I.

ON THE BEING OF THINGS.

CHAPTER II.

OF THE QUALITY OF THINGS.

CHAPTER III.

OF THE REAL AND REALITY.

CHAPTER IV.

OF BECOMING AND CHANGE.

CHAPTER V.

OF THE NATURE OF PHYSICAL ACTION.

CHAPTER VI.

THE UNITY OF THINGS.

CHAPTER VII.

CONCLUSION.

BOOK II.

Cosmology.

CHAPTER I.

OF THE SUBJECTIVITY OF OUR PERCEPTION OF SPACE.

CHAPTER II.

DEDUCTIONS OF SPACE.

CHAPTER III.

OF TIME.

Table of Contents.

CHAPTER IV.

OF MOTION.

CHAPTER V.

THE THEORETICAL CONSTRUCTION OF MATERIALITY.

CHAPTER VI.

THE SIMPLE ELEMENTS OF MATTER.

CHAPTER VII.

THE LAWS OF THE ACTIVITIES OF THINGS.

CHAPTER VIII.

THE FORMS OF THE COURSE OF NATURE.

BOOK III.

Psychology.

CHAPTER I.

THE METAPHYSICAL CONCEPTION OF THE SOUL.

CHAPTER II.

SENSATIONS AND THE COURSE OF IDEAS.

CHAPTER III.

ON THE MENTAL ACT OF 'RELATION.'

CHAPTER IV.

THE FORMATION OF OUR IDEAS OF SPACE.

CHAPTER V.

THE PHYSICAL BASIS OF MENTAL ACTIVITY.

Table of Contents.

BOOK I.

ON THE CONNEXION OF THINGS.

———

INTRODUCTION.

1. REAL is a term which we apply to things that are in opposition to those that are not; to events that happen in distinction from those that do not happen; to actually existing relations in contrast with those that do not exist. To this usage of speech I have already had occasion to appeal. I recall it now in order to give a summary indication of the object of the following enquiries. It is not the world of the thinkable, with the inexhaustible multiplicity of its inner relations—relations which are eternally valid—that here occupies us. Our considerations are expressly directed to this other region, of which the less palpable connexion with that realm of ideas, ever since the attention of Plato was first fastened upon it, has remained the constantly recurring question of Philosophy. It is a region that has been described in opposite terms. It has been called a world of appearance, of mere phenomena—and that in a depreciatory sense —by men who contrasted the variable multiplicity of its contents with the imperturbable repose and clearness of the world of ideas. To others it presented itself as the true reality. In its unfailing movement, and in the innumerable activities pervading it, they deemed themselves to have a more valuable possession than could be found in the solemn shadow-land of unchangeable ideas. This diversity of appellation rests on a deep antithesis of conception, which will attract our notice throughout all philosophy. My only reason for mentioning it here is that the two views, while wholly different in their estimates of value, serve equally to bring to light the centre round which metaphysical enquiries, so far as their essence is concerned, will always move; i. e. the fact of change. While predicable only by metaphor of anything that is merely object of thought, change com-

pletely dominates the whole range of reality. Its various forms—
becoming and decay, action and suffering, motion and development
—are, as a matter of fact and history, the constant occasions of those
enquiries which, as forming a doctrine of the flux of things in
opposition to the permanent being of ideas, have from antiquity been
united under the name of Metaphysic.

II. It is not that which explains itself but that which perplexes us
that moves to enquiry. Metaphysic would never have come into
being if the course of events, in that form in which it was presented
by immediate perception, had not conflicted with expectations, the
fulfilment of which men deemed themselves entitled to demand from
whatever was to be reckoned as truly existing or truly taking place.
These expectations might be accounted for in various ways. They
might be held to be innate to the intelligent spirit. If that were true
of them, it would follow that, in the form of necessary assumptions
as to the mode of existence and connexion of anything that can
possibly be or happen, they determine our judgment upon every
occurrence with which observation presents us. Or they might be
taken to consist in requirements arising in the heart out of its needs,
hopes, and wishes; in which case their fulfilment by the external
world, as soon 'as attention was recalled to it, would be no less
strongly demanded. Or finally it might be held that, without carrying
any intellectual necessity in their own right, they had arisen out of the
de facto constitution of experience as confirmed habits of apprehension,
suggesting that in every later perception the same features were to be
met with as had been found in the earlier. The history of philosophy
may convince us of the equally strong vivacity and assurance, with
which these different views have asserted themselves. The tendency
of the present day, however, is to deny the possession of innate cognition, to refuse to the demands of the heart every title to a share in
the determination of truth, to seek in experience alone the source of
that certain knowledge which we would fain acquire in regard to the
connexion of things.

III. Philosophy has been too painfully taught by the course of its
history how the neglect of experience avenges itself, for any fresh
reminder of its indispensableness to be required. Taken by itself,
however, and apart from every presupposition not furnished by itself,
experience is not competent to yield the knowledge which we seek.
For our wish is not merely to enumerate and describe what has
happened or is happening. We also want to be able to predict what
under definite circumstances will happen. But experience cannot

show us the future ; and cannot even help us to conjecture what it will be unless we are certain beforehand that the course of the world is bound to follow consistently, beyond the limits of previous observation, the plan of which the beginning is presented to us within those limits.

An assurance, however, of the validity of this supposition is what experience cannot afford us. Grant as much as you please that observation in its ceaseless progress had up to a certain moment only lighted on cases of conformity to the rules which we had inferred from a careful use of earlier perceptions : still the proposition that this accumulation of confirmatory instances, which has so far gone on without any exception being met with, has increased the probability of a like confirmation in the future, is one that can only be maintained on the strength of a previous tacit admission of the assumption, that the same order which governed the past course of the world will also determine the shape to be taken by its future. This one supposition, accordingly, of there being a universal inner connexion of all reality as such which alone enables us to argue from the structure of any one section of reality to that of the rest, is the foundation of every attempt to arrive at knowledge by means of experience, and is not derivable from experience itself. Whoever casts doubt on the supposition, not only loses the prospect of being able to calculate anything future with certainty, but robs himself at the same time of the only basis on which to found the more modest hope of being able under definite circumstances to consider the occurrence of one event as more probable than that of another.

IV. There have been philosophers of sceptical tendency who have shown themselves well aware of this. Having once given up the claim to be possessors of any such innate truth as would also be the truth of things, they have also consistently disclaimed any pretension from a given reality to infer a continuation of that reality which was not given with it. Nothing in fact was left, according to them, in the way of knowledge but the processes of pure Mathematics, in which ideas are connected without any claim being made that they hold good of reality, or history and the description of what is or has been. A science of nature, which should undertake from the facts of the present to predict the necessity of a future result, they held to be impossible. It was only in practical life that those who so thought relied with as much confidence as their opponents on the trustworthiness of those physical principles, which within the school they maintained to be quite without justification. The present professors of natural science, who by their noisy glorification of experience compel every meta-

physical enquiry at the outset to this preliminary self-defence, appear
to be only saved by a happy inconsistency from the necessity of a like
disclaimer. With laudable modesty they question in many individual
cases whether they have yet discovered the true law which governs
some group of processes under investigation: but they have no doubt
in the abstract as to the presence of laws which connect all parts of
the world's course in such a way that, if once complete knowledge had
been attained, infallible inferences might be made from one to the
other. Now experience, even if it be granted that in its nature it is
capable of ever proving the correctness of this assumption, certainly
cannot be held to have yet done so. There still lie before us vast
regions of nature, as to which, since we know nothing of any connexion of their events according to law, the assertion that they are
throughout pervaded by a continuous system of law cannot rest on
the evidence of experience, but must be ventured on the ground of a
conviction which makes the systematic connexion of all reality a
primary certainty.

V. There are various ways of trying to compromise the difficulty.
Sometimes the admission is made that the science of nature is only
an experiment in which we try how far we can go with the arbitrary
assumption of a law regulating the course of things; that only the
favourable result which experience yields to the experiment convinces
us of the correctness of the assumption made. Upon this we can in
fact only repeat the remark already made, and perhaps it will not be
useless actually to repeat it. If a question is raised as to the nature
of the connexion between two processes, of which the mutual dependence is not deducible from any previously known truth, it is
usual no doubt to arrive at the required law by help of an hypothesis,
of which the proof lies in the fact that no exception can be found to
its application. But in truth an hypothesis thus accredited is intrinsically after all nothing more than a formula of thought in which we
have found a short expression for the common procedure which *has
been* observable in all instances, hitherto noticed, of the connexion in
question. The character of a law is only imparted to this expression
by the further thought, which experience cannot add, but which *we*
add—the thought that in the future members of this endless series of
instances the same relation *will* hold good which, as a matter of experience, we have only found to hold good between the past members
of the series.

It is again only by a repetition of what I have already said that I
can reply to the further expansion of the view referred to. It may

readily be allowed that the observation of the same connexion between two occurrences, when constantly repeated without an instance to the contrary, gives an ever increasing probability to the assumption of a law connecting them and renders their coincidence explicable only on this assumption. But on what after all does the growing power of this surmise rest? If to begin with we left it an open question whether there is any such thing as law at all in the course of things, we should no longer be entitled to wish to find an explanation for a succession of events, and in consequence to favour the assumption which makes it explicable. For every explanation is in the last resort nothing but the reduction of a mere coincidence between two facts to an inner relation of mutual dependence according to a universal law. Every need of explanation, therefore, and the right to demand it, rests on the primary certainty of conviction that nothing can in truth be or happen which has not the ground of its possibility in a connected universe of things, and the ground of its necessary realisation at a definite place and time in particular facts of this universe. If we once drop this primary conviction, nothing any longer requires explanation and nothing admits of it ; for that mutual dependence would no longer exist which the explanation consists in pointing out. Or, to employ a different expression : if we did not start from the assumption that the course of things was bound by a chain of law, then and for that reason it would not be a whit more improbable that the same processes should always occur in a uniform, and yet perfectly accidental, connexion, than that there should be the wildest variety of the most manifold combinations. And just because of this the mere fact of a constantly repeated coincidence would be no proof of the presence of a universal law, by the help of which a further forecast might become possible as to the yet unobserved cases that lie in the future. It is not till the connexion of manifold facts according to law is established as a universal principle that any standard can exist for distinguishing a possible from an impossible, a probability from an improbability. Not till then can the one case which has been observed to occur, to the exclusion of the multitude of equally possible cases, warrant us in assuming the persistency of a special relation, which in accordance with the universal reign of law yields this one result and excludes other results that are in themselves equally possible.

All experience accordingly, so far as it believes itself to discover a relation of mutual dependence between things according to law, is in this only confirming the supposition, previously admitted as correct, of there being such a relation. If the supposition is still left in doubt, experience can never prove it. And the actual procedure of physical enquiry is in complete harmony with this state of the case. Even where the processes observed seem to contradict every thought of a uniting law, the investigator never takes himself to have found in these experiences a disproof of the supposition stated, such as would render further effort useless. He merely laments that a confirmation of it is not forthcoming, but never despairs of arriving at such a confirmation by further research.

VI. If then we enquire not so much into ostensible principles, which are generally drawn up for contentious purposes, as into those which without being put into words are continually affirmed by practice, we may take the prevalent spirit of the natural sciences to be represented by the confession that the certainty of there being a relation of mutual dependence between things according to law is independent of experience. Nay, it is common in these sciences to take that relation for granted in the particular form of a relation according to universal law with an exclusiveness which philosophy cannot accept off-hand. But in this admission that there are laws the investigator of nature still believes that all he has done has been to admit a general point of view. The question what the laws of reality are, which in fact includes every object of further enquiry, he reserves as one that is to be dealt with exclusively by the elaboration of experience. He denies the necessity or possibility of any metaphysical enquiry which in this region might aspire to add anything to the results that experience may give. Against such claims the only adequate defence of Metaphysic would consist in the complete execution of its aims; for it would only be in detail that it could be made intelligible how the manipulation, which experience must undergo in order to yield any result, is impossible, unless by the aid of various definite intermediary ideas, which contain much that does not arise out of the mere general idea of conformity to law, as such, and of which, on the other hand, the certainty cannot in turn be founded on empirical evidence.

For the present this brief hint on the subject may be taken to suffice —the more so as it is to be immediately followed by a comprehensive concession to our opponents. In our view Metaphysic ought not to repeat the attempt, which by its inevitable failure has brought the science into disrepute. It is not its business to undertake a demonstration of the special laws which the course of things in its various directions actually follows. On the contrary, while confining itself to an enquiry into the universal conditions, which everything that is to be counted as existing or happening at all must, according to it, be expected to fulfil, it must allow that what does in reality exist or happen is a thing which it cannot know of itself but can only come to know by experience. But it is only from this final knowledge of fact that those determinate laws of procedure could be derived, by which this particular reality satisfies those most general requirements which hold good for every conceivable reality. Metaphysic accordingly will only be able to unfold certain ideal forms (if that expression may be allowed), to which the relations between the elements of everything real must conform. It can supply none of those definite proportions, constant or variable, by the assignment of which it might give to those forms the special mathematical construction necessary to their applicability to a real world that is throughout determined in respect of quality, magnitude, number, and sequence. All this Metaphysic leaves to experience. It will still, however, continue to demand that the results at which experience arrives should admit of being so interpreted as to fit these ideal forms and to be intelligible as cases of their application; and to treat as fictions or as unexplained facts those which remain in contradiction with them.

VII. There would be nothing then to forbid us from identifying Metaphysic with the final elaboration of the facts with which the sciences of experiment and observation make it acquainted—but an elaboration distinguished from such sciences by the pursuit of other aims than those towards which they are directed with such laudable and unremitting energy. Natural science, while employing the conceptions of certain elements and forces most effectually for the acquisition of knowledge, foregoes the attempt to penetrate to the proper nature of those elements and forces. In a few cases important discoveries, leading to rapid progress in further insight, have been made by application of the calculus to certain assumed processes, at any possible construction of which science itself has been unable to arrive. We therefore do no injustice to science in taking its object to consist in a practical command over phenomena; in other words, the capability, however acquired, of inferring from given conditions of the present to that which either will follow them, or must have preceded them, or must take place contemporaneously with them in parts of the universe inaccessible to observation. That for the acquisition of such command, merely supposing a mutual dependence of phenomena according to some law or other, the careful comparison of phenomena should to a great extent suffice, without any acquaintance with the true nature of what underlies them, is a state of things intelligible in itself and of which the history of science gives ample evidence.

That the same process should always suffice for the purpose is not so easy to believe. On the contrary, it seems likely that after reaching a certain limit in the extent and depth of its enquiries, natural science will feel the need, in order to the possibility of further progress, of reverting to the task of defining exhaustively those centres of relation, to which it had previously been able to attach its calculations while leaving their nature undetermined. In that case it will either originate a new Metaphysic of its own or it will adopt some existing system. So far as I can judge, it is now very actively engaged in doing the former. Its efforts in that direction we observe with great interest but with mixed feelings. The enviable advantage of having acquired by many-sided investigation an original knowledge of facts, for which no appropriation of other men's knowledge can form a perfect substitute, secures a favourable judgment in advance for these experiments of naturalists: and there is the more reason that this should be so, since the philosophical instinct, which is able to ensure their success, is not the special property of a caste, but an impulse of the human spirit which finds expression for itself with equal intensity and inventiveness among those of every scientific and practical calling. But there is a drawback even here. It arises from the involuntary limitation of the range of thought to the horizon of the accustomed occupation, to external nature, and from the unhesitating transference of methods which served the primary ends of natural science correctly enough, to the treatment of questions bearing on the ulterior relations of the facts of which mastery has been obtained, and on their less palpable dependence upon principles to which reference has been studiously avoided in the ascertainment of the facts themselves.

Of course it is not my intention to indicate here the several points at which, as it seems to me, these dangers have not been avoided. I content myself with referring on the one hand to the inconsiderate habit of not merely regarding the whole spiritual life from the same ultimate points of view as the processes of external nature, but of applying to it the same special analogies as have determined our conception of those processes; and secondly to the inclination to count any chance hypothesis of which the object is one that admits of being presented to the mind, or, failing of this, of being merely indicated in words, good enough to serve as a foundation for a wholly new and paradoxical theory of the world. I do not ignore the many valuable results that are due to this mobility of imagination. I know that man must make trial of many thoughts in order to reach the truth, and that a happy conjecture is apt to carry us further and more quickly on our

way than the slow step of methodical consideration. Still there can be
no advantage in making attempts of which the intrinsic impossibility
and absurdity would be apparent if, instead of looking solely at the
single problem of which the solution is being undertaken, we carried
our view to the entire complex of questions to which the required
solution must be equally applicable. I do not therefore deny that the
metaphysical enterprises of recent physical investigators, along with
the great interest which they are undoubtedly calculated to excite,
make pretty much the same impression on me, though with a somewhat different colouring, as was made on the votaries of exact science
by the philosophy of nature current in a not very remote past.

Our business, however, is not with such individual impressions. I
only gave a passing expression to them in order to throw light on the
purpose of the following dissertation. The qualification of being
conducted according to the method of natural science, by which it is
now the fashion for every enquiry to recommend itself, is one which
I purposely disclaim for my treatise. Its object is indeed among
other things to contribute what it can to the solution of the difficult
problem of providing a philosophical foundation for natural science ;
but this is not its only object. It is rather meant to respond to the
interest which the thinking spirit takes, not merely in the calculations
by which the sequence of phenomena on phenomena may be foretold,
but in ascertaining the impalpable real basis of the possibility of all
phenomena, and of the necessity of their concatenation. This interest,
reaching beyond the region on which natural science spends its
labour, must necessarily take its departure from other points of view
than those with which natural science is familiar, nor would I disguise
the fact that the ultimate points of view to which in the sequel it will
lead us will not be in direct harmony with the accustomed views of
natural science.

VIII. There is a reproach, however, to which we lay ourselves
open in thus stating the problem of Metaphysic. It is not merely that
experience is vaunted as the single actual source of our ascertained
knowledge. Everything which cannot be learnt from it is held to be
completely unknowable : everything which in opposition to the observable succession of phenomena we are apt to cover by that comprehensive designation, the essence of things. The efforts, therefore,
to which we propose to devote ourselves will be followed with the
pitying repudiation bestowed on all attempts at desirable but impracticable undertakings. Beyond the general confidence that there
is such a thing as a connexion of things according to law, the human

spirit, it is held, has no source of knowledge, which might serve the purpose of completing or correcting experience. It would be a mere eccentricity to refuse to admit that a confession of the inscrutability of the essence of things, *in a certain sense*, must at last be elicited from every philosophy; but what if the more exact determination of this sense, and the justification of the whole assertion of such inscrutability, should be just the problem of Metaphysic, which only promises to enquire, but does not fix beforehand the limits within which its enquiry may be successful? And it is clear that the assertion in question, if prefixed to all enquiry, is one that to a certain extent contradicts itself. So long as it speaks of an essence of things, it speaks of something and presupposes the reality of something as to the existence of which according to its own showing experience can teach nothing. As soon as it maintains the unknowability of this essence, it implies a conviction as to the position in which the thinking spirit stands to the essence, which, since it cannot be the result of experience, must be derived from a previously recognized certainty in regard to that which the nature of our thought compels us to oppose, as the essence of things, to the series of phenomena. But it is just these tacit presuppositions, which retain their power over us all the time that we are disputing our capacity for knowledge, that stand in need of that explanation, criticism, and limitation, which Metaphysic deems its proper business. Nor have we any right to take for granted that the business is a very easy one, and that it may be properly discharged by some remarks well-accredited in general opinion, to be prefixed by way of introduction to those interpretations of experience from which alone a profitable result is looked for. When we assume nothing but conformity to law in the course of things, this expression, simple itself, seems simple in its signification: but the notions attached to it turn out to be various and far-reaching enough, as soon as it has to be employed in precisely that interpretation of experience which is opposed to Metaphysic.

I will not enlarge on the point that every physical enquiry employs the logical principles of Identity and Excluded Middle for the attainment of its results : both are reckoned as a matter of course among the methods which every investigation follows. But meanwhile it is forgotten that these principles could not be valid for the connected series of phenomena without holding good also of the completely unknown basis from which the phenomena issue. Yet many facts give sufficient occasion for the surmise that they apply to things themselves and their states in some different sense from that in which they apply to the judgments which are suggested to us in thinking about these states. We show as little scruple in availing ourselves of mathematical truths, in order to advance from deduction to deduction. It is tacitly assumed that the unknown essence of things, for one manifestation of which we borrow from experience a definite numerical value, will never out of its residuary and still unknown nature supply to the consequence which is to be looked for under some condition an incalculable coefficient, which would prevent the correspondence of our mathematical prediction with the actual course of events.

Nor is this all. Besides these presumptions which are at any rate general in their character and which are all that can be noticed at the outset, in the actual interpretation of experience there are implied many unproven judgments of a more special sort, which can only be noticed in the sequel. For logical laws hold good primarily of nothing but the thinkable content of conceptions, mathematical laws of nothing but pure quantities. If both are to be applied to that which moves and changes, works and suffers, in space and time, they stand in constant need of fresh ideas as to the nature of the real, which as connecting links make it possible to subordinate to the terms of those laws this new region of their application. It is vain for us therefore to speak of a science founded on experience that shall be perfectly free from presuppositions. While this science thinks scorn of seeking support from Metaphysic and disclaims all knowledge of the essence of things, it is everywhere penetrated by unmethodised assumptions in regard to this very essence, and is in the habit of improvising developments, as each separate question suggests them, of those principles which it does not deem it worth while to subject to any systematic consideration.

IX. In making these remarks I have no object in view but such as may properly be served by an introduction. I wish to prepossess that natural feeling of probability, which in the last instance is the judge of all our philosophical undertakings, in favour of the project of putting together in a systematic way the propositions in regard to the nature and connexion of what is real, which, independently of experience and in answer to the questions with which experience challenges us, we believe ourselves to have no option but to maintain. I expressly disclaim, however, the desire to justify this belief, from which as a matter of fact we are none of us exempt, by an antecedent theory of cognition. I am convinced that too much labour is at present spent in this direction, with results proportionate to the groundlessness of the claims which such theories make. There is something convenient and seductive in the plan of withdrawing attention from the solution of definite questions and applying oneself to general questions in regard to cognitive capacities, of which any one *could* avail himself who set seriously about it. In fact, however, the history of science shows that those who resolutely set themselves to mastering certain problems generally found that their cognisance of the available appliances and of the use of them grew keener in the process; while on the other hand the pretentious occupation with theories of cognition has seldom led to any solid result. It has not itself created those methods which it entertains itself with exhibiting but not employing. On the contrary, it is the actual problems that have compelled the discovery of the methods by which they may be solved. The constant whetting of the knife is tedious, if it is not proposed to cut anything with it.

I know that such an expression of opinion is in unheard-of opposition to the tendency of our time. I could not, however, repress the conviction that there is an intrinsic unsoundness in the efforts made to found a Metaphysic on a psychological analysis of our cognition. The numerous dissertations directed to this end may be compared to the tuning of instruments before a concert, only that they are not so necessary or useful. In the one case it is known what the harmony is which it is sought to produce: in the other case the mental activities which are believed to have been discovered are compared with a canon which the discoverers profess that they have still to find out. In the last resort, however, every one allows that as to the truth of our cognition and its capability of truth no verdict can be compassed which is independent of that cognition itself. It must itself determine the limits of its competence. In order to be able to do this—in order to decide how far it may trust itself to judge of the nature of the real, it must first arrive at a clear notion of the propositions which it is properly obliged—obliged in thorough agreement with itself—to assert of this real. It is by these assumptions, which are simply necessary to Reason, that the conception of the real which is supposed to be in question is determined ; and it is only their content that can justify Reason, when the question is raised, in forming any judgment with regard to its further relation to this its object— either that is in maintaining the unknowability of its concrete nature, or in coming to the conclusion as the only one compatible with the reconciliation of all its thoughts, that the conception of things which it generates has no independent object, or in persistently retaining a belief in such an object in some sense which reason itself determines—a belief which, because of such a nature, neither requires nor admits further proof. On the other hand it strikes me as quite unjustifiable to treat the most obscure of all questions, that of the psychological origin of knowledge and the play of conditions which co-operate in producing it, as a preliminary question to be easily dealt with, of which the issue might settle decisively the validity or invalidity collectively or severally of the utterances of reason. On the contrary the psychological history of the origin of an error only conveys a proof that it is an error on supposition that we are previously acquainted with the truth and can thus be sure that the originating condition of the error involved a necessary aberration from that truth.

Thus the doctrine which I would allege rests not on any conviction which has previously to be admitted as to the psychological roots of our knowledge, but simply on an easily recognisable fact, of which the admission is implied by the very act of disputing it. Every one, evade it as he will, must in the last instance judge of every proposition submitted to him and of every fact with which experience presents him upon grounds of which the constraining force presses itself upon him with an immediate assurance. I say, 'in the last instance,' for even when he undertakes to examine this self-evidence, his final affirmation or denial of it must always rest on the like self-evidence as belonging to his collected reasons for deciding on the matter. In regard to that which this self-supported reason must affirm, now that by the space of centuries it has, in sequence on experience, reflected on itself, a comprehensive consciousness may be obtained or at least sought. But how all this takes place in us, and how it comes about that those fundamental truths which are necessities of our thought acquire their self-evidence—these are points on which enlightenment, if possible at all, can only be looked for in a remote future. But whenever it may come, it can only come after the first question has been answered. The process of our cognition and its relations to objects must, whether we like it or no, be subject to those judgments which our reason passes as necessities of thought upon *every* real process and on the effect of *every* element of reality upon every other. These declarations are not in the least at war with the high interest which we take in psychology as a proper region of enquiry. They only amount to a repetition of the assertion which every speculative philosophy must uphold, that while Psychology cannot be the foundation of Metaphysic, Metaphysic must be the foundation of Psychology.

X. It is time, however, for some more precise statements as to the line which it is proposed to take in the following enquiry. In referring to the supposition of a universal relation of mutual dependence between all things real as the common foundation of all scientific investigation, I at the same time indicated a doubt with reference to the exclusive form to which in the present stage of scientific culture it is the fashion to reduce this relation—the form of conformity to universal law. This form is neither the only one nor the oldest under which the human spirit has presented to itself the connexion of things. It was emphatically not as instances of a universal rule but as parts of a whole that men first conceived things: as related to each other not primarily by permanent laws but by the unchangeable purport of a plan, of which the realisation required from the several elements not always and everywhere an identical procedure, but a changeable one. In this conviction originated the dazzling forms of the idealistic constructions of the universe. Starting from a supreme idea, into the depths of which they claimed to have penetrated by immediate intuition, the authors of these schemes thought to deduce the manifold variety of phenomena in that order in which the phenomena were to contribute to the realisation of the supposed plan. It was not the discovery of laws that was their object, but the establishment of the several ends which the development of things had gradually to attain and of which each determined all habits of existence and behaviour within the limits of that section of the universe which it governed. The barrenness of these schemes is easily accounted for. They failed in that in which men always will fail, in the exact and exhaustive definition of that supreme thought, which they held in honour. Now any shortcoming in this outset of the theory must be a source of constantly increasing defect in its development, as it descends to particulars. If ever a happy instinct led it to results that could be accepted, it was only an aesthetic satisfaction that such guesses yielded, not any certainty that could meet doubt by proof. Yet the general conviction from which the speculations in question set out does not yield in any way, either. as less certain or as less admissible, to the supposition of universal conformity to law, which in our time is deemed alone worthy of acceptance. For my part therefore—and I wish there to be no uncertainty on the point—I should reckon this theory of the universe, if it could be carried out in detail, as the completion of philosophy; and though I cannot but deem it incapable of being thus carried out, I yet do not scruple to allow to the conviction, that its fundamental thought is virtually correct, all the influence which it is still possible for it to retain on the formation of my views.

But from among the objects of the enquiry before us, this theory, at least as carrying any immediate certainty, remains excluded. For we are not to employ ourselves upon the world of ideas itself, with its constituents arranged in an order that holds good eternally and is eternally complete, but upon the given world, in which the process of realisation of the ideas is supposed to be visible. Now it is not once for all nor in a systematic order that this real world unfolds ectypes of the ideas. In that case it would scarcely be possible to say in what respect the series of the ectypes is distinguishable from that of the archetypes. But the world of reality presents innumerable things and occurrences distributed in space and time. It is by shifting relations of these that the content of the ideas is realised in manifold instances and with degrees of completeness or incompleteness—is so realised only again to disappear. However then we may think on the obscure question of the position in which the ideas stand to the world of phenomena and of the regulation of this world by them, it is certain that as soon as their realisation becomes dependent on the changing connexion between a number of points brought into relation, there must arise a system of universal laws, in accordance with which in all like cases of recurrence a like result necessarily follows, in unlike cases an unlike result, and a certain end is attained in one case, missed in another. Accordingly, even the idealistic theory of the world, which believes reality to be governed by ends that belong to a plan, if it would render the process of realisation of these ends intelligible, necessarily generates the conception of a universal connexion of things according to law as a derived principle, though it may refuse it the dignity of an ultimate principle. It will find no difficulty in admitting further that the human spirit does not possess any immediate revelation as to an end and direction of the collective movement of the universe, in which according to its own supposition that spirit is a vanishing point. Having for its vocation, however, to work at its limited place in the service of the whole according to the same universal laws which hold good for all the several elements of the whole, the human spirit will more easily possess an immediate consciousness of this necessity by which it like everything else is determined.

Considerations of this sort settle nothing objectively: but they suffice to justify the abstract limitation of our present problem. Metaphysic has merely to show what the universal conditions are which must be satisfied by anything of which we can say without contradicting ourselves that it is or that it happens. The question remains open whether these laws, which we hope to master, form the ultimate object which our knowledge can reach, or whether we may succeed in deducing them from a highest thought, as conditions of its realisation which this thought imposes on itself.

XI. In order to the discovery of the truths we are in search of it would be desirable to be in possession of a clue that could be relied on. The remarks we have just made at once prevent us from availing ourselves of a resource in which confidence was placed by the philosophers of a still recent period. The followers of the idealistic systems to which I last referred imagined that in their dialectic method they had security for the completeness and certainty of the formulae in which they unfolded the true content of the universe. They directed their attention but slightly to the riddles of experience. To a much greater degree they had allowed themselves to be affected by the concentrated impression of all the imperfections by which the world outrages at once our knowledge our moral judgment, and the wishes of our hearts. In opposition to that impression there arose in their minds with great vivacity but, as was not denied, in complete obscurity the forecast of a true being, which was to be free from these shortcomings and at the same time to solve the difficult problem of rendering the presence of the shortcomings intelligible. This forecast, into which they had gathered all the needs and aspirations of the human spirit, they sought by the application of their method to unfold into its complete content. In their own language they sought to raise that into conception [1] which at the outset had been apprehended only in the incomplete form of imagination [2].

I do not propose to revert to the criticism of this method, on the logical peculiarity of which I have enlarged elsewhere. It is enough here to remark that in accordance with the spirit of the theories in which it was turned to account, it has led only to the assignment of certain universal forms of appearance which cannot be absent in a world that is to be a complete ectype of the supreme idea. It has not led to the discovery of any principles available for the solution of questions relating to the mutual qualification of the several elements, by which in any case the realisation of those forms is completely or incompletely attained. The method might conceivably be transformed so as to serve this other end, for its essential tendency, which is to clear up obscure ideas, will give occasion everywhere for its use.

[1] [Begriff.]   [2] [Vorstellung.]

But in this transformation it would lose the most potent part of that which formerly gave it its peculiar charm. Its attraction consisted in this, that it sought in a series of intuitions, which it unfolded one out of the other, to convey an immediate insight into the very inner movement which forms the life of the universe, excluding that labour of discursive thought which seeks to arrive at certainty in roundabout ways and by use of the most various subsidiary methods of proof. As making such claims, the method can at bottom only be a form of that process of exhibiting already discovered truths which unfolds them in the order which after much labour of thought in other directions comes to be recognised as the proper and natural system of those truths. If however the method is to be employed at the same time as a form of *discovering* truth, the process, questionable at best, only admits of being in some measure carried out in relation to those universal and stable forms of events and phenomena, which we have reason for regarding as an objective development of the world's content or of its idea. In regard to the universal laws, by which the realisation of all these forms is uniformly governed, we certainly cannot assume that they constitute a system in which an indisputable principle opens out into a continuous series of developments. We cannot in this case ascribe the development to the reality[1] as objective, but only to our thoughts about the reality[1] as subjective. The Dialectic method would therefore have to submit to conversion into that simpler dialectic, or, to speak more plainly, into that mere process of consideration in which the elementary thoughts that we entertain as to the nature and interconnection of the real are compared with each other and with all the conditions which warrant a judgment as to their correctness, and in which it is sought to replace the contradictions and shortcomings that thereupon appear by better definitions. Nothing is more natural and familiar than this mode of procedure, but it is also obvious that it does not of itself determine beforehand either the point of departure for the considerations of which it consists or in detail the kind of progress which shall be made in it.

XII. Other attempts at the discovery of a clue have started from

---

[1] ['Sache' in this work means whatever a name can stand for, is coextensive with 'Vorstellbarer Inhalt' (a content which can be presented in an idea), Logic, sect. 34², and therefore has 'objectivity' (Objectivität), Logic, sect. 3; on the other hand it is much wider than 'Ding' (a thing), which has not only 'Objectivität' but also 'Wirklichkeit' (concrete external reality); cp. Logic, sect. 3. There is no exact English equivalent for 'Sache' in this sense.]

a conception of classification. There lies a natural charm in the assumption that not only will the content of the universe be found to form an ordered and rounded whole according to some symmetrical method, but also that the reason, of which it is the vocation to know it, possesses for this purpose innate modes of conception in organised and completed array. The latter part of this notion, at any rate, was the source of Kant's attempt by a completion of Aristotle's doctrine of Categories to find the sum of truths that are necessities of our thought. In the sense which Aristotle himself attached to his Categories, as a collection of the most universal predicates, under which every term that we can employ of intelligible import may be subsumed, they have never admitted of serious philosophical application. At most they have served to recall the points of view from which questions may be put in regard to the objects of enquiry that present themselves. The answers to those questions always lay elsewhere— not in conceptions at all, but in fundamental judgments directing the application of the conception in this way or that. Kant's reformed table of Categories suffers primarily from the same defect; but he sought to get rid of it by passing in fact from it to the 'principles of Understanding' which, as he held, were merely contracted in the Categories into the shape of conceptions and could therefore be again elicited from them. The attempt is a work of genius, but against the reasoning on which it is founded and the consequences drawn from it many scruples suggest themselves. Kant found fault with Aristotle for having set up his Categories without a principle to warrant their completeness. On the other hand, plenty of people have been forthcoming to point out the excellence of the principles of division which Aristotle is supposed to have followed. I do not look for any result from the controversy on this point. Given a plurality of unknown extent, if it is proposed to resolve it not merely by way of dichotomy into M and non-M but ultimately into members of a purely positive sort, M, N, O, P, Q, there can be no security in the way of method for the completeness of this disjunctive process. From the nature of the case we must always go on to think of a residuary member R, of which nothing is known but that it is different from all the preceding members. Any one who boasts of the completeness of the division is merely saying that for his part he cannot add a fresh member R. Whoever denies the completeness affirms that a further member R has occurred to him which with equal right belongs to the series. Aristotle may have had the most admirable principles of division; but they do not prove that he has noticed all the members which properly fall under them. But the same remark holds equally good against Kant. It may be conceded to him that it is only in the form of the judgment that the acts of thought are performed by means of which we affirm anything of the real. If it is admitted further as a consequence of this that there will be as many different primary propositions of this kind as there are essentially different logical forms of judgment, still the admission that these different forms of judgment have been exhaustively discovered cannot be insisted on as a matter, properly speaking, of methodological necessity. The admission will be made as soon as we feel ourselves satisfied and have nothing to add to the classification ; and if this agreement were universal, the matter would be practically settled, for every inventory must be taken as complete, if those who are interested in its completeness can find nothing more to add to it. But that kind of theoretical security for an unconditional completeness, which Kant was in quest of, is something intrinsically impossible.

These however are logical considerations, which are not very decisive here. It is more important to point out that the very admission from which we started is one that cannot be made. The logical forms of judgment are applied to every possible subject-matter, to the merely thinkable as well as to the real, to the doubtful and the impossible as well as to the certain and the possible. We cannot therefore be the least sure that all the different forms, which are indispensable to thought for this its wide-reaching employment, are also of equal importance for its more limited application to the real. So far however as their significance in fact extends also to this latter region, it is a significance which could not be gathered in its full determination from that general form in which it was equally applicable to the non-real. The categorical form of judgment leaves it quite an open question, whether the subject of the judgment to which it adds a predicate is a simple 'nominal essence[1]' remaining identical with itself, or a whole which possesses each of its parts, or a substance capable of experiencing a succession of states. The hypothetical form of judgment does not distinguish whether the condition contained in its antecedent clause is the reason of a consequence, or the cause of an effect, or the determining end from which the fact stated in the consequent proceeds as a necessary condition of its fulfilment. But these different conceptions, which are here presented in a like form, are of different importance for the treatment of the real. The metaphysical significance

[1] ['Einfacher Denkinhalt.']

C 2

of the Categories is, therefore, even according to Kant's view, only
a matter of happy conjecture, and rests upon material considerations,
which are unconnected with the forms of judgment, and to which
the systematisation of those logical forms has merely given external
occasion.   It is only these incidentally suggested thoughts that have
given to the Categories in Kant's hands a semblance of importance
and productiveness, which these playthings of philosophy, the object
of so much curiosity, cannot properly claim.   This roundabout road
of first establishing a formal method affords us no better security
than we should have if we set straight to work at the thing—at the
matter of our enquiry.

XIII.   We are encouraged to this direct course by the recollection
that it is not a case of taking possession for the first time of an
unknown land.   Thanks to the zealous efforts of centuries the objects
we have to deal with have long been set forth in distinct order, and
the questions about them collected which need an answer.   Nor had
the philosophy which has prepared the way for us itself to break
wholly new ground.   In regard to the main divisions of our subject
it had little to do but to repeat what everyone learns anew from his
own experience of the world.   Nature and spirit are two regions so
different as at first sight to admit of no comparison, and demanding
two separate modes of treatment, each devoted to the essential
character by which the two regions are alike self-involved and separate from each other.   But on the other hand they are destined to
such constant action upon each other as parts of one universe, that
they constrain us at the same time to the quest for those universal
forms of an order of things which they both have to satisfy alike in
themselves and in the connexion with each other.   It might seem as
if this last-mentioned branch of its enquiry must be the one to which
early science would be last brought.   As a matter of history, however,
it has taken it in hand as soon as the other two branches, and has
long devoted itself to it with greater particularity than, considering
the small progress made in the other branches, it could find conducive
to success.   But whatever may be the case historically, now at least
when we try to weigh the amount of tenable result which has been
won from such protracted labour, we are justified in beginning with
that which is first in the order of things though not in the order
of our knowledge ; I mean with Ontology, which, as a doctrine
of the being and relations of all reality, had precedence given to
it over Cosmology and Psychology—the two branches of enquiry
which follow the reality into its opposite distinctive forms.   It is

to this division of the subject that with slight additions or omissions,
Metaphysic under every form of treatment has to all intents and
purposes returned. The variety in the choice of terms occasioned
by peculiar points of view adopted antecedently to the consideration
of the natural division of the subject, has indeed been very great.
But to take any further account of these variations of terminology,
before entering on the real matter in hand, seems to me as useless as
the attempt to determine more exactly that limitation of the problems
before us which metaphysicians have had before them in promising
to treat only of rational cosmology and psychology, as opposed in a
very intelligible manner to the further knowledge which only experience can convey.

XIV. No period of human life is conceivable in which man did
not yet feel himself in opposition to an external world around him.
Long in doubt about himself, he found around him a multitude of
perceptibly divided objects, and he could not live long without
having many impressions forced upon him as to their nature and
connexion. For none of the every-day business that is undertaken
for the satisfaction of wants could go on without the unspoken conviction that our wishes and thoughts have not by themselves the
power to make any alteration in the state of the outer world, but that
this world consists in a system of mutually determinable things, in
which any alteration of one part that we may succeed in effecting is
sure of a definite propagation of effects on other parts. Moreover
no such undertaking could be carried out without coming on some
resistance, and thus giving rise to the recognition of an unaccountable
independence exercised by things in withstanding a change of state.
All these thoughts as well as those which might readily be added on
a continuation of these reflections, were primarily present only in the
form of unconsciously determining principles which regulated actions
and expectations in real life. It is in the same form that with almost
identical repetition they still arise in each individual, constituting the
natural Ontology with which we all in real life meet the demand for
judgments on events. The reflective attempt to form these assumptions into conscious principles only ensued when attention was called
to the need of escaping contradictions with which they became em-
barrassed when they came to be applied without care for the consequences to a wider range of knowledge.

It was thus that Philosophy, with its ontological enquiries, arose.
In the order of their development these enquiries have not indeed
been independent of the natural order in which one question suggests

another. Still owing to accidental circumstances they have often
drifted into devious tracks ; have assumed and again given up
very various tendencies. There is no need, however, in a treatise
which aims at gathering the product of these labours, to repeat this
chequered history. It may fasten directly on the natural conception
of the universe which we noticed just now—that conception which
finds the course of the world only intelligible of a multiplicity of persistent things, of variable relations between them, and of events
arising out of these changes of mutual relation. For it is just this
view of the universe, of which the essential purport may be thus sum-
marised, which renews itself with constant identity in every age.
Outside the schools we all accommodate ourselves to it. Not to us
merely, but to all past labourers in the field of philosophy, it has
presented itself as the point of departure, as that which had either to
be confirmed or controverted. Unlike the divergent theories of speculative men, therefore, it deserves to be reckoned as itself one of the
natural phenomena which, in the character of regular elements of
the universe, enchain the attention of philosophy. For the present
however all that we need to borrow from history is the general conviction that of the simple thoughts which make up this view there is
none that is exempt from the need of having its actual and possible
import scientifically ascertained in order to its being harmonised with
all the rest in a tenable whole. No lengthy prolegomena are needed
to determine the course which must be entered on for this purpose.
We cannot speak of occurrences in relations without previously thinking of the things between which they are supposed to take place or to
subsist. Of these things, however—manifold and unlike as we take
them to be—we at the same time affirm, along with a distinction in
the individual being of each, a likeness in respect of that form of
reality which makes them things. It is with the simple idea of this
being that we have to begin. The line to be followed in the sequel may
be left for the present unfixed. Everything cannot be said at once.
That natural view of the world from which we take our departure,
simple as it seems at first sight, yet contains various interwoven
threads ; and no one of these can be pursued without at the same time
touching others which there is not time at the outset to follow out on
their own account and which must be reserved to a more convenient
season. For our earlier considerations, therefore, we must ask the
indulgence of not being disturbed by objections of which due account
shall be taken in the sequel.

CHAPTER I.

On the Being of Things.

1. ONE of the oldest thoughts in Philosophy is that of the opposition between true being and untrue being. Illusions of the senses, causing what is unreal to be taken for what is real, led to a perception of the distinction between that which only appears to us and that which is independent of us. The observation of things taught men to recognise a conditional existence or a result of combination in that which to begin with seemed simple and self-dependent. Continuous becoming was found where only unmoving persistent identity had been thought visible. Thus there was occasioned a clear consciousness of that which had been understood by 'true being,' and which was found wanting in the objects of these observations. Independence not only of us but of everything other than itself, simplicity and unchanging persistence in its own nature, had always been reckoned its signs. Its signs, we say, but still only its signs ; for these characteristics, though they suffice to exclude that of which they are not predicable from the region of true being, do not define that being itself. Independence of our own impressions in regard to it is what we ascribe to every truth. It holds good in itself, though no one thinks it. Independence of everything beside itself we affirm not indeed of every truth, but of many truths which neither need nor admit of proof. Simplicity exclusive of all combination belongs to every single sensation of sweetness or redness; and motionless self-subsistence, inaccessible to any change, is the proper character of that world of ideas which we oppose to reality on the ground that while we can say of the ideas that they eternally *hold good* we cannot say that they *are.* It follows that in the characteristics stated of Being not only is something wanting which has been thought though not expressed but the missing something is the most essential element of that which we are in quest of. We still want to know what exactly that Being itself is to which those terms may be applied by way of

distinguishing the true Being from the apparent, or what that reality consists in by which an independent simple and persistent Being distinguishes itself from the unreal image in thought of the same independent simple and persistent content.

2. To this question a very simple answer may be attempted. It seems quite a matter of course that the thinking faculty should not be able by any of its own resources, by any thought, to penetrate and exhaust the essential property of real Being, in which thought of itself recognises an opposition to all merely intelligible existence. The most that we can claim, it will be said, is that real Being yields us a living experience of itself in a manner quite different from thinking, and such experiences being once given, a ground of cognition with reference to them thereupon admits of being stated, which is necessary not indeed for the purpose of inferring that presence of real Being which is matter of immediate experience but for maintaining the truth of this experience against every doubt. Upon this view no pretence is made of explaining by means of conceptions the difference of real Being from the conception of the same, but immediate sensation [1] has always been looked upon as the ground of cognition which is our warrant for the presence of real Being. Even after the habit has been formed of putting trust in proofs and credible communications, we shall still seek to set aside any doubt that may have arisen by rousing ourselves to see and hear whether the things exist and the occurrences take place of which information has been given us ; nor does any proof prove the reality of its conclusion unless, apart from the correctness of its logical concatenation, not merely the truth of its original premisses, as matter of thought, but the reality of its content is established—a reality which in the last resort is given only by sensuous perception. It may be that even sensation sometimes deceives and presents us with what is unreal instead of with what is real. Still in those cases where it does not deceive, it is the only possible evidence of reality. It may in like manner be questioned whether sensation gives us insight into the real *as it is*. Still *of the fact that* something which really is underlies it, sensation does not seem to allow a doubt.

3. The two objections just noticed to the value of sensation cannot here be discussed in full, but with the second there is a difficulty connected which we have to consider at once. The content of simple sensations cannot be so separated from the sensitive act as that detached images of the two, complete in themselves, should remain

[1] ['Sinnlichen Empfindung.']

after the separation. We can neither present redness, sweetness, and warmth to ourselves as they would be if they were not felt, nor the feeling of them as it would be if it were not a feeling of any of these particular qualities. The variety, however, of the sensible qualities, and the definiteness of each single quality as presented to the mind's eye, facilitate the attempt which we all make to separate in thought what is really indivisible. The particular matter which we feel, at any rate, appears to us independent of our feeling, as if it were something of which the self-existent nature was only recognised and discovered by the act of feeling.

But we do not succeed so easily in detaching the other element—that real being, of which, as the being of this sensible content, it was the business of actual sensation in opposition to the mere recollection or idea of it to give us assurance. It cannot be already given in this simplest affirmation or position which we ascribed to the sensible contents, and by which each of them is what it is and distinguishes itself from other contents. Through this affirmation that which is affirmed only comes to hold good as an element in the world of the thinkable. It is not real merely because it is in this sense something, as opposed to nothing void of all determination. In virtue of such affirmation Red is eternally Red and allied to Yellow, not allied to what is warm or sweet. But this identity with itself and difference from something else holds good of the Red of which there is no actual sensation as of that of which there is actual sensation. Yet it is only in the case of the latter that sensation is supposed to testify to real existence. Apart from that simplest affirmation, however, the various sensible qualities in abstraction from the sensitive act which apprehends them have nothing in common. If therefore we assert of them, so far as they are felt, a real Being different from this affirmation, this Being is not anything which as attaching to the nature of the felt quality would merely be recognised and discovered by the sensitive act. On the contrary, it lies wholly in the simple fact of being felt, which forms the sole distinction between the actual sensation of the quality that is present to sense and the mere idea of quality which is not so present. Thus it would appear that the notion with which we started must be given up; for sensation is not a mere ground of cognition of a real Being which is still something different from it and of which the proper nature has still to be stated; and the being which on the evidence of sensation we ascribe to things consists in absolutely nothing else than the fact of their being felt.

4. This assertion, however, can only be hazarded when certain

points of advanced speculation have been reached, which we shall
arrive at later. The primary conception of the world is quite remote
from any such inference. According to it sensation is certainly the
only 'causa cognoscendi' which convinces us of Being, and just
because it is the only one, there easily arises the mistake of supposing
that what it alone can show consists only of it ; whereas in fact Being
is, notwithstanding, independent of our recognition of it, and all
things, of which we learn the reality, it is true, only from sensation,
will continue to be, though our attention is diverted from them and
they vanish from our consciousness. Nothing indeed appears more
self-evident than this doctrine. We all do homage to it. Yet the
question must recur, what remains to be understood by the Being of
things, when we have got rid of the sole condition under which it is
cognisable by us. It was as objects of our feeling that things were
presented to us. In this alone consisted as far as we could see
what we called their Being. What can be left of Being when we
abstract from our feeling? What exactly is it that we suppose ourselves to have predicated of things, in saying that they *are* without
being felt ? Or what is it that for the things themselves, by way of
proof, confirmation, and significance of their being, takes the place of
that sensation which for us formed the proof, confirmation, and significance of their being.

The proper meaning of these questions will become clearer, if I
pass to the answers which the natural theory of the world gives to
them ; for it must not be supposed that this theory makes no effort to
remedy the shortcoming which we have noticed. Its simplest way
of doing so consists in the reflection that on the disappearance of our
own sensation that of others takes its place. The men whom we
leave behind will remain in intercourse with others. Places and
objects, from which we are removed, will be seen by others as
hitherto by us. This constitutes their persistency in Being, while
they have vanished from our senses. Everyone, I think, will find
traces in himself of this primary way of presenting the case. Yet it
helps us rather to put off the question than to answer it. It is sure
to repeat itself at once in another form. Being was said to be
independent of any consciousness on the part of a sentient subject.
What then if consciousness is extinguished out of the entire universe
and there is no longer any one who could have cognisance of the
things that are supposed to exist ? In that case, we answer, they will
continue to stand in those relations to each other in which they stood
when they were objects of perception. Each will have its place in

space or will change it. Each will continue to exercise influences on others or to be affected by their influence. These reciprocal agencies will constitute that in which the things possess their being independently of all observation. Beyond this view of the matter the natural theory of things scarcely ever goes. In what respect it is unsatisfactory and in what it is right we have now to attempt to consider.

6. There is one point on which it is held to be defective, but unfairly, because its defect consists merely in its inability to answer an improper question, which we have simply to get out of the habit of putting. The question arises in this way. All those relations, in which we just now supposed the reality of things to consist, may be thought of equally as real and as unreal. But they must be actually real and not merely thought of as real, if they are to form the Being of things and not merely the idea of this Being. In what then, we ask, consists this reality of that which is in itself merely thinkable, and how does it arise? That this question is unanswerable and self-contradictory needs no elaborate proof. In what properly consists the fact—how it comes about or is made—that there is something and not nothing, that something and not nothing takes place; this it is eternally impossible to say. For in fact, whatever the form of the question in which this curiosity might find expression, it is clear that we should always presuppose in it as antecedent to that reality of which we seek an explanation, a prior connected reality, in which from definite principles definite consequences necessarily flow, and among them the reality that has to be explained. And the origin of this latter reality would not be like that of a truth which arises as a consequence out of other truths but which yet always subsisted along with them in eternal validity. The origin in question would be expressly one in which a reality, *that was previously itself unreal*, arises out of another reality. Everything accordingly which we find in the given reality—the occurrence of events, the change in the action of things upon each other, the existence of centres of relation between which such action may take place—all this we must assume to begin with in order to render the origin of reality intelligible.

This obvious circle has been avoided by the common view. Nor can it be charged with having itself fallen into another circle in reducing the real Being of things to the reality of those relations the maintenance of which it supposed to constitute what was meant by this Being. For it could not be intended to analyse this most general conception of reality, of which the significance can only be conveyed in the living experience of feeling. All that could be meant by definitions of Being in the common theory was an indication of that which within this given miracle of reality is to be understood as the Being of the Things in distinction from other instances of the same reality, from the existence of the relations themselves and from the occurrence of events. Whether the common theory has succeeded in this latter object is what remains to be asked.

8. Philosophy has been very unanimous in denying that it has. How, it is asked, are we to understand those relations, in the subsistence of which we would fain find the Being of the Things? If they are merely a result of arbitrary combinations in which we present things to our minds, we should equally fail in our object whether the things ordered themselves according to this caprice of ours or whether they did not. In the former case we should not find the Being independent of ourselves which we were in search of. If the latter were the true state of the case, it would make it still more plain that there must be something involved in the Being of things which our definition of this Being failed to include—the something in virtue of which they are qualified to exist on their own account, not changing with and because of our changeable conception of their Being. We cannot be satisfied therefore without supposing that the relations, of which we assume the existence, exist between the things themselves, so as to be discoverable by our thought but not created by or dependent on it. The more, however, we insist on this objective reality of relations, the more unmistakeable we make the dependence of the Being of everything on the Being of everything else. No thing can have its place among the other things, if these are not there to receive it among them. None can work or suffer, before the others are there to exchange impressions with it. To put the matter generally; in order to there being such a thing as an action of one thing upon another, it would seem that the centres of relation between which it is to take place must be established in independent reality. A Being in things, resting wholly on itself and in virtue of this independence rendering the relations possible by which things are to be connected, must precede in thought every relation that is to be taken for real. This is the *pure* Being, of which Philosophy has so often gone in quest. It is opposed by Philosophy, as being of the same significance for all things, to the empirical Being which, originating in the various relations that have come into play between things, is different for every second thing from what it is for the third, and which Philosophy hopes somehow to deduce as a supervening result from the pure Being.

7. I propose to show that expectation directed to this metaphysical use of the conception of pure Being is a delusion, and that the natural theory of the world, in which nothing is heard of it, is on this point nearer the truth than this first notion of Speculation. Every conception, which is to admit of any profitable application, must allow of a clear distinction between that which is meant by it and that which is not meant by it. So long as we looked for the Being of things in the reality of relations in which the things stand to each other, we possessed in these relations something by the affirmation of which the Being of that which is, distinguishes itself from the non-Being of that which is not. The more we remove from the conception of Being every thought of a relation, in the affirmation of which it might consist, the more completely the possibility of this distinction disappears. For not to be at any place, not to have any position in the complex of other things, not to undergo any operation from anything nor to display itself by the exercise of any activity upon anything; to be thus void of relation is just that in which we should find the nonentity of a thing if it was our purpose to define it. It is not to the purpose to object that it was not this nonentity but Being that was meant by the definition. It is not doubted that the latter was the object of our definition, but the object is not attained, so long as the same definition includes the opposite of that which we intended to include in it.

No doubt an effort will be made to rebut this objection in its turn. It will be urged that if, starting from the comparison of the multiform Being of experience, we omit all the relations on which its distinction rests, that which remains as pure Being is not the mere privation of relations but that of which this very unrelatedness serves only as a predicate, and which, resting on itself and independent, is distinguished by this hardly to be indicated but still positive trait from that which is not. Now it is true that our usage is not to employ these and like expressions of that which is not or of the nothing, but the usage is not strictly justifiable so long as we apply the expressions to this pure Being. They only have an intelligible sense because we already live in the thought of manifold relations, and within the sphere of these the true Being has opportunity of showing by a definite order of procedure what is the meaning of its independence and self-subsistence. Once drop this implication, and all the above expressions, in the complete emptiness of meaning to which they thereupon sink, are unquestionably as applicable to Nothing as they are to Being, for in fact independence of everything else, self-subsistence and complete absence of relation are not less predicable of the one than of the other.

8. We may expect here the impatient rejoinder—'There still remains the eternal difference that the unrelated Being is while the unrelated non-Being is not : all that comes of your super-subtle investigation is a contradiction of your own previous admission. For the meaning of Being, in the sense of reality and in opposition to not-being, is as you say undefinable and only to be learnt by actual living. The cognition thus gained necessarily and rightfully presupposes the conception of pure Being, as the positive element in the experienced Being. We have not therefore the problem of distinguishing Being from not-Being any longer before us. That is settled for us in the experience of life. Our problem merely is within real Being by negation of all relations to isolate the *pure* Being, which must be there to begin with in order to the possibility of entrance into any relations whatever. In forming this conception of pure Being therefore, Thought is quite within its right, although for that which it looks upon as the positive import of the conception it can only offer a name, of which the intelligibility may be fairly reckoned on, not a description.'

Now by way of reply to these objections I must remind the reader that what I disputed was not at all the legitimacy of the formation of the idea in question but only the allowability of the metaphysical use which it is sought to make of it. The point of this distinction I will endeavour first to illustrate by examples. Bodies move in space with various velocities and in various directions. No doubt we are justified as a matter of thought in fixing arbitrarily and one-sidedly now on one common element, now on another, in these various instances, and thus in forming the conception of direction without reference to velocity, that of velocity apart from direction, that of motion as the conception of a change of place, which leaves direction and velocity unnoticed. There is nothing whatever illegitimate in the formation of any of these abstractions. Nor is it incompatible with the nature of the abstractions that instances of each of them should be so connected in thought as to yield further knowledge. None of them, however, immediately and by itself allows of an application to reality without being first restored to combination with the rest from which our Thought, in arbitrary exercise of its right of abstraction, had detached them. There will never be a velocity without direction ; never a direction *ab* in the proper sense of the term without a velocity leading from *a* to *b*, not from *b* to *a*. There will never be a motion that is a mere change of place, as yet without direction and velocity and waiting to assume these two qualifications later on. That which we are here seeking to convey is essentially, if not altogether, the familiar truth that general ideas are not applicable to the real world in their generality, but only become so applicable when each of their marks, that has been left undetermined, has been limited to a completely individual determinateness, or, to use an expression more suited to the case before us, when to each partial conception necessary to the complete definition there has been again supplied in case it expresses a relation, the element to which the relation attaches.

9. We take the case to be just the same with the conception of pure Being. It is an abstraction formed in a perfectly legitimate way, which aims at embracing the common element that is to be found in many cases of Being and that distinguishes them from not-Being. We do not value this abstraction the less because the simplicity of what it contains is such that a verbal indication of this common element, as distinct from any systematic construction of it, is all that is possible. Still, like those to which we compared it, it does not admit, as it stands, of application to anything real. Just as an abstract motion cannot take place, just as it never occurs but in the form of velocity in a definite direction, so pure Being cannot in reality be an antecedent or substance of such a kind as that empirical existence with its manifold determinations should be in any sort a secondary emanation from it, either as its consequence or as its modification. It has no reality except as latent in these particular cases of it, in each of these definite forms of existence. It is merely in the system of our conceptions that these supervene upon it as subsequent and subordinate kinds. There was a correct feeling of this in what I call the natural theory of the world. It was quite aware of the intellectual possibility of detaching the affirmation that is the same in all cases from the differences of the manifold relations which are affirmed by it in the different cases of Being, just as the uniform idea of quantity can be detached from the different numbers and spaces which are subordinate to it. But it rightly held to the view that the pure Being thus constituted has not reality as pure but only in the various instances in which it is a latent element; just as is the case with quantity, which never occurs as pure Quantity but only as this or that definite Quantum of something.

10. The length of this enquiry, which leads to a result seemingly so simple, must be justified by the sequel. It may be useful, I think, to repeat the same thought once again in another form. There are

other terms which have been applied to pure Being, in the desire to
make that which admits of no explanatory analysis at least more
intelligible by a variety of signs. Thus it is usual to speak of it as an
unconditional and irrevocable Position [1] or Putting. It will be readily
noticed that as so applied, each of these terms is used with an extension of meaning in which it ceases to represent any complete
thought. They alike tend to give a sensuous expression to the idea
in question by recalling the import in which they are properly used;
and when that on which their proper meaning rests has again to
be expressly denied the result is obscurity and confusion. We
cannot speak of a putting or Position in the proper sense of the
term without stating what it is that is put. And not only so, this
must be put somewhere, in some place, in some situation which is
the result of the putting and distinguishes the putting that has taken
place from one that has not taken place. Any one who applied this
term to pure Being would therefore very soon find himself pushed
back again to a statement of relations, in order to give to this ' Position' or pure Being the meaning necessary to its distinction from the
not-putting, the pure non-Being. The notion which it is commonly
attempted to substitute for this—that of an act of placing pure and
simple, which leaves out of sight every relation constituted by the act
—remains an abstraction which expresses only the purpose of the
person thinking to think of Being and not of not-Being, while on the
other hand it carefully obliterates the conditions under which this
purpose can attain its end and not the precise opposite of this end.
Nor would it be of any avail to be always reverting to the proposition
that after all it is by this act of putting that there is constituted the
very intelligible though not further analysable idea of an objectivity
which can be ascribed only to that which is, not to nothing. For,
apart from every other consideration, if we in fact not merely performed the act of mere putting, as such, but by it put a definite
content, without however adding what sort of procedure or what
relations were to result to the object from this act of putting, the
consequence would merely be that the thing put would be presented
to our consciousness as an essence which signifies something and
distinguishes itself from something else, but not as one that is in
opposition to that which is not. Real Being, as distinct from the
mere truth of the thinkable, can never be arrived at by this bare act

---

[1] ['Position oder Setzung.' It seems unavoidable that the English word ' Position' should be used, though it has of course no active meaning such as belongs to
' Position' and ' Setzung.']

of putting, but only by the addition in thought of those relations, to be placed in which forms just the prerogative which reality has over cogitability.

The other general signification, which the expressions 'Position' and 'putting' have assumed, illustrates the same state of the case. We cannot affirm simply something, we can only affirm a proposition —not a subject, but only a predicate as belonging to a subject. Now it is psychologically very intelligible that from every act of affirmation we should look for a result, which stands objectively and permanently before thought, while all negation implies the opposite expectation, that something will vanish which previously thus stood before it. It is quite natural therefore that we should fall into the delusion of imagining that in the purpose and good will to affirm there lies a creative force, which if it is directed to no definite predicate but exercised in abstraction would create that universal and pure Being which underlies all determinate Being. In fact however the affirmation does not bring into Being the predicate which forms its object, and it could just as well, though for psychological reasons not so naturally, assert the not-Being of things as their Being. The Being of things, therefore, which is in question, cannot be found in the affirmation of them merely as such but only in the affirmation of their Being. We are thus brought back to the necessity of first determining the sense of this Being in order to the presence of a possible object of the affirmation, and this determination we have, so far at least, found no means of carrying out except by presupposition of relations, in the reality of which the Being of that which is consists in antithesis to the not-Being of that which is not.

11. There is a further reason for avoiding the expression which I have just been examining. 'Position' and 'putting forth' are alike according to their verbal form terms for actions[1]. Now it may seem trifling, but I count it important all the same, to exercise a precaution in the choice of philosophical expressions and not to employ words which almost unavoidably carry with them an association which has a disturbing influence on the treatment of the matter expressed. In the case before us the prejudicial effects apprehended have not remained in abeyance. It has not indeed been believed possible to achieve a putting forth which should create Being : but there was always associated with the application of the word the notion that it has been by a corresponding act, from whomsoever proceeding, that this Being so unaccountably presented to us has originated and that we then

[1] [v. note on p. 32.]

penetrate to its true idea when we repeat in thought this history of its origin. We shall find the importance of this error, if we revert to the reproach brought against the natural theory of the world. It is objected that in looking for the Being of every thing in its relations to other things, it leaves no unconditioned element of reality—none that would not have others for its presupposition. If a can only exist in relation to b, then, it is said, b must be there beforehand; if b exists only in relation to c, then c must be its antecedent. And if perchance there were a last element s dependent not on any further elements but on the first a, this, it will be urged, would only make still more apparent the untenability of a construction of reality which after all has to make the being of a itself the presupposition of this Being. But this whole embarrassment could only be incurred by one, whose problem it was to *make* a world; nor would he incur it, unless a limitation on his mode of operation interfered with the making of many things at the same time and compelled him to let an interval of time elapse in passing from the establishment of the one element to that of the other: for undoubtedly, if Being consists only in the reality of relations, a could not stand by itself and therefore could not exist till the creating hand had completed the condition of its being by the after-creation of b. But what could justify us in importing into the notion of this productive activity this habit of our own thinking faculty, which does, it is true, in presenting relations to itself pass from one point of relation to another? Why should we not rather assume that the things as well as the relations between them were made in a single act, so that none of them needed to wait, as it were hung in the air during a certain interval, for the supplementary fulfilment of the conditions of its reality? We will not attempt however further to depict a process, which cannot be held to be among the objects of possible investigation. It is not our business to discover in what way the reality of things has been brought about, but only to show what it is that it must be thought of and recognised as being when once in some way that we cannot conceive it has come to be. We have not to make a world but so to order our conceptions as that they may correspond without contradiction to the state of the given world as it stands. Such a contradiction we may be inclined to think is involved in the thought of a creative 'Position,' which could only put forth things that really are under the condition of their being mutually related, yet on the other hand could only put them forth one after the other. But there is no contradiction in the recognition of a present world of reality, of which the collective elements are as a matter of fact so conditioned by the tension of mutual relatedness that only in this can the meaning of their Being and its distinction from not-Being be recognised.

12. The foregoing remarks contain an objection to the metaphysical doctrine of Herbart, which requires some further explanation. It need not be said that Herbart never entertained the unphilosophical notion that the irrevocable 'position,' in which he found the true Being of things, was an activity still to be exercised. He too looked on it as a fact to be recognised. As to how the fact came to be so it was in his eyes the more certain that nothing could be said as, being unconditioned and unchangeable according to his understanding of those terms, it excluded every question in regard to origin and source. But a certain ambiguity seems to me to lie in the usage of this expression of an irrevocable 'position.'

There are two *demands* which may no doubt be insisted on. In the first place, assuming that we are in undoubted possession of the true conception of Being, we should be bound to be on our guard in its application against attaching it to qualities which on more exact consideration would be found to contradict it. Nothing can then compel us on this assumption to revoke the affirmation or 'position,' as an act performed by ourselves, by which we recognised the presence in some particular case of that 'position,' not to be performed by us, in which true Being consists. If on the other hand instead of being in possession of the correct conception of Being, we are only just endeavouring to form it, intending at a later stage to look about for cases of its application, in that case we have so to construct it as to express completely what we meant, and necessarily meant, to convey by it. Nothing therefore ought to be able to compel us again to revoke the recognition that in the characteristics found by us there is apprehended the true nature of that position which we have not to make but to accept as the Being presented to us. Here are two sorts of requirement or necessity, but in neither case have we to do with anything except an obligation incumbent on our procedure in thinking. The proposition—Being consists in so and so, and the proposition—this is a case of Being, ought alike to be so formed as that we shall not have to revoke either as premature or incorrect. But as to the nature of Being itself nothing whatever is settled by either requirement and it is not self-evident that the 'position' which constitutes Being and which is not one that waits to be performed by us, is in itself as irrevocable as our thoughts about it should be. The common view of the world does not as a matter of fact, at least at the beginning, make this claim for Being. The fixedness of Being, which it ascribes to things, only amounts to this, that they serve as relatively persistent points on which phenomena fasten and from which occurrences issue. But according to this view if once reason had been found to say of a thing, ' It has been,' it would in spite of this revocation of its further persistence still be held that, so long as it has been, it has had full enjoyment of the genuine and true Being, beside which there is no other specifically different Being.

The question whether such a view is right or wrong I reserve for the present. Herbart decided completely against it. True Being according to him is only conceived with irrevocable correctness, if it is apprehended as itself a wholly irrevocable ' position.' This necessary requirement, however, with him involved the other—the requirement that every relation of the one thing to another, which could be held necessary to the Being of the Thing, should be excluded, and that what we call the true Being should be found only in the pure ' position,' void of relation, which we have not to exercise but to recognise. No doubt it is our duty to seek such a cognition of the real as will not have again to be given up. But I cannot draw the deduction that the object of that cognition must itself be permanent, and therefore I cannot ascribe self-evident truth to this conviction of Herbart's. It is a Metaphysical doctrine in regard to which I shall have more frequent opportunity later on of expressing agreement and hesitation, and which I would now only subject to consideration with reference to the one point, with which we are specially occupied. In order to preserve the connexion of our thoughts, I once again recall the point that the conception of a pure, completely unrelated Being turned out to be correctly formed indeed, but perfectly inapplicable. We were able to accept it only as an expression or indication of that most general affirmation, which is certainly present in every Being, and distinguishes it from not-Being. But we maintained that it is never merely by itself, but only as having definite relations for its object, that this affirmation constitutes the Being of the real ; that thus pure Being neither itself is, nor as naked ' Position' of an unrelated content forms the reality of that content, nor is rightly entitled to the name of Being at all.

13. On the question how determinate or empirical Being issues from pure Being, the earlier theories, which started from the independence of pure Being, pronounced in a merely figurative and incomplete manner. The wished for clearness we find in Herbart. According to his doctrine pure Being does not lie behind in a mythical past. Each individual thing enjoys it continuously, for each thing is in virtue of a 'position' which is alien to all relations and needs them not. It is just the complete indifference of things to all relations, and it alone, that makes it possible for them to enter into various relations towards each other, of which in consequence of this indifference none can in any way add to or detract from the Being of the things. From this commerce between them, which does not touch their essence, arises the chequered variety of the course of the given world.

I cannot persuade myself that this is an admissible way of presenting the case. Granting that there really is such a thing as an element *a* in the enjoyment of this unrelated 'Position' of being unaffected by others and not reacting upon them, it does not indeed contradict the conception of this Being that ideas of relation should afterwards be connected with it. But in reality it is impossible for that to enter into relations which was previously unrelated. For *a* could not enter into relations *in general*. At each moment it could only enter into the definite relation *m* towards the definite element *b*, to the exclusion of every other relation μ towards the same element. There must therefore be some reason in operation which in each individual case allows and brings about the realisation only of *m*, not that of a chance μ. But since *a* is indifferent towards every relation, there cannot be contained in its own nature either the reason for this definite *m*, nor even the reason why it should enter into a relation, that did not previously obtain, with *b* and not rather with *c*. That which decided the point can therefore only be looked for in some earlier relation *l*, which however indifferent it might be to *a* and *b*, in fact subsisted between them. If *a* and *b* had been persistently confined each to its own pure Being, without as yet belonging at all to this empirical reality and its thousandfold order of relations, they would never have issued from their ontological seclusion and been wrought into the web of this universe. For this entry could only have taken place into some region in space, at some point of time, and in a direction somewhither ; and all this would imply a determinate place outside the world, which the things must have left in a determinate direction. Therefore, while thus seemingly put outside the world into the void of pure Being, the Things would have already stood, not outside all relations to the world, but only in other and looser relations instead of in the closer ones, which are supposed to be established later. And just as it would be impossible for them to enter into relations if previously unrelated, so it would be impossible for them wholly to escape again from the web of relations in which they had once become involved.

It may indeed be urged with some plausibility that, since we take the relations of things to be manifold and variable, Being can attach to no single one of them, and therefore to none at all: that therefore it cannot be Being which the Thing loses, if we suppose all its relations successively to disappear. But this argument would only be a repetition of the confusion between the constancy of a general idea and the reality of its individual instances. Colour, for instance, is not necessarily green or red, but it is no colour at all if it is none of these different kinds. Were it conceivably possible that all relations of a thing should disappear without in their disappearance giving rise to new ones—a point of which I reserve the consideration—we could not look upon this as the return of the thing into its pure Being, but only as its lapse into nonentity. A transition, therefore, from a state of unrelatedness into relation, or *vice versa*, is unintelligible to us. All that is intelligible is a transition from one form of relation to another. And an assumption which would find the true Being of Things in their being put forth without relations, seems at the same time to make the conception of these things unavailable for the Metaphysical explanation of the universe, while it was only to render such explanation possible that the supposition that there are Things was made at all.

14. There is yet one way out of the difficulty to be considered. ' In itself,' it may be said, ' pure Being is foreign to all relations, and no Thing, in order to be, has any need whatever of relations. But just because everything is indifferent to them, there is nothing to prevent the assumption that the entry of all things into relations has long ago taken effect. No thing has been left actually to enjoy its pure Being without these relations that are indifferent to it, and it is in this shape of relatedness that the sum of things forms the basis of the world's changeable course.' Or, to adopt what is surely a more correct statement—' It has not been at any particular time in the past that this entry into relations has taken place, which, as we pointed out, is unthinkable. Every thing has stood in relations from eternity. None has ever enjoyed the pure Being which would have been possible for its nature.' In this latter transformation, however, the thought would essentially coincide with that which we alleged in opposition to it. It would amount simply to this, that there *might be* a pure Being, in which Things, isolated and each resting on itself, without any mutual relation, would yet *be*; that there is no such Being, however, but in its stead only that manifoldly determined empirical Being, in each several form of which pure Being is latently present. Between the view thus put and our own there would no longer be any difference, except the first part of the statement, supposing it to be adhered to. A Being, which might be but is not, would for us be no Being at all. The conception of it would only purport to be that of a possibility of thought, not the conception of that reality of which alone Metaphysic professes to treat. We should certainly persist in denying that this pure Being so much as *could be* elsewhere than in our thoughts. We take the notion of such Being to be merely an abstraction which in the process of thinking, and in it only, separates the common affirmation of whatever is real from the particular forms of reality, as applied to which alone the affirmation is itself a reality.

CHAPTER II.

Of the Quality of Things.

15. ACCORDING to the natural theory of the world, as we have so far followed it, the Being of Things is only to be found in the reality of certain relations between one and another. There are two directions therefore in which we are impelled to further enquiry. We may ask in the first place, what is the peculiar nature of these relations, in the affirmation of which Being is supposed to lie? In that case its definition would assign a number of conditions, which whatever is to be a Thing must satisfy. We feel, secondly, with equal strength the need of trying to find first in the conception of the Thing the subject which would be capable of entering into the presupposed relations. The order of these questions does not seem to me other than interchangeable, nor is it indeed possible to keep the answers to them entirely apart. It may be taken as a pardonable liberty of treatment if I give precedence to the second of the mutually implied forms of the problem. It too admits of a double signification. For if we speak of the essence of Things, we mean this expression to convey sometimes that by which Things are distinguished and each is what it is, sometimes that in virtue of which they all are *Things* in opposition to that which is not a Thing. These two questions again are obviously very closely connected, and it might seem that the mention of the first was for us superfluous. For it cannot be the business of ontology to describe the peculiar qualities by which the manifold Things that exist are really distinguished from each other. It could only have to indicate generally what that is on the possible varieties of which it may be possible for distinctions of Things to rest. But this function it seems to fulfil in investigating the common structure of that which constitutes a Thing as such ; for this necessarily includes the idea and nature of that by particularisation of which every individual Thing is able to be what it is and to draw limits between itself and other Things. The sequel of our discussion may however justify our procedure in allowing ourselves to be driven to undertake an answer to this second question by a preliminary attempt at answering the first.

10. What the occasions may be which psychologically give rise in us to the idea of the Thing, is a question by which the objects of our present enquiry are wholly unaffected. The idea having once arisen, and it being impossible for us in our natural view of the world to get rid of it, all that concerns us is to know what we mean by it, and whether we have reason, taking it as it is, for retaining it or for giving it up. As we have seen, sensation is our only warrant for the certainty that something is. It no doubt at the same time warrants the certainty of our own Being as well as that of something other than ourselves. It is necessary, however, in this preliminary consideration to forget the reference to the feeling subject, just as the natural view of the world at first forgets it likewise and loses itself completely in the sensible qualities, of which the revelation before our eyes is at the supposed stage of that view accepted by it as a self-evident fact. It is only in sensation therefore that it can look, whether for the certainty of there being something, or, beyond this, for the qualities of that which is. Yet from its very earliest stage it is far from taking these sensible qualities as identical with that which it regards as the true Being in them. Not till a later stage of reflection is it attempted to maintain that what we take to be the perception of a thing is never more than a plurality of contemporary sensations, held together by nothing but the identity of the place at which they are presented to us, and the unity of our consciousness which binds them together in its intuition. The natural theory of the world never so judges. Undoubtedly it takes a thing to be sweet, red, and warm, but not to be sweetness, redness, and warmth alone. Although it is in these sensible qualities that we find all that we experience of its essence, still this essence does not admit of being exhaustively analysed into them. In order to convey what is in our minds when we predicate such qualities of a Thing, the terms which connote them must, in grammatical language, be construed into objects of that ' *is*,' understood in a transitive sense, which according to the usage of language is only intransitive. The other ways of putting the same proposition, such as ' the thing tastes sweet,' or ' it looks red,' help to show how in the midst of these predicates, as their subject or their active point of departure, the Thing is thought of and its unity not identified with their multiplicity. This idea, however far it may be from being wrought out into clear consciousness, in every case lies at the bottom of our practical procedure where we act aggressively upon the external world,

seeking to get a hold on things, to fashion them, to overcome their resistance according to our purposes.

I need not dwell on the occasions—readily suggesting themselves to the reader—which confirm us in this conception, while at the same time they urgently demand a transformation of it which will make good its defects. Such are the change in the properties in which the nature of a determinate thing previously seemed to consist, and the observation that none belongs to the thing absolutely, but each only under conditions, with the removal of which it disappears. The more necessary the distinction in consequence becomes between the thing itself and its changeable modes of appearance, the more pressing becomes the question, what it is that constitutes the thing itself, in abstraction from its properties. But I do not propose to dwell on the more obvious answers to this question any more than on the occasions which suggest it. Such are the statements that the Thing itself is that which is permanent in the change of these properties, that it is the uniting bond of their multiplicity, the fixed point to which changing states attach themselves and from which effects issue. All this is no doubt really involved in our ordinary conception of the Thing, but all this tells us merely how the true Thing behaves, not what it is. All that these propositions do is to formulate the functions obligatory on that which claims to be recognised as a Thing. They do not state what we want to know, viz. what the Thing must be in order to be able to perform these required functions. I reserve here the question whether and how far we may perhaps in the sequel be compelled, by lack of success in our attempts, to content ourselves with this statement of postulates. The object of ontological thinking is in the first instance to make the discovery on which the possibility of fulfilling the ontological problem depends—to discover the nature of that to which the required unity, permanence, and stability belong.

17. It is admitted that sensation is the single source from which we not only derive assurance of the reality of some Being, but which by the multiplicity of its distinguishable phenomena, homogeneous and heterogeneous, first suggests and gives clearness to the idea of a particular essence [1] which distinguishes itself from some other particular essence. It is quite inevitable therefore that we should attempt to think of the required essence [2] of things after the analogy of this sensible material, so far at any rate as is compatible with the simultaneous

[1] ['Die Vorstellung eines Was, das von einem andern Was sich unterscheidet.']
[2] ['Was.']

problem of avoiding everything which would disqualify sensations for adequately expressing this essence[1].

This attempt has been resolutely made in the ontology of Herbart. To insist on the mere unity, stability, and permanence of Things, was a common-place with every philosophy which spoke of Things at all. It was then left to the imagination to add in thought some content to which these formal characteristics might be applicable. Herbart defines the content. A perfectly simple and positive quality, he holds, is the essence of every single thing, i. e. of every single one among those real essences, to the combinations of which in endless variety we are compelled by a chain of thought, of which the reader can easily supply the missing links, to reduce the seemingly independent 'Things' of ordinary perception. Now if Herbart allows that these simple qualities of Things remain completely unknown to us; that nothing comes to our knowledge but appearances flowing from them as a remote consequence, then any advantage that might otherwise be derived from his view would disappear unless we ventured to look for it in this, that his unknown by being brought under the conception and general character of quality would at least obtain an ontological qualification, by which it would be distinguished from a mere postulate, as being a concrete fulfilment of such postulate.

If however we try to interpret to ourselves what is gained by this subordination, we must certainly confess that Quality in its proper sense is presented to us exclusively in sensations, and in no other instances. Everything else which in a looser way of speaking we so call consists in determinate relations, which we gather up, it is true, in adjectival expressions and treat as properties of their subjects, but of which the proper sense can only be apprehended by a discursive comparison of manifold related elements, not in an intuition. There would be nothing in this, however, to prevent us from generalising the conception of Quality in the manner at which, to meet Herbart's view, we should have to aim. Our own senses offer us impressions which do not admit of comparison. The colour we see is completely heterogeneous to the sound we hear or the flavour we taste. Just as with us, then, the sensations of the eye form a world of their own, into which those of the ear have no entry, so we are prepared to hold of the whole series of our senses that it is not a finished one, and to ascribe to other spirits sensations which remain eternally unknown to us, but of which, notwithstanding, we imagine that to those who are capable of them they would exhibit themselves with the same

[1] ['Wesen.']

character of being vividly and definitely pictured, with which to us the sensations of colour, for instance, appear as revelations of themselves.

It is always difficult in the case of the simplest ideas by the help of words about them to represent the characteristic trait, scarcely expressible but by the ideas themselves, in virtue of which they satisfy certain strongly felt needs of thought. Still I trust to be sufficiently intelligible if I find in the character, just mentioned, of being presentable as a mental picture or image immediately without the help of a discursive process, the reason of our preference for apprehending the essence of a thing under the form of a simple quality. Just as the colour red stands before our consciousness, caring, so to speak, to exhibit nothing but itself, pointing to nothing beyond itself as the condition of its being understood, not constituting a demand that something should exist which has still to be found out, but a complete fulfilment ; so it is thought that the super-sensible Quality of the Thing, simple and self-contained, would reveal its essence, not as something still to be sought for further back, but as finally found and present. And even when further reflection might be supposed to have shaken our faith in the possibility of satisfying this craving for an intuitive knowledge and limited us to laying down mere forms of thinking which determine what the essence of things is not ; even then we constantly revert to this longing for the immediate presentability of this essence, which after all can only be satisfied with the likeness of the *quaesitum* to a sensible quality. We may have to forego intuition ; but we feel its absence as an abiding imperfection of our knowledge.

18. That the demand in question must really be abandoned is not in dispute. Whatever eternal simple and super-sensible Quality we may choose to think of as the essence of the Thing, it will be said that, as a Quality, it always remains in need of a subject, to which it may belong. It may form a How, but not the What of the Thing. It will be something which the Thing has, not which it is.

This objection, familiar as it is to us all, with the new relation which it asserts between Subject and Quality, rests meanwhile on two grounds of which the first does not suffice to render impossible the previously assumed identity of the Thing with its simple quality. In our thought and in its verbal expression, the Qualities—red, sweet, warm—appear as generalities, which await many more precise determinations, in the way of shade, of intensity, of extension, and of form, from something which belongs to the nature of the individual case in which they are sensible, and thus not to the qualities themselves. We thus present them to ourselves in an adjectival form, as not themselves amounting to reality but as capable of being employed by the real, which lies outside them, through special adjustment to clothe its essence; as a store of predicable materials, from which each thing may choose those suitable to the expression of its peculiar nature. Then of course the question is renewed as to the actual essence which with this nature of its own lies behind this surface of Quality.

But we must be on our guard against repeating in this connexion a question which in another form we have already disclaimed. We gave up all pretension of being able to find out how things are made and we confessed that the peculiar affirmation or 'position,' by which the real is eternally distinguishable from the thinkable, may indeed be indicated by us—but that we cannot follow its construction as a process that is taking place. But it is precisely this objection that may now be brought up against us, that we are illegitimately attempting to construe that idea of the Thing, which must comprehend the simple supra-sensible Quality along with its reality, into the history of a process by which the two constituent ideas which make up the idea of the Thing—or rather the objects of these ideas—have come to coincide. For if we maintain the above objection in its full force— [the objection founded on the distinction between the Quality of the Thing and the Thing itself] and refuse to keep reverting to the supposition that some still more subtle quality constitutes the Thing itself, while a quality of the kind just objected to merely serves as a predicate of the Thing, the result will be that we shall have on the one side a Quality still only generally conceived, unlimited, and unformed, as it presents itself merely in thought and therefore still unreal; on the other side a 'position' which is still without any content, a reality which is as yet no one's reality. It would be a hopeless enterprise to try to show how these two—such a quality and such a 'position'— combine, not in our thought to produce an idea of the Thing, but in reality to produce the Thing itself.

This however was not what was meant by the view, which sought to identify the essence of the Thing with its simple supra-sensible Quality. It was emphatically not in the form of a still undetermined generality —not as the redness or sweetness which we think of, but obviously only in that complete determination, in which red or sweet can be the object of an actually present sensation—it was only in this form that the Quality, united with the 'position' spoken of, was thought of as identical with the essential Being (the τί ἐστι) of Things. It was not supposed that there had ever been a process by which the realities signified by these two constituent ideas had come to be united, or by which the complete determinateness of the Quality as forming the essential Being of the Thing, had been elaborated as a secondary modification out of the previous indeterminateness of a general Quality. It is true, that in our usage of terms there unavoidably attaches to the word 'Quality' a notion of dependence, of its requiring the support of a subject beyond it; and it is this notion which occasions 'Quality' to be treated as synonymous with the German 'Eigenschaft[1].' But in truth this impression of its dependence issues only from the general abstraction of Quality, which we form in thought, and is improperly transferred to those completely determined qualities, which form the content of real feelings and constitute the occasions of these abstractions.

19. But, true as this defence of the view referred to may be, we still gain nothing by it. Undoubtedly, if a quality in the complete determinateness which we supposed, simple and unblended with anything else, formed an unchangeable object of our perception, we should have no reason to look for anything else behind it, for a subject to which it attached. But if we just now took this in the sense that this quality might in that case pass directly for the Thing itself, we must now subjoin the counter-remark that in that case, if nothing else were given, we should have no occasion at all to form the conception of a Thing and to identify that quality with it. For the impulse to form the conception and the second of the reasons which forbid the identification of the simple quality with the Thing, lie in the given change. The fact that those qualities which form the immediate objects of our perception, neither persist without change nor change without a principle of change, but always in their transition follow some law of consecutiveness, has led to the attempt to think of the Thing as the persistent subject of this change and of the felt qualities merely as predicates of which one gives place to the other. Whether this attempt is justified at all—whether an entirely different interpretation of the facts of experience ought not to be substituted for it—is a question which we reserve as premature. For the present our business is only to consider in what more definite form this assumption of Things, in case it is to be retained, must be presented to thought, if it is to render that service to our cognition for the sake of which it is made; if, i. e., it is to make the fact of change thinkable without contradiction.

[1] [lit. 'Property.']

And in regard to this point I can only maintain that speculative philosophy, while trying to find a unity of essence under change, was wrong
in believing that this unity was to be found in a simplicity, which in
its nature is incapable of being a unity or of forming the persistent
essence of the changeable.   Change of a thing is only to be found
where an essence a, which previously was in the state a^1, remains
identical with itself while passing into the state a^2.   In this connexion I
still leave quite on one side the difficulties which lie in the conception,
apparently so simple, of a state.   For the present it may suffice to
remark that we are obliged by the notion we attach to the term 'state'
to say not that the essence is identically like [1] itself, but only that it is
identical with itself, in its various states.   For no one will deny that a,
if it finds itself in the state a^1, cannot be taken to be exactly like a^2,
without again cancelling the difference of the states, which has been
assumed.   All that we gain by the distinction, however, is, to begin
with, two words.   For the question still remains : In what sense can
that at different moments remain identical with itself, which yet in one
of these moments is not identically like itself as it was in the other?
It is scarcely necessary to remark how entirely unprofitable the
answers are which in the ordinary course of thought are commonly
given to this question ; such as, The essence always remains the
same with itself, only the phenomenon changes ; the matter remains
the same, the form alters ; essential properties persist, but many unessential ones come and go ; the Thing itself abides, only its states
are variable.   All these expressions presuppose what we want to
know.   We have here pairs of related points, of which one term corresponds in each case to the Thing a, the other is one of its states
a^1, a^2.   How can the first member a of these pairs be identical with
itself, if the several second members are not identical with each
other, and if, notwithstanding, the relation between the two members
of each pair is to be maintained, in the sense that the second member,
which is the Form, the Phenomenon, the State, is to be Form, Phenomenon, or State of the first member?

So long as we are dealing with the compounded visible things of

---

[1] ['Gleichheit,' used here, and in §§ 59 and 268, with a strict insistance on all
that is involved in its meaning of equality; viz. on the qualitative likeness, without
which comparison by measurement is impossible.   Thus in the places referred to
the terms which are 'gleich' are a and a, and neither 'equal' nor 'like' translates
'gleich' adequately; it includes both.   'Identity' was used in Logic, § 335 ff.,
but will not do here, because of the contrast with the continued identity, 'Identität,'
imputed to a *thing.*]

common perception, the pressure of this difficulty is but slight. In such cases we look upon a connected plurality of Predicates *pqr*, as the essence of a thing. This coherent stock may not only assume and again cast off variable additions, *s* and *t*, but it may in itself by the internal transposition of its components in *qrp, rpq, prq*, experience something which we might call its own alteration in opposition to the mere variation of those external relations. Or finally it may be the form of combination that remains the same, while the elements themselves, *p q* and *r*, vary within certain limits. In these cases the imagination still finds the two sides of its object before it, and can ascribe to one of them the identity[1], to the other the difference[2]. What justifies it in understanding the fluctuations of that which does not remain exactly like itself as a series of states of the Identical, is a question which is left to take care of itself. The difficulty involved in it comes plainly into view if we pass from the apparent things of perception to those which we might in truth regard as independent elements in the order of the Universe, and we think of each of these as determined by a simple quality, *a*. The simple, if it alters at all, alters altogether, and in the transition from *a* to *b*, there remains nothing over to which the essence would withdraw, as to the kernel that remains the same in the process of change. Only a succession, *abc*, of different essences — one passing away, the other coming into being — would be left, and with this disappearance of all continuity between the different appearances there would disappear the only reason which led us to regard them as resting on subject Things.

20. This inference cannot be invalidated by an objection which readily suggests itself and which I have here other reasons for noticing. It is to the instance of sensations that we must constantly revert, if we would explain to ourselves what supra-sensible Qualities really mean to us when we combine them with sensations under the common idea of Quality. Let us then take a simple Red colour, *a*, in which we find no mixture with other colours, still less a combination of other colours, as representing the manner in which the simple quality, *a*, of an essence would appear to us, if it were perceivable by the senses. It will then be argued as follow : If this Red passes into an equally simple Yellow, there still undoubtedly remains a common element, which we feel in both colours, though it is inseparable from *a* and *b*, the universal *C* of colour. Neither the redness in the red, nor that which makes the yellow what it is, has any existence either in

---

[1] ['Identität.']   • [2] ['Ungleichheit.']

fact or in thought apart from the luminous appearance in which the nature of colour consists, nor has this appearance any existence of its own other than in the redness or yellowness. On the contrary its whole nature shows itself now in one colour, now in the other. In the same way the essence of the thing will now be the perfectly simple *a*, now the equally simple *b*, without this implying a disappearance of the common *C*, the presence of which entitles us to regard *a* and *b* merely as its varying states or predicates. It would be idle to meet this argument by saying that the common element *C* of colour is only a product of our intellectual process of comparison ; nay, not even such a product, but merely the name for the demand, simply unrealisable, which we make upon our intellect to possess itself of this common element presumed to be present in red and yellow, in detachment from both colours. For the fact, it might be replied, would still remain that we should not make this impracticable demand, if it were not felt in the perception of red and yellow, 'There is something there, which we look for though we do not find it as anything perceivable or separate, this common *C*, for which we have made the name colour.'

Now since we readily forego the pretension of apprehending the essence of things in the way of actual intuition, and confine ourselves to enquiring for the form of thought under which we have to conceive its unknown nature, we might certainly continue to look upon the comparison just stated as conveying the true image of the matter in hand, i. e. the image of that relation, in which the simple essence stands to its changeable states. We might at the same time regard this analogy of our sensations as a proof of the fact that the demand which we make upon the nature of things for an identity within the difference does not, as such, transgress the limits of the actually possible. In more detail the case might be put thus : What may be the look of that persistent *C*, which maintains itself in the change of the simple qualities of the Thing, of this it is true we have no knowledge, and we as little expect to know it as we insist on seeing the general colour *C*, which maintains itself in the transition from Red to Yellow. The mere fact, however, that in order to render this transition possible the continuous existence of this universal is not merely demanded without evidence by our thought, but is immediately testified to by sensation as plainly present though not separable from particular sensible objects—this proves to us that the continuance of a common element in a series of different and absolutely simple members is at any rate something possible, and not a combination of words to which no real instance could correspond.

21. The above will, I hope, have made plain the meaning of this rejoinder. I should wish ultimately to show that it is inapplicable, but before I attempt this, I may be allowed to avail myself of it for the purpose of more exactly defining certain points so as to save the necessity of enlarged explanations further on. When in our comparison we chose to pass from the simple quality red to another equally simple, to point to yellow as this second quality seemed a selection which might be made without hesitation. But sour or sweet might equally have presented themselves. It was only the former transition, however, (from red to yellow) which left something actually in common between the different members ; while the second on the contrary (from red to sweet) would have left no other community than that which belongs to our subjective feeling as directed to those members. Our selection therefore was natural, for we knew what the point was at which we wished to arrive and allowed ourselves to be directed by this reference. The fact however that the other order of procedure is one which we can equally present to ourselves reminds us that the transition from one simple quality to another is not in every case possible without loss of the common element C. This however is no valid objection. It will be at once replied that in speaking of change it has always been understood that its course was thus limited to certain definite directions. No one who takes the essence of a thing to admit of change can think of it as changeable without measure and without principle. To do so would be again to abolish the very reason that compelled us to assign the succession of varying phenomena to a real subject in the Thing ; for that reason lay merely in the consecutiveness with which definite transitions take place while others remain excluded. The only sense therefore that has ever attached to the conception of change, the only sense in which it will be the object of our further consideration, is that in which it indicates transformations or movements of a thing within a limited sphere of qualities. Beyond this will be another equally limited sphere of qualities, forming the range within which another essence undergoes change, but it is understood that in change the thing never passes over from one sphere into the other. As regards the more precise definition of these spheres, our comparison with colours can only serve as a figure or illustration. As colour shifts to and fro from one of its hues to another, without ever approximating to sounds or passing into them, it serves well as a sensible image of that limitation of range which we have in view. But this does not settle the question whether the various forms $a^1 a^2 a^3 \ldots$, into which

the essence *a* might change now and again, are kinds of a common *C* only in the same sense in which the colours are so, or whether they are really connected with each other in some different form, which logical subordination under the same generic idea does not adequately symbolise.

22. It is time, however, to show the unsatisfactoriness of this attempt to justify a belief in the capacity for change on the part of a Thing, of which the essence was confined to a perfectly simple Quality. *If* our imagination ranges through the multiplicity of sensible qualities, it finds certain groups of these within which it succeeds in arresting a common element *C*, while beyond them it fails to do so. This was the point of departure of our previous argument. Passing from this consideration of an intellectual process to consideration of the Thing, we said; '*if* the essence of a thing changes, the limitation within itself of such a sphere of states affords it the possibility of completing its change within the sphere without loss of its abiding nature *C*. Only if it passed beyond these limits would all continuity disappear and a new essence take its place.' Very well; but what correspondence is there between these two ' if's ' which we allowed to follow each other as if completely homogeneous ? The former refers to a movement of our intellect. Meanwhile the object presented to the intellect stands before it completely unmoved. The general colour, of which we think, is not sometimes Red, sometimes Yellow, but is always simultaneously present in each of these colours and in each of the other hues, which we class together as equally external primary species of colour. In the Thing, however, the supposed *C* cannot be made so simply to stand towards the manifold $a^1 a^2 a^3$ in the relation of a universal kind to its species. Even were it the case that in respect of their nature $a^1 a^2 a^3$ admit of being regarded as species of *C*, still, *if* the thing changes, they are not contained in it, as in a universal *C*, with the eternal simultaneity of species that exist one along with the other. They succeed each other, and the essence *a*, if it is a^1, for that reason excludes from itself a^2 and a^3. Thus it is just this that remains to be asked, how that second *if* can be understood ; how we are to conceive the state of the case by which it comes about that the thing moves—moves, if you like, within a circumscribed sphere of qualities $a^1 a^2 a^3 \ldots$, but still within it does move, and so passes from one to the other of the qualities as that, being in the one, it excludes the others ; how it is that it so moves while yet these qualities are the species of a universal *C*, eternally simultaneous and only differing as parts of a system. And, be it observed, we are at present not enquiring for a cause which produces this motion, but only how the essence a is to be thought of, in case the motion takes place. This question we cannot answer without coming to the conclusion that the change is not reconcilable with the assumption of a simple quality, constituting this essence. At the moment when a has the form a^1 and in consequence excludes the forms a^2 and a^3, it cannot without reservation be identified with a C, which includes $a^1 a^2 a^3$ equally in itself. It would have to be C^1 in order to be a^1, C^2 in order to be a^2, and the same course of changes which we wished to combine with a persistent simple quality would find its way backwards into this quality itself.

23. I could not avoid the appearance of idle subtlety if I pursued this course of thought without having shown that it is forced upon us. Why, it will be asked, do we trouble ourselves, out of obstinate partiality for the common view, to give a shape to the idea of the Thing in which it may include the capacity of change? Why do we not follow the enlightened view of men of science which finds no difficulty in explaining the multiplicity of phenomena by the help of changeable relations between unchangeable elements? There is the more reason for the question since this supposition not only forms the basis of the actual procedure of natural science but is precisely that for which Herbart has enforced respect on the part of every metaphysical enquirer.

Let us pursue it then in the definite form which this philosopher has given to it. According to him, not only as a matter of fact do elements, which undergo no change in the course of nature, underlie phenomena, but according to their idea the real essences, the true things which we have to substitute for the apparent things of perception, are unchangeably identical with themselves, each resting on itself, standing in need of no relation to each other in order to their Being, but for that reason the more capable of entering into every kind of relation to each other. Of their simple qualities we have no knowledge, but undoubtedly we are entitled to think of them as different from each other and even as opposed in various degrees without being obliged in consequence to transfer any such predicates, supposing them to be found by our comparison, to the qualities themselves as belonging to their essence; as if, that is, some of the qualities were actively negated by others, and some were presupposed by and because of others. This admission made, let us suppose that two essences, A and B, come into that relation M to each other which Herbart describes as their being *together*. I postpone my remarks about the proper sense of this 'together.' All that we now know of it is that it is the condition under which what Herbart considers to be the indifference of essences towards each other ceases. Supposing them then to be '*together*,' it might happen that A and B without detriment to their simplicity might yet be representable by the compound equivalent expressions $a+\gamma$ and $\beta-\gamma$. In that case the continuance of this state of being 'together' would require the simultaneous subsistence of $+\gamma$ and $-\gamma$; i.e. the continuance of two opposites, which if we put them together in thought, seem necessarily to cancel each other. But they cannot really do so. Neither are the simple essences A and B according to their nature accessible to a change, nor are the opposite elements which our Thought, in its comparing process, might distinguish in them, actually separable from the rest, in combination with which they belong to two absolutely simple and indivisible Qualities.

' But, if this be so, nothing happens at all and everything remains as it is!' This is the exclamation which Herbart expects to hear, but he adds that we only use such language because we are in full sail for the abyss which should have been avoided. I must however repeat it. What has taken place has been this. We, the thinkers, have imagined that from the contact of opposites there arose some danger for the continuance of the real essences. We have then reminded ourselves that their nature is inaccessible to this danger. Thus it has been we who have maintained the *conception* of the real essence in its integrity against the falsification which would have invaded it in every attempt to account its object capable of being affected by any disturbance from without. This has taken place in our thought, but in the essence itself nothing has in fact happened. The name of self-maintenance, which Herbart gives to this behaviour on the part of the Things, can at this stage of his theory as yet mean nothing but the completely undisturbed continuance of that which in its nature is inaccessible to every disturbance that might threaten it. An activity issuing from the essences, a function exercised by them, it indicates as little as a real event which might occur to them. And just for this reason the multiplicity of kinds and modes, in which Herbart would have it that this self-maintenance takes effect, cannot really exist for it. The undisturbed continuance is always the same, and except the variation of the external relations, through which the so-called ' being together' of the essences is brought about and again annulled, nothing new whatever in consequence of this being 'together' happens in the universe.

24. Quite different from this sense of self-maintenance, which

Herbart himself expressly allows in the Metaphysic, is that other sense in which he applies the same conception in the Psychology. Only the investigator of Nature could have satisfied himself with the conclusion just referred to. For him the only concern is to ascertain the external processes, on which for us the change in the qualitatively different properties of things as a matter of fact depends. It is no part of his task to enquire in what way these processes, supposing them to take place, bring it about that there is such a thing as an appearance to us. If it is the belief of the students of Natural Science that the theory, which regards all those processes as mere changes in the relations of elements themselves unchangeable, is adequate for its purpose—though in the sequel I shall have to deny that according to this way of presenting the case any but an incomplete view even of the course of external nature is possible—yet for the present I am ready to allow that there may be apparent success upon this method in the attempt to eliminate all changes on the part of the real itself from the course of the outer world.

But this only renders the admission of change a yet more inevitable necessity, if we bear in mind that the *entire* order of the universe which forms the object of Metaphysical enquiry includes the origin of the phenomenon in us no less than the external processes which are its *de facto* conditions. Thus, if the physical investigator explains the qualitative change of things as mere appearance, the metaphysician has to consider how an appearance is possible. Herbart is quite right—and I do not for the present trouble myself with the reproaches which might be brought against this point of his doctrine—in assuming the simple real essence of the soul as the indispensable subject, for which alone an appearance can arise. Whereas in regard to no other real essence do we know in what its self-maintenance consists, this, according to him, is clear in regard to the soul. Each of its primary acts of self-maintenance, he holds, has the form of an idea, i.e. of a simple sensation. Between these aboriginal processes there take place a multitude of actions and reactions, from which is supposed to result, in a manner which we need not here pursue in detail, the varied whole of the inner life. These acts of self-maintenance on the part of the soul, however—consisting at one time in a sensation, at another in the hearing of a sound; now in the perception of a flavour, now in that of warmth—are manifestly no longer simple continuations of the imperturbable essence of the soul. Taking a direction in kind and form according to the kind and form of the threatening disturbance, they are functions, activities, or reactions of the soul, which are not possible to an unchangeable but only to a changeable Being. For it is not in a merely threatened disturbance but only in one which has actually taken effect that the ground can lie of the definite reaction, which ensues at every moment to the exclusion of many others that, as far as the nature of the soul goes, are equally possible for it. In order to be able to meet the threatened disturbance *a* by an act of self-maintenance *a*, the other disturbance *b* by another act *β*, the soul must take some note of the fact that at the given moment it is *a* and not *b*, or *b* and not *a*, that demands the exercise of its activity. It must therefore itself suffer in both cases, and differently in one case from the other. This change on its own part—I say change, for it would be useless to seek to deny that various kinds of suffering are inconceivable without various kinds of change on the part of the subject suffering—cannot be replaced by the mere change in the relations between the soul unchanged in itself, and other elements. Any such relation would only be a fact for a second observer, which might awaken in him the appearance of a change taking place in the observed soul, which in reality does not take place: but even for this observer the appearance could only arise, if he on his own part at least actually possessed that capability of change which in the observed soul he holds to be a mere appearance.

It is therefore quite impossible entirely to banish the inner liability to change on the part of the real from an explanation of the course of the universe. If it were feasible to exclude it from a theory of the outer world, it would belong the more inevitably to the essence of that real Being, for which this outer world is an object of perception. But, once admitted in this position, it cannot be a self-evident impossibility for the real elements, which we regard as the vehicles of natural operations. That, on the contrary, it is a necessity even for these, we shall try to show later on.

Our consideration of the question, however, so far rests on a certain supposition; on the necessity, in order to render the fact of appearance intelligible, of conceiving a simple real subject, the soul. There is no need for me here to justify this assumption against the objections which are specially directed against it. It is no object of our enquiry, so far, to decide whether the conception of Things is tenable at all; whether it does not require to be superseded by another conception. I repeat; it is only *in case* Things are to be taken to exist and to serve to make the world intelligible, that we then enquire in what way they must be thought of. And to that question we have given the answer that Essence, Thing or Substance, can only be that which admits of Change. Only the predicates of Things are unchangeable. They vary indeed in their applicability to Things, but each of them remains eternally the same with itself. It is only the Things that change, as they admit of and reject now one predicate, now another. This thought indeed is not new. It has already been expressly stated by Aristotle. For us, however, it necessarily raises at once questions that are new.

CHAPTER III.

Of the Real and Reality.

25. The changes which we see going on, and the consecutiveness which we believe to be discoverable in them, compelled us to assume the existence of Things, as the sustainers or causes of this continuity. The next step was, if possible, to ascend from that which needs explanation to the unconditioned, in regard to which only recognition is possible. For this purpose we tried to think of the Thing as unchangeably the same with itself, and, impressed with the need of assimilating the idea of it as much as possible to what is contained in sensation, since sensation alone actually gives us an independent something instead of merely requiring it, we took its nature to consist in a simple quality. We convinced ourselves, however, that an unchangeable and simple quality is not thinkable as a subject of changeable states or appearances, and thus we are compelled to give up the claim to any such immediate cognition as might reveal the essence of Things to us in a simple perception. I do not mean to imply by this that we should have hoped really to attain this perception. But we indulged the thought that, for such a spirit as might be capable of it, there would be nothing in the essence of Things incompatible with their being thus apprehended. This conviction in its turn we have now to abandon. In its very nature that which is to be a Thing in the sense of being a subject of change would repel the possibility of being presented as an unmoving object of any intuition. A new form has therefore to be sought for that which is to be accounted the essence of any Thing; and in order to find it we again take our departure from that natural theory of the world which without doubt has tried answers of its own to all these questions that are constantly reasserting themselves with fresh insistance.

26. In regard to the common objects of perception we answer the question, What are they? in two ways, of which one soon reduces itself to the other. Products of art, which exhibit a purpose on the part of a maker, we denote by reference to the end for which they are intended, setting aside the variety of forms in which they fulfil that end. The changeable products of nature, in the structure of which a governing purpose is more or less obscure to us, we characterise according to the kind and order of phenomena into which they develope of themselves or which could be elicited from them by external conditions. In both cases by the essence of the thing that we are in quest of we understand the properties and modes of procedure, by which the Thing is distinguished from other things. The other series of answers, on the contrary, exhibits as this essence the material out of which the things are made, overlooking the various kinds of behaviour and existence to which in the case of each thing the particular formation of this material gives rise. Yet after all this second mode of answering the question ultimately passes over into the former. It satisfies only so long as it consists in a reduction of a compound to more simple components. Supposing us to have discovered this simple matter, how then do we answer the question, What after all is the simple matter itself? What for instance is the Quicksilver, of which we will suppose ourselves to have discovered that something else consists of it? So long as our concern was to reduce this other thing to it, it was taken for something simple. But itself in its simplicity, what is it? We find it fluid at our ordinary temperatures, fixed at lower temperatures, vaporous at higher ones; but we could not say what it is in itself, supposing it not to be acted on by any of these external conditions or by any of the other conditions, under which its phenomenal properties change in yet other ways.

We can in fact only answer, that it is in itself the unassignable something, which under one condition appears as a^1, under another as a^2, under a third as a^3, and of which we assume that, if these conditions succeed each other in reverse order, it will pass again from a^3 into a^2 and a^1, without ever being converted into β^1, β^2 or β^3—forms which in a like mutual connexion exhibit the various phenomena of another thing, say Silver. Thus, it may be stated as a general truth, that our idea of that which makes a Thing what it is consists only in the thought of a certain regularity with which it changes to and fro within a limited circle of states whether spontaneously or under visible external conditions, without passing out of this circle, and without ever having an existence on its own account and apart from any one of the forms which within this circle it can assume. This way of presenting the case, while fully sufficient for the needs of ordinary judgment, has given occasion to various further metaphysical experiments.

27. If attention is directed to the qualities by which one Thing distinguishes itself from another, its essence in this sense cannot any longer be thought of as object of a simple perception, but only in the logical form of a conception, which expresses the permanently uniform observance of law in the succession of various states or in the combination of manifold predicates. From this point a very natural course of thought leads us to two ways of apprehending the Thing. We may define it first by the collective marks, which at a given moment it exhibits, in their *de facto* condition. This gives us a statement of what the essence is, τὸ τί ἐστι according to Aristotle's expression. But it would be conceivable that, like two curves which have an infinitely small part of their course in common, so two different things, *A* and *B*, should coincide in the momentary condition of their marks, but should afterwards diverge into paths of development as different as were the paths that brought them to the state of coincidence. In that case the essence of each will be held only to be correctly apprehended, if the given condition of each is interpreted as the result of that which it previously was, and at the same time as the germ of that which it will be. This seems the natural point of departure from which Aristotle arrived at the formula τί ἦν εἶναι. He did not complete it by the other equally valuable τί ἔσται εἶναι, though the notion that might have been so expressed was not alien to his way of thinking. In practice, it must be admitted, these determinations of the idea of the Thing, which theoretically are of interest, cannot be carried through. Even the actual present condition of a Thing would not admit of exhaustive analysis, without our thinking of the mutual connexion between the manifold phenomena which it exhibits, as already specifically ordered according to the same law which would appear still more plainly upon a consideration of the various states, past and to be expected, of the Thing. The second formula therefore only gives general expression to the intention of constantly gaining a deeper view of the essence of the Things, in a progression which admits of indefinite continuance, while a fuller regard is for ever being paid to the multiplicity of the different ways in which the Thing behaves under different conditions, to its connexion with the rest of the world, and lastly—according to a direction of enquiry very natural, though still out of place in this part of Metaphysics—to the final purpose of which the fulfilment is the Thing's vocation in the universe. As a means of setting aside the difficulties, which beset us at this point, the expressions referred to have not in fact been used, nor do they seem at all available for the purpose.

28. We proceed to particularise some of these. Had we succeeded in making the essential idea of a thing so completely our own, that all modes of procedure of the .thing under all conditions would flow from the idea self-evidently as its necessary consequences, we should after all in so doing have only attained an intellectual image of that by which as by its *essentia* the Thing is distinguished from everything else. The old question would repeat itself, what it is which makes the thing itself more than this its image in thought, or what makes the object of our idea of the thing more than thinkable, and gives it a place as a real thing in the world. Just as the Quality demanded a Subject to which it might attach, so still more does the idea, less independent than the quality, seem to require a fixed kernel to give its matter that reality which, as the material contained in an idea, it does not possess. If we have once forbidden ourselves to look for the essence of the Thing in a simple uniform quality that may be grasped in perception ; if we resolved rather to find an expression for it in the law which governs the succession of its phenomena ; then that which we are in quest of has to fulfil for all things the same indistinguishable function. Itself without constituent qualities it has to give reality to the varying qualities constituent of things. We are thus brought to the notion of a material of reality, a Real pure and simple, which in itself is neither this nor that, but the principle of reality for everything.

The history of Philosophy might recount numerous forms under which this notion has been renewed ; but it is needless to treat them here in detail. The natural requirements of the case have always led, when once this path has been entered on, to the same general determinations as Plato assigned to this ὕλη. The consideration that observation presents us with an indefinite number of mutually independent Things, permanent or transitory, caused this primary matter of all things to be regarded by the imagination as divisible, in order that there might be a piece of it in each single thing, sufficient to stiffen the thing's ideal content into reality. But this conception of divisibility in its turn had to be to a certain extent withdrawn. For it would imply that before its division the matter has possessed a continuity, and this would be unthinkable without the assumption of its having properties of some kind, by which it would have been possible for this material of reality to be distinguished from other thinkable materials. But thus understood, as already definitely qualified, it would not have disposed of the metaphysical question which it was meant to solve. For the question was not, what quality of primary matter as a matter-of-fact formed the basis of the individual things that fashion themselves out of it, but what it is that is needed to help any and every thinkable quality to be more than thinkable, to be real. If therefore the imagination did notwithstanding, as we do not doubt that it did, present this ultimate Real to itself mainly as a continuous and divisible substance, this delineation of it, occasioned by reference to the observation of natural objects, strictly speaking went beyond that which in this connexion it was intended to postulate. All that had to be supposed was the presence in every single thing, however many things there might be, of such a kernel of reality, wholly void of properties. There were therefore according to this notion an indefinite number of instances of this conception of the real, but they did not stand in any connexion with each other any more than in any other case many instances of a general idea, merely because they are all subordinate to that idea, stand in any actual connexion with each other. But I will not continue this line of remark; for the obscurity of this whole conception is not to be got rid of by criticism, but by pointing out its entire uselessness.

29. It is manifest that a representation which has its value in the treatment of ordinary objects of experience, has been applied to a metaphysical question, which it is wholly insufficient to answer. In sensuous perception we are presented with materials, which assume under our hands such forms as we will, or are transformed by operations of nature into things of the most various appearance. But a little attention informs us that they are but relatively formless and undetermined. The possibility of assuming new forms and of manifold transmutation they all owe to the perfectly determinate properties which they possess, and by which they offer definite points of contact to the conditions operating on them. The wax, which to the ancients represented the primary matter on which the ideas were supposed to be impressed in order to their realisation, would not take this impression, and would not retain the form impressed on it but for the peculiar unelastic ductility and the cohesion of its minute parts, and any finer material which we might be inclined to substitute for it, though it might possess a still more many-sided plasticity, would at the same time be still less capable of preserving the form communicated to it.

It is therefore a complete delusion to hope by this way of ascent to arrive at something which, without any qualification on its own part, should still bear this character of pure receptivity, necessary to the Real we are in quest of. After all we should only arrive at a barren matter R, which would be equally incapable of receiving a definite shape, and of duly retaining it when received. For that which was without any nature of its own different from everything else, could not be acted on by any condition p at all, nor by any condition p otherwise than by another q. No position of circumstances therefore would ever occur under which that indeterminate subject R could be any more compelled or entitled to assume a certain form π rather than any other we like, κ. If we supposed however this unthinkable event to come about and R to be brought into the form π, there would be nothing to move it to the retention of this form to the exclusion of any other, κ, since every other would be equally possible and equally indifferent to it. In this absence of any resistance, which could only rest on some nature of R's own, every possibility of an ordered course of the world would disappear. In every moment of time everything that was thinkable at all would have an equal claim to reality, and there would be none of that predominance of one condition over another which is indispensable to account for any one state of things or to bring about a determinate change of any state of things. But not only would any origin or preservation of individual forms be reduced to nothing by the complete absence of qualities on the part of the Real. The relation itself, which at each moment must be supposed to obtain between it and the content to which it gives reality, would from a metaphysical point of view be unmeaning. Words no doubt may be found by which to indicate it metaphorically. We speak of the properties which constitute the whole essence of a Thing, as inhering in the unqualified substance of the Real, or as attaching to it, or as sustained by it. But all these figurative expressions with the use of which language cannot dispense, are in contradiction with the presupposed emptiness and formlessness of the matter. Nothing can sustain anything, or allow it to attach to or depend upon itself, which does not by its own form and powers afford this other points of contact and support. Or, to speak without a figure, it is impossible to see what inner relation could be meant, if we ascribed to a certain Real a property π or a group of properties π as its own. R would be as void of relation to the property or group of properties, as alien to it, as any other R[1].

30. These shortcomings on the part of the conception of the Real would make themselves acutely felt as soon as an attempt was made,

not merely to set it up in isolated abstraction, but to turn it to account for the actual explanation of the course of things. It would then become evident that nothing could be built on it which had any likeness to a Static or Mechanic of change. But it will be objected that we are fighting here against ghosts raised by ourselves, so long as we speak of processes by which the connexion of the real with the qualities it contains is supposed for the first time to have come about. This, however, it will be said, is what has never been meant. Even the ancients, who originated the conception of matter in question, we find were aware that at no place or time did the naked and unformed matter exist by itself. It had existed from eternity in union with the Forms, by means of which the different Things, now this, now that, had been fashioned out of it. In the plainest way it was stated that, taken by itself, it was rather without being, a μὴ ὄν, and that Being first arose out of its indefeasible union with the qualitative content supplied by the Ideas. This may be fairly urged, and in this explanation we might perfectly acquiesce, if it were one that really admitted of being taken at its word. If it were so taken, it would amount simply to a confession that what the theory understood and looked for under the designation of the Real is nothing more than the 'Position,' throughout inseparable from the constituent qualities of Being, by which these qualities not merely are thought of but *are ;* and that consequently it would be improper for this ' Position,' which only in thought can be detached as the uniform mode of putting forth from that which is put forth by it, to be regarded in a substantive character as itself a something, a Real, the truly existing Thing; improper that, compared with it, everything which on other grounds we took to form the essence of the Thing, should be forced into the secondary position of an unessential appendage.

The doctrines, however, which speak of the real material of Being, are far from conveying this unreserved admission even in the explanation adduced. On the contrary, they continue to interpret the distinction between the principle that gives reality and the real itself as if it represented something actual. When they ascribe to the matter, which has no independent existence, successive changes of form, they do not merely mean by this that the inexplicable ' Position' passes from the content π to the other content κ. In that case all that would be attained would be a succession, regulated or unregulated, of states of fact without inner connexion. Their object rather is to be able to treat the matter R as the really permanent connecting member which experiences π and κ, or exchanges the one for the

other, as states of itself, and which, in virtue of its own nature, forbids
the assumption of other phenomena ϕ and ψ, or the realisation of
another order of succession. Without this last addition the conception
of the Real R would not, upon this view any more than upon other,
have any value. For I repeat, it is only under the obligation of explaining a particular consecutiveness in the course of the world,
which does not allow any and every thinkable variation in the
state of facts, that we are constrained, instead of resting in the
phenomena, to look for something behind them under the name
of the Real, however that is to be conceived. A flux of absolute
becoming without any principle, once allowed, demands no explanation and needs no assumption to be made which could lead to such an
explanation, intrinsically impossible, as the one given. The doctrines
in question, therefore, under the guidance of this natural need which
they think to satisfy by the supposition of the Real pure and simple,
do not in fact make the admission which they seem to make. Although their 'matter' R nowhere exists in its nakedness, this is, so to
speak, only a fact in the world's history, which need not follow from
the idea of R. Although as a matter of fact everywhere imprisoned
in variously qualified forms, still in all those forms R continues to
exist as the single self-subsistent independent Being and imparts its
own reality to the content which changes in dependence on it. Thus
the matter, considered by itself and in detachment from the forms in
which it appears, is still not properly, as it is called, a μὴ ὄν, but
according to the proper sense even of the doctrines which so designate
it, merely an οὐκ ὄν, if weight may be laid on the selection of these
expressions. And against this permanent residuum of the doctrine of
the ὕλη the objections already made retain their force. It is impossible
to transfer the responsibility of providing for the reality of the determinate content to a Real without content, understood in a substantive
sense, for none of the connecting thoughts are possible which would
be needed in order to bring this Real into the desired relation with
the qualities assigned to it.

81. I cannot therefore believe that interpreters, as they went deeper
into this ancient notion of an empty Real as such, of an existing
nothing which yet purports to be the ground of reality to all definite
Being, would find in it a proportionately deeper truth. To us it is
only an example of an error of thought, which is made too often and
too easily not to deserve an often-repeated notice. If we ask whence
the colour of a body proceeds, we usually think at first of a pigment
which we suppose to communicate the colour to it. And in this we

are often right; for in compound things it may easily be that a property, which seems to be spread over the whole of them, attaches
only to a single constituent. But we are wrong already in as far as
our phrase implies that the pigment communicates its colour to the
whole body. Nothing of the sort really happens, but a combination
of physical effects brings it about that in our sensation the impression
of colour produced by the pigment completely disguises the other
impression, which would have been produced by the other constituents
of the body, that have throughout remained colourless. But when we
repeat our question, it appears that the same answer cannot always
be repeated. The pigment cannot owe its colour to a new pigment.
Sooner or later the colouring must be admitted as the immediate
result of the properties which a body possesses on its own account
as its proper nature, and does not borrow from anything else.

Our procedure has been just the same with reference to the things
and their reality. We desired to know whence their common property of reality is derived, and in imagination introduced into each of
them a grain of the stuff of reality which we supposed to communicate to the properties gathered about it the fixedness and consistency
of a Thing. What actual behaviour, however, or what process this
expression of 'communication' so easily used, is to signify, remained
more than we could say. In fact, just as little as a pigment would
really convey its colouring to anything else, could the mere presence
of the Real convey the reality, which is emphatically held to be
peculiar to it, to an essence in the way of qualities, which, we are to
suppose, have somehow grouped themselves around it. Indeed, the
metaphysical representation is in much worse case than that which
we made use of in the example just instanced. For of the pigment
we did not dream that it was itself not merely colourless, but in its
nature completely indifferent to the various colours that may be
thought of, and that it proceeded to assume one of them as if the
colours, before they were properties of a thing, already possessed a
reality which enabled them to enter into a relation to bodies and to
let themselves be assumed by bodies. In this case we were aware
that the Redness, which we ascribe to the pigment, is the immediate
result of its own nature under definite circumstances; that it could
not exist, that nothing could have it, until these circumstances acted
on this nature, and that it would change if the body, instead of being
what it is, were another equally determinate body. But in our metaphysical language, when we spoke of the properties in opposition to
the real essence of things, we in fact spoke as if the thinkable quali-

ties, by which one thing is distinguished from another, before they really existed as qualities of a Thing might already possess a reality which should enable them to enter into a definite relation to an empty Real—a relation by which, without having any foundation more than all other qualities in the nature of this Real, it was possible for them to become *its* properties.

I leave this comparison, however, to be pursued on another occasion. Apart from figure, our mistake was this. We demanded to know what it is on which that Being of Things which makes them Things rests. By way of answer we invented the Substantive conception of the Real pure and simple, and believed that by it we had represented a real object, or rather the ultimate Real itself. In fact however *real* is an adjectival or predicative conception, a title belonging to everything that in some manner not yet explained behaves as a Thing—changes, that is to say, in a regular order, remains identical with itself in its various states, acts and suffers; for it is this that we assumed to be the case with Things, *supposing that there are Things*. The question was, on what ground this actual behaviour rests. It is a question that cannot be settled by thinking of our whole requirement as satisfied in general by the assumption of a Real as such, of which after all, as has been shown, we could not point out how in each single case it explains the reality which itself is never presented to us as universal and homogeneous, but only as a sum of innumerable different individual cases.

The conception of the Real therefore is liable to a criticism similar to though somewhat different from that which is called for by the conception of pure Being. This latter we found correctly formed, but inapplicable, so long as the definite relations are not made good again, which had been suppressed in it by the process of abstraction. Of the conception of the Real on the contrary it may be maintained that it is untruly formed. That which is conceived in this conception everywhere presupposes the subject to which it may belong, and cannot itself be subject. For this reason it cannot be spoken of in substantive form as the Real, but only applied adjectivally to all that is real. It would be well if the usage of language favoured this way of speaking, more lengthy though it is, in order to keep the thought constantly alive that it is not through the presence of a Real in them that Things become or are real, but that primarily they are only called real if they exhibit that mode of behaviour which we denominate reality. In regard to this we have stated what we mean by it. The mode under which it may be thinkable has still to be ascertained.

32. With a view to answering the above question we are naturally led to the opposite path to that hitherto pursued. Let us see how far it will take us. The two incomplete ideas, by the union of which we form the conception of the Thing—that of the content by which it is distinguished from other things and that of its reality—cannot be any longer taken to represent two actually separable elements of its Being. The Reality must simply be the form in which the content actually exists, and can be nothing apart from it. But the requirement that this should be so meets at once with a serious objection. So long as we could answer the question *What* the Thing is by calling it a simple quality, we had a uniform content, apprehensible in intuition, before us, to which it seemed, to begin with at least, that the ' Position ' of reality might be applied without contradiction. We have now decided that this essence is only to be found in a law, according to which the changeable states, properties or phenomena, $a^1 a^2 a^3$ of the thing, are connected with each other. But how could a law be that which, if simply endowed with reality, would constitute a thing? How could it be gifted with those modes of behaviour which we demand of whatever claims to be a Thing?

This question involves real difficulty, but it also expresses doubts which merely arise from a scarcely avoidable imperfection in our linguistic usage. The first of these doubts is analogous to that which we raised against the simple Quality as essence of the Thing, and which we found to have no justification. As long as we thought of the Quality in the way presented to us in language by adjectives, as a generality abstracted from many instances, distinct indeed from other qualities but undetermined in respect of intensity, extent and limitation ; so long it could not be accepted as the essence of a Thing. After all the determinateness still lacking to it had been made good, it might have been so accepted, if the necessary requirement of capability of change had not prevented this. In like manner the conception of law is at the outset understood in a similar general sense. Abstracted from a comparison between the modes of behaviour of different things, it represents primarily the rule, according to which from a definite general class of conditions a definite class of results is derived. The rule indeed is such that there is a permanent proportion according to which definite changes in the results correspond to definite changes in the conditions ; but the cases in which the law will hold good, and the determined values of the conditions which give rise in each of these cases to equally determined values on the part of the effects—these are not contained in the law itself or contained in it only as possibilities which are thought of along with it, but of which it asserts none as a fact. In this shape a law cannot be that of which the immediate reality, even if it were thinkable, would form a Thing. But this is not what is meant by the theories which employ such an expression [which identify thing and law]. What they have in view, to put it shortly, is not a general law but an instance of its application. This latter expression, however, needs further explanation and limitation.

33. If in the ordinary general expression of a law, for all quantities left indefinite, we substitute definite values, it is not our habit, it is true, to call the individual instance thus obtained any longer a law at all, because unless we revert to the general form of which it is an application it is no longer fitted to serve as a ground of judgment upon other like cases, and this assistance in reasoning is the chief service which in ordinary thinking we expect from a law. Intrinsically, however, there is no such real difference between the individual instance and the universal as would forbid us from subsuming the former under the name of Law. On the contrary, it is itself what it is in respect of its whole nature only in consequence of the law, and conversely the law has no other reality but in the case of its application. It is therefore a legitimate extension of the usage of terms, if we apply the name of a law to the definite state of facts itself, which includes a plurality of relations between elements which are combined according to the dictates of the general law. It may be the general law of a series of quantities that each sequent member is the nth power of the preceding one. It is not, however, in this general form that the law forms a series. We have no series till we introduce in place of n a definite value, and at the same time to give to some one of the members, say the first, a definite quantitative value. Applying this to our present case, the general law would correspond only to the abstract conception of a *Thing as such;* the actual series on the other hand, which this laws governs, to the conception of some individual Thing. And it is only in this latter sense as corresponding to the actual series that it can be intended to represent a law as being the essence to which 'Position' as a Thing belongs.

Upon this illustration two remarks have to be added. In our parallel the definite series appears as an example of a general law, of which innumerable other examples are equally possible. It may turn out in the sequel that this thought has an equally necessary place in the metaphysical treatment of things; but at this point it is still foreign to our enquiry. It does not belong to that essence of a thing of which we are here in quest, that the law which orders its content should apply also to the content of other things. On the contrary, it is completely individual and.single of its kind, distinguishing this thing from all other things. On this point we are often in error, misled by the universal tendency to construct reality out of the abstractions, which the reality itself has alone enabled us to form. The course, which investigation cannot avoid taking, thoroughly accustoms us to look on general laws as the *Prius*, to which the manifold facts of the real world must afterwards, as a matter of course, subordinate themselves as instances. We might, however, easily remind ourselves that as a matter of fact all general laws arise in our minds from the comparison of individual cases. These are the real Prius, and the general law which we develope from them is primarily only a product of our thought. Its validity in reference to many cases is established by the experiences from the comparison of which it has arisen, and is established just so far as these confirm it. Had our comparison, instead of being between one thing and other things, been a comparison of a thing with itself in various states—and that is the sort of comparison to which alone our present course of enquiry would properly lead—then it would by no means have been self-evident that the consecutiveness and conformity to law, which we had found to obtain between the successive states of the one thing, must be transferable to the relations between any other elements whatever they might be, and thus to the states and nature of another thing. We should have no right therefore to regard the essence of the Thing as an instance of a universal law to which it was subject. At the same time it is obvious that this law of the succession of states in a single thing, wholly individual as it is, if it were apprehended in thought, would continue logically to present itself to us as an idea, of which there might be many precisely similar copies. It is quite possible to attempt to make plurals even of the idea of the universe and of the supreme Being. It is considerations in a different region, not logical but material, that alone exclude the possibility of there being such plurals; and it is these alone which in our Metaphysic can in the sequel decide for or against the multiplicity of precisely similar things, for or against the validity of universal laws which they have to obey. To make my meaning clearer, I will supplement the previous illustration of a numerical series by another. We may compare the essence of a thing to a melody. It is not disputed that the successive sounds of a melody are governed by a law of æsthetic consecutive-

ness, but this law is at the same time recognised as one perfectly
individual. There is no sense in regarding a particular melody as a
kind, or instance of the application, of a general melody. Leaving
to the reader's reflection the task, which might be a long one, of
making good the shortcomings from which this illustration, like the
previous one, suffers, I proceed to the second supplementary remark
which I have to make.

If we develope a general law from the comparison of different
things under different circumstances, two points are left undetermined—one, the specific nature of the things, the other, the particular character of the conditions under which the things will behave
in one way or in another. Let both points be determined, and we
arrive at that result, identical with itself and unchangeable, which we
represented by comparison with a definite series of quantities, but
which cannot answer our purpose—the purpose of apprehending that
essence of the Thing which remains uniform in change. We have
therefore, as already remarked, only to carry out the comparison of a
thing with itself in its various states. The consecutiveness and conformity to law, that would thus appear, would be the individual law or
essence of the Thing in opposition to the changeable conditions that
have now to be left undetermined. One more misunderstanding I
should like to get rid of in conclusion. It is no part of our present
question whether and how this comparison and the discovery of the
abiding law is possible for us with reference to any particular thing.
Our problem merely is to find the form of thought in which its
essence could be adequately apprehended supposing there to be no
hindrance in the nature of our cognition and in its position towards
Things to the performance of the process. The same reserve is made
by every other metaphysical view. Even the man who looks for the
essence of the Thing in a simple Quality does not expect to know
that Quality and therefore satisfies himself with establishing the
general form in which it would appear to him, but denies himself
the prospect of ever looking on this appearance.

84. So much for those objections to the notion of a law as constituting the essence of the Thing, which admit of being set aside by
an explanation of our meaning. In fact, *if* we thought of the
'Position' which conveys reality as lighting upon this individual law,
it would form just that permanent yet changeable essence[1] of a Thing
which we are in search of. The reader, however, will find little satisfaction in all this. The question keeps recurring whether after all

---

[1] ['Das beständige und dennoch veränderliche Was.']

that 'Position' of reality, applied to this content, can in fact exhaustively constitute the essence of a real Thing; whether we have not constantly to search afresh for the something which, while following this law, would convey to it—convey to what is in itself a merely thinkable mode of procedure—reality? In presence of this constantly recurring doubt I have no course but to repeat the answer which I believe to be certainly true. Let us, in the first place, recall the fact that in what we are now asking for there is something intrinsically unthinkable. We are not satisfied with the doctrine that the Thing *is* an individual law. We believe that we gain something by assuming of it that in its own nature it is something more and other than this, and that its conformity to this law, by which it distinguishes itself from everything else, is merely its mode of procedure.

Can we however form any notion of what constitutes the process which we indicate by this familiar name of conformity to law? If this nucleus of reality, which we deem it necessary to seek for, possessed a definite nature, alien to that which the law enjoins, how could it nevertheless come to adjust itself to the law? And if we would assume that there are sundry conditions of which the operation upon it might compel it to such obedience, would this compulsion be itself intelligible, unless its own nature gave it the law that upon these conditions supervening it should obey that other law supposed to be quite alien to its nature? In any case that which we call conformity to law on the part of a Thing would be nothing else than the proper being and behaviour of the Thing itself. On the other side: What exactly are we to take the laws to be before they are conformed to? What sort of reality, other than that of the Things, could belong to them, such as they must certainly have if it is to be possible for a nature of Things, assumed hitherto to lie beyond them, to adjust itself to them? There is only one answer possible to these questions. It is not the case that the things follow a mode of procedure which would in any possible form be actually separable from them. Their procedure is whatever it may be, and by it they yield the result which we afterwards, upon reflective comparison, conceive as their mode of procedure and thereupon endow in our thought with priority to the Things themselves, as if it were the pattern after which they had guided themselves. If we would avoid this conclusion by denying to the required nucleus of the Thing any nature of its own, we should be brought back to that conception of the absolute Real, *R*, which we have already found so useless. Even if this real Nothing were itself thinkable, it would certainly not be capable of distributing the reality, which it is supposed to have of its own, over the content which forms the essence of a determinate Thing. It could not therefore represent our *quaesitum*, the something of which we require a so-called conformity to a determinate mode of procedure. There is therefore, it is clear, nothing left for us but to attempt to defend the proposition, that the real Thing is nothing but the realised individual law of its procedure.

35. I shall be less wearisome if I connect my further reflections on the subject with an historical antithesis of theories. Idealism and Realism have always been looked upon as two opposite poles of the movement of philosophical thought, each having different though closely connected significations, according as the enquiry into what really is, or the reference to that which is to be valued and striven after in life, was the more prominent. The opposition was in the first instance occasioned by the question which now occupies us. In the inexhaustible multiplicity of perceivable phenomena Plato noticed the recurrence of certain uniform Predicates, forming the permanent store from which, in endless variety of combination, all things derive their particular essence or the nature by which one distinguishes itself from the other and each is what it is. And just as the simple elements, so the real combinations of these which the course of nature exhibited, were no multiplicity without a Principle, but were subject on their own part to permanent types, within which they moved. Further, the series of relations, into which the different things might enter with each other—ultimately even the multiplicity of that world which our own action might and should institute—testified no less to this inner order of all reality. The case was not such as the Sophists, his predecessors in philosophy, had tried to make it out to be. It was not the case that a stream of Becoming, with no check upon its waves, flowed on into ever new forms, unheard of before, without obligation to return again to a state the same with or like to that from which it set out. On the contrary, everything which it was to be possible for Reality to bring about was confined within fixed limits. Only an immeasurable multiplicity of places, of times, and of combinations remained open to it, in which it repeated with variations this content of the Ideal world.

The full value of this metaphysical conception I shall have to bring out later. For the present I wish to call attention to the misleading path, never actually avoided, into which it has drawn men astray. It was just the multiplicity in space and time of scattered successive and intersecting phenomena—the course of things—that properly constituted the true reality, the primary object given us to be perceived and known. That world of Ideas, on the other hand, which comprehended the permanent element in this changing multiplicity and the recurrent forms in the transmutation of the manifold, was in contrast with it something secondary, having had its origin in the comparisons instituted by our thought, and, so far as of this origin, neither real nor calculated to produce in turn any reality out of itself. However great the value of the observation that Reality is such as to enable us by the connexion of those ideas of ours to arrive at a correspondence with its course; still it was wrong to take this world of ideas for anything else than a system of abstractions or intellectual forms, which only have reality so far as they can be considered the modes of procedure of the things themselves, but which could in no sense be opposed to the course of things as a *Prius* to which this course adjusts itself, completely or incompletely, as something secondary.

In order to make my meaning quite clear, I must emphasize the proposition that the only reality given us, the true reality, includes as an inseparable part of itself this varying flow of phenomena in space and time, this course of Things that happen. This ceaselessly advancing melody of event—it and nothing else—is the metaphysical place in which the connectedness of the world of Ideas, the multiplicity of its harmonious relations, not only is found by us but alone has its reality. Within this reality single products and single occurrences might be legitimately regarded as transitory instances, upon which the world of ideas impressed itself and from which it again withdrew: for before and after and beside them the living Idea remained active and present in innumerable other instances, and while changing its forms never disappeared from reality. But the whole of reality, the whole of this world, known and unknown together, could not properly be separated from the world of Ideas as though it were possible for the latter to exist and hold good on its own account before realising itself in the given world, and as though there might have been innumerable equivalent instances — innumerable other worlds—besides this, in which the antecedent system of pure Ideas might equally have realised itself. Just as the truth about the individual Thing is not that there is first the conception of the Thing which ordains how it is to be, and that afterwards there comes the mere unintelligible fact, which obeys this conception, but that the conception is nothing more than the life of the real itself; so none of the Ideas is an antecedent pattern, to be imitated by what is. Rather, each Idea is the imitation essayed by Thought of one of the

traits in which the eternally real expresses itself. If the individual
Ideas appear to us as generalities, to which innumerable instances
correspond, we have to ascribe this also to the nature of that supreme
Idea, into which we gather the individual Ideas. The very meaning
of there being such an Idea is that a stream of phenomena does not
whirl on into the immeasurable with no identity in successive
moments, without ever returning to what it was before and without
relationship between its manifold elements. The generality of the
Ideas therefore is implied in the systematic character of what fills
the universe, in the inner design of the pattern, of which the unbroken reality and realisation constitute the world. It is completely
misinterpreted as an outline-sketch of what might be in impeachment
of what is—of a possibility which, in order to arrive at reality, would
require the help of a second Cosmos, of a real and of movements of
the real that are no part of itself.

86. I shall have frequent opportunity in the sequel of dwelling
again on this system of thought; nor in fact can I hope to make it
perfectly clear till I shall have handled in detail the manifold difficulties which oppose a return to it. I say expressly—a return to it;
for to me it seems the simplest and most primary truth, while to representatives of the present intricate phase of scientific opinion it
usually appears a rash and obscure imagination. Psychologically it
is almost an unavoidable necessity that the general laws, which we
have obtained from comparison of phenomena, should present themselves to us as an independent and ordaining *Prius*, which precedes
the cases of its application. For in relation to the movement of our
cognition they are really so. But if by their help we calculate a future
*result beforehand from the given present conditions, we forget that
what comes first in our reflection as a major premiss is yet only the
expression of the past and of that nature of its own which Reality in
the past revealed to us. So accustomed are we to this misunderstanding, so mastered by the habit of first setting what is in truth the
essence of the Real over against the Real, as an external ideal for it
to strive after, and of then fruitlessly seeking for means to unite what
has been improperly separated, that every assertion of the original
unity of that which has been thus sundered appears detrimental to the
scientific accuracy to which we aspire. True, the need of blending
Ideal and Real, as the phrase is, has at all times been keenly felt;
but it seems to me that the attempts to fulfil this problem have sometimes promoted the error which they combated. In demanding a
special act of speculation in order to achieve this great result, they

maintain the belief in a gulf, not really there, which it needs a bold leap to pass.

For the present, however, I propose to drop these general considerations, and, if possible, to get rid of the obscurity and apparent inadmissibility of the result just arrived at. One improvement is directly suggested by what has been said. We cannot express our Thesis, as we did just now, in the form : ' The Thing is the realised individual law of its behaviour.' This expression, if we weigh its terms, would contain all the false notions against which we were anxious to guard. Instead of the ' *realised* law' it would clearly be better to speak of the law never realised, but that always has been real. But no verbal expression that we could find would serve the purpose of excluding the suggested notion which we wish to be expressly excluded. For in speaking of a law, we did not mean one which, though real as a law, had still to wait to be followed, but one followed eternally ; and so followed that the law with the following of it was not a mere fact or an event that takes place, but a self-completing activity. And this activity, once more, we look upon not in the nature of a behaviour separable from the essence which so behaves, but as forming the essence itself—the essence not being a dead point behind the activity, but identical with it. But however fain we might be to speak of a real Law, of a living active Idea, in order the better to express our thought, language would always compel us to put two words together, on which the ordinary course of thinking has stamped two incompatible and contradictory meanings. We therefore have to give up the pretension of remaining in complete accord with the usage of speech.

CHAPTER IV.

Of Becoming and Change.

37. WHEN I first ventured, many years ago, on a statement of metaphysical convictions, I gathered up the essence of the thoughts, with which we were just then occupied, in the following proposition : 'It is not in virtue of a substance contained in them that Things are ; they are, when they are qualified to produce an appearance of there being a substance in them.' I was found fault with at the time on two grounds. It was said that the proposition was materially untrue, and that in respect of form the two members of the proposition appeared not to correspond as antitheses. The latter objection would have been unimportant, if true : but I have not been able to convince myself of its truth, or of the material incorrectness of my expression. According to a very common usage the name ' Substance ' was employed to indicate a rigid real nucleus, which was taken, as a self-evident truth, to possess the stability of Reality—a stability which could not be admitted as belonging to the things that change and differ from each other without special justification being demanded of its possibility. From such nuclei the Reality was supposed to spread itself over the different properties by which one thing distinguishes itself from another. It was thus by its means, as if it was a coagulative agent, which served to set what was in itself the unstable fluid of the qualitative content, that this content was supposed to acquire the form and steadfastness that belong to the Thing. It was matter of indifference whether this peculiar crystallisation was thought of as an occurrence that had once taken place and had given an origin in time to Things, or whether the solidifying operation of the substance was regarded as an eternal process, carried on in things equally eternal and without origin in time as an essential characteristic of their nature. In either case the causal relation remained the same. It was by means of a substance empty in itself that Reality, with its fixedness in the course of changes, was supposed to be lent to the determinate content.

I believe myself to have shown that no one of the thoughts involved in this view is possible. In going on, however, to supplement the conclusion that it is not in virtue of a substance that Things are, by the further proposition that, if they are qualified to produce an appearance of the substance being in them, then they are, I did not intend any correspondence between this and the other member of the antithesis in the sense of opposing to the rejected construction of that which makes a Thing a Thing another like construction. What I intended was to substitute for every such *construction* (which is an impossibility) that which alone is possible, the *definition* of what constitutes the Thing. The notion which it was sought to convey could only be this, that when we speak of something that makes a Thing, as such ('die Dingheit'), we mean the form of real existence belonging to a content, of which the behaviour presents to us the appearance of a substance being present in it; the truth being that the holding-ground which under this designation of substance we suppose to be supplied to Things is merely the manner of holding itself exhibited by that which we seek to support in this impossible way.

88. There was no great difficulty in showing the unthinkableness of the supposed real-in-itself. The denial is easy, but is the affirmation of a tenable view equally easy? Setting aside the auxiliary conception just excluded, have we other and better means—are we left with means that still satisfy us—of explaining the functions which we cannot but continue still to expect of Things, if the assumption of their existence is to satisfy the demands for the sake of which it was made? On this question doubts will arise even for a man who resolves to adopt by way of experiment the result of the previous considerations. I repeat: A world of unmoved ideal contents, if it were thinkable without presupposing motion at least on the part of him to whom it was object of observation, would contain nothing to occasion a quest for Things behind this given multiplicity. Nor is it the mere variety of these phenomena, but only the regularity of some kind perceived or surmised in it, that compels us to the assumption of persistent principles by which the manifold is connected.

Common opinion, under a mistake soon refuted, had thought to find these subjects of change in the Things perceivable by the senses. For these we substituted supra-sensible essences of perfectly simple quality. But the very simplicity of these would have made any alternative but Being or not-Being impossible for them, and would thus have excluded change. Yet change must really take place somewhere, if only to render possible the appearance of change some-

where else. Then we gave up seeking the permanent element of
Things in a state of facts always identical with itself, and credited
ourselves with finding it in the very heart of change, as the uniform
import of a Law, which connects a multiplicity of states into one
rounded whole. Even thus, however, it seemed that only an expression had been gained for that in virtue of which each Thing is
what it is, and distinguishes itself from what it is not. As to the
question how an essence so constituted can partake of existence in
the form of a Thing, there remained a doubt which, being insufficiently
silenced, evoked the attempt to represent the real-in-itself as the un-
yielding stem to which all qualities, with their variation, were related
as the changeable foliage. The attempt has failed, and leaves us still
in presence of the same doubt. The first point to be met is this : *If*
we think of change as taking place, then the law which comprehends
its various phases as members of the same series will serve to
represent the constant character of the Thing which persists throughout the change ; but how can we think the change itself, which we
thus presuppose ? How think its limitation to these connected
members of a series ? And then we shall have to ask : Would the
regularity in the succession of the several states a^1, a^2, a^3 . . . really
amount to that which, conceived as persistence of a Thing, we believe
it necessary to seek for in order to the explanation of phenomena ?
These questions will be the object of our next consideration.

89. Under the name 'change,' in the first place, there lurks a
difficulty, which we must bring into view. It conveys the notion that
the new real, as other than something else, is only the continuation of
a previous reality. It tends to avoid the notion of a naked coming
into being, which would imply the origin of something real out of a
complete absence of reality. Yet after all it is only the distinctive
nature of the new that can anyhow be thought of as contained in the
previously existing. The reality of the new, on the other hand, is not
contained in the reality of the old. It presupposes the removal of that
reality as the beginning of its own. It thus beyond a doubt *becomes*
(comes into being) in that sense of the term which it is sought to
avoid. It is just this that constitutes the distinction between the
object of Metaphysic and that world of ideas, in which the content of
a truth *a* is indeed founded on that of another *b*, but, far from arising
out of the annihilation of *b*, holds good along with it in eternal
validity.

If now we enquire, how this becoming, involved in every change, is
to be thought of, what we want to know, as we naturally suppose, is

not a process by which it comes about. The necessity would be too obvious of again assuming the unintelligible becoming in this process by which we would make it intelligible. Nor can even the notion of becoming be represented as made up of simpler notions without the same mistake. In each of its forms, origination and decay, it is easy to find a unity of Being and not-Being. But the precise sense in which the wide-reaching term 'Unity' would have in this connexion to be taken, would not be that of coincidence, but only that of transition from the one to the other, and thus would already include the essential character of becoming. There is no alternative but to give up the attempt at definition of the notion as well as at construction of the thing, and to recognise Becoming, like Being, as a given perceivable fact of the cosmos.

Only on one side is it more than object of barren curiosity. It may appear to contain a contradiction of the law of Identity, or at least of the deductions thought to be derivable from this law. No doubt this law in the abstract sense, which I previously stated[1], holds good of every object that can be presented to thought. a will never cease to $= a$ till it ceases to be. That which is, never is anything that is not, so long as it is at all. On the same principle that which becomes, originates, passes away, is only something that becomes so long as it is becoming, only something that originates so long as it originates, only something that passes away so long as it passes away. There does not therefore follow from the law of Identity anything whatever in regard to the reality of any m. Let m be what it will, it will be $= m$, *in case* it is and so long as it is. But whether it is, and whether, once being, it must always be, is a point on which the principle of Identity does not directly decide at all. Yet such an inference from it is attempted. Because the conception of Being, like every other conception, has an unchangeable import, it is thought that the reality, which the conception indicates, must belong as unchangeably to that to which it once belongs. The doctrines of the irremoveability and indiscerptibility of everything that truly is are thus constantly recurrent products of the movement of metaphysical thought.

But this inference is limited without clear justification to the subsistence of the Things on which the course of nature is supposed to rest. That relations and states of Things come into Being and pass away is admitted without scruple as a self-evident truth. It is true that without this admission the content of our experience could not be presented to the mind at all. If, however, it were the principle of

[1] [Logic, § 55.]

Identity that required the indestructibility of Things, the same principle would also require the unchangeableness of all relations and states. For of everything, not merely of the special form of reality, it demands permanent equality with itself. This consideration might lead us to repeat the old attempts at a denial of all Becoming, or—since it cannot be denied—to undertake the self-contradictory task of explaining at least the becoming of the appearance of an unreal becoming. But if we refuse to draw this inference from the principle of Identity, then that persistency in the Being of Things, which we hitherto tacitly presupposed, needs in its turn to be established on special metaphysical grounds, and the question arises whether the difficult task of reconciling it with the undeniable fact of change cannot be altogether avoided by adopting an entirely opposite point of view.

40. This question has in fact already been often enough answered in the affirmative. Theories have been advanced in the history of Thought, which would allow of no fixed Being and reduced everything to ceaseless Becoming. They issued, however,—as the enthusiasm with which they were generally propounded was enough to suggest— from more complex motives than we can here examine. We must limit ourselves to following the more restricted range of thoughts within which we have so far moved. Still, we too have seen reason to hold that it is an impossible division of labour to refer the maintenance of the unity which we seek for in succession to the rigid unalterableness of real elements, and the production of succession merely to the fluctuation of external relations between these elements. Change must find its way to the inside of Being. We therefore agree with the last-mentioned theorists in thinking it worth while to attempt the resolution of all Being into Becoming, and in the interpretation of its permanence, wherever it appears, as merely a particular form of Becoming; as a constantly repeated origination and decay of Things exactly alike, not as a continuance of the same Thing unmoved. But it would be useless to speak of Becoming without at the same time adding a more precise definition. Neither do we find in experience an origination without limit of everything from everything, nor, if we did find it, would its nature permit it to be the object of scientific enquiry, or serve as a principle of any explanation. Even those theorists who found enthusiastic delight in the sense of the unrestrained mobility enjoyed by the Becoming which they held in honour as contrasted with the lifeless rigidity of Being—even they, though they have set such value on the inexhaustible variety of

Becoming, and on its marvellous complications, have yet never held its eternal flux to be accidental or without direction. Even in Heraclitus we meet with plain reference to inexorable laws which govern it. It is only, then, as involving this representation of a definite tendency that the conception of Becoming merits further metaphysical examination.

41. The thought just stated first had clear expression given it by Aristotle in his antithesis of δύναμις and ἐνέργεια. The undirected stream of event he encloses, so to speak, within banks, and determines what is possible and what is impossible in it. For what he wishes to convey is not merely the modest truth, that anything which is to be real must be possible. It is of this possibility rather that he maintains that it cannot be understood as a mere possibility of thought, but must itself be understood as a reality. A Thing exists δυνάμει when the conditions are really formed beforehand for its admission as an element of reality at some later period, while that alone can exist ἐνεργείᾳ, of which a δύναμις is contained in something else already existing ἐνεργείᾳ. Thus all Becoming is characterised throughout by a fixed law, which only allows the origination of real from real, nay more, of the determinate from the determinate. We have here the first form of a principle of Sufficient Reason, transferred from the connected world of Ideas to the world of events. The first conscious assertion of a truth, which human thought has made unconscious use of from the beginning, is always to be looked on with respect as a philosophical achievement, even if it does not offer the further fruits which one would fain gather from it. Barren in detail, however, these two Aristotelian conceptions certainly are, however valuable the general principle which they indicate. They would only be applicable on two conditions; if they were followed by some specific rule as to what sequent can be contained δυνάμει in what antecedent, and if it could be shown what is that *C* which must supervene in order to give reality to the possible transition from δύναμις into ἐνέργεια.

To find a solution of the first problem has been the effort of centuries, and it is still unfound. On the second point a clearer explanation might have been wished for. The examples of which Aristotle avails himself include two cases which it is worth while to distinguish. If the stones lying about are δυνάμει the house, or the block of marble δυνάμει the statue, both stones and marble await the exertion of activity from without, to make that out of them ἐνεργείᾳ which indeed admits of being made out of them but into which they do not develope themselves. They are possibilities of something future because they are available for that something if made use of by a form-giving motion. On the other hand, if the soul is the activity of the living body, it is in another sense that the body is δυνάμει the soul. It does not wait to have the end to which it is to shape itself determined from without, as the stone waits for external handling to be worked into a house or into a statue. On the contrary it involves in itself the necessary C, the active impulse which presses forward to the realisation of that single end, of which the conditions are involved in it to the exclusion of all other ends. Each case is metaphysically important. The first is in point where we have to deal with the connexion between different elements of which one acts on the other and with the conveyance of a motion to something which as yet is without the motion. The second case apart from anything else involves the question, on which we propose to employ ourselves in the immediate sequel: granted that a thing a, instead of awaiting from without the determination of that which it is to become, contains in its own nature the principle of a and the principle of exclusion of every β, how comes it about that this is not the end of the matter but that the a of which the principle is present proceeds to come into actual being, and ceases to exist merely in principle?

42. I shall most easily explain at once the meaning of this question and the reason for propounding it, by adducing a simple answer, which we might be tempted to employ by way of setting the question aside as superfluous. It is self-evident, we might say, that a proceeds from a because a conditions this a and nothing but this a, not any β. Now it is obvious that this answer is only a repetition of the questionable supposition which we just made. The very point we wanted to ascertain was, what process it is in the thing that *in reality* compels the conditioned to issue from that which conditions it, as necessarily as in our thought the consciousness of the truth of the proposition which asserts the condition carries with it the certainty of the truth of that which asserts the conditioned. We do not in this case any more than elsewhere cherish the unreasonable object of finding out the means by which in any case a realised condition succeeds further in realising its consequence. But to point to it as a self-evident truth that one fact should in reality call another into being, if to the eye of thought they are related as reason and consequence, is no settlement of our question. I reserve for the present the enquiry into the manner in which we think in any case of the intelligible nature of a consequence F as contained in the nature of its reason G[1]. Whatever

[1] [G and F refer to the German words used here 'Grund' and 'Folge.']

this relation may be, the mere fact that it obtains does not suffice to make the idea of F arise out of G even in our consciousness. Were it so, every truth would be immediately apparent to us. No round-about road of enquiry would be needed for its discovery, nor should we even have a motive to seek for it. The universe of all truths connected in the way of reason and consequent would stand before our consciousness, so long as we thought at all, in constant clearness. But this is not the case. Even in us the idea of the consequence F arises out of that of its reason G only because the nature of our soul, with the peculiar unity which characterises it, is so conditioned by particular accompanying circumstances, p, that it cannot rest in the idea of G and, supposing no other circumstances, q, to condition it otherwise, cannot but pass on account of its own essence to the idea of F—to that and no other. In the absence of those accompanying conditions, p, which consist in the whole situation of our soul for the moment, the impulse to this movement is absent likewise; and for that reason innumerable ideas pass away in our consciousness without evoking images of the innumerable consequences, F, of which the content is in principle involved in what these ideas contain. If instead of the conditions, p, those other circumstances, q, are present —consisting equally in the general situation of the soul for the moment—then the movement may indeed arise but it does not necessarily issue in the idea of F. It may at any moment experience a diversion from this goal. This is the usual reason of the distraction and wandering of our thoughts. It is never directly by the logical affinity and concatenation of their thinkable objects that their course is determined but by the psychological connexion of our ideas, so far as these are the momentary states of our own nature. Of the connexion of reason and consequence in Things we never recognise more than just so much as the like connexion on the part of our own states enables us to see of it.

It is not enough therefore to appeal to the principle, that the content of G in itself, logically or necessarily, conditions that of F, and that therefore in reality also F will ensue upon G. The question rather is why the Things trouble themselves about this connexion between necessities of thought; why they do not allow the principle G which they contain to be for ever a barren principle, but actually procure for it the consequence F which it requires; in other words, what addition of a complementary C must be supposed in order that the Things in their real being may pass from G to F just as our thought—not always or unconditionally—passes from the knowledge of G to the knowledge of F.

43. We are thus brought back to a proposition which I shall often in the sequel have occasion to repeat: namely that the error lies just in this, in first setting up in thought an abstract series of principles and consequences as a law-giving power, to which it is supposed that every world that may possibly be created must be subject, and in then adding that, as a matter of self-evidence, the real process of becoming can and must *in concreto* strike only into those paths which that abstract system of law has marked out beforehand. It will never be intelligible whence the conformity of Things to rules of intellectual necessity should arise, unless their own nature itself consists in such conformity. Or, to put the matter more correctly, as I stated in detail above (**34**); it is just this real nature of things that is the First in Being—nay the only Being. Those necessary laws are images in thought of this nature, secondary repetitions of its original procedure. It is only for our cognition that they appear as antecedent patterns which the Things resemble. It is therefore of no avail to appeal to the indefeasible necessity, by which Heraclitus thought the waves of Becoming to be directed. Standing outside the range of Becoming, this 'Ανάγκη would have had no control over its course. It became inevitable that Becoming should be recognised as containing the principle of its direction in itself, as soon as we admitted the necessity of substituting its mobility for the stationariness of things. Now if we attempt to find the necessity in the Becoming, one thing is clear. Between the extinction of the reality of m and the origin of the new reality of μ, no gap, no completely void chasm can be fixed. For the mere removal of m would in itself be exactly equivalent to the removal of anything else, p or q, that we like to imagine. Any other new reality therefore, π or κ, would have just as much or as little right to follow on the abolished m, as that μ; and it would be impossible that definite consequents should flow from definite antecedents. It is impossible therefore that the course of nature should consist in successive abolitions of one and originations of another reality. Every effort to conceive the order of events in nature as a mere succession of phenomena according to law, can only be justified on the ground that it may be temporarily desirable for methodological reasons to forego the search for an inner connexion. As a theory of the true constitution of reality it is impossible.

But the theory of Becoming might with perfect justification admit all this and only complain of a misinterpretation of its meaning.

Just as motion, it will be said, cannot be generated by stringing together moments of rest in the places *a, b, c,* so Becoming cannot be apprehended by supposing a succession of realities *a, b, c,* of which each is detached from the rest and looked upon as a self-contained and—for however brief an interval—motionless Being. On the contrary, to each single one of these members the same conception of Becoming must be applied as to the series, and just as the definitely directed velocity, with which the moving object without stopping traverses its momentary place *a,* necessarily carries it over into the place *b* and again through it into another, so the inner Becoming of the real *a,* as rightly apprehended, is the principle of its transition into *b* and into *b* only. For this is self-evident : that, just as it is not Being that is, but Things that are, so it is not Becoming that becomes, but the particular becoming thing ; and that consequently there is no lack of variety in the qualities *a, b, c,* which at each moment mark out in advance the direction in which the Becoming is to be continued.

I do not doubt that this defence would have expressed the mind of Heraclitus, with whose more living thought that modern invention of the schools which explains Becoming as a mere succession of pheno-mena stands in unfavourable contrast. And we might go further in the same spirit. ' You,' we might say, ' who treat a motionless content as existing, have certainly no occasion to contemplate its change ; but for all that we have nothing but your own assurance for it that the " Position " by which you suppose *a* to have been once constituted will endure for ever. In reality you can assign no reason why such should be the case with it, unless you look upon the *a* of one moment as the condition of *a* in the next moment and thus after all make *a* become *a.* But in the nature of reality there may be contained the springs of movement which are lacking to mere thought. If we think of an *a,* of which the essence consists only in the motion to *b,* we are indeed as little able to state how this *a* and its efflux is made, as you would be to state how your *a* and its rest is made. But your conception has no advantage over ours. For the motion, which (as extended to Things themselves) you find fault with, you after all have to allow in regard to the external relations of your Things, where you are as little able to construct it as in the inner nature of Things. To us, however, if admitted (within Things) as a characteristic of the real, it affords the possibility of explaining not merely the manifold changes in the course of nature but also as a special case that persistency in it which you are fond of putting in the foreground, without going into particu-lars, as something intelligible of itself, but which at bottom you present to yourselves merely as an obstruction to your own thoughts. Your law of Identity, moreover, would be equally suited by our assumption. We could not indeed suppose *a* to become *b* and *c* in three successive moments, unless it were precisely *b* in the second moment and *c* in the third—thus at each moment exactly what it is. More than this— more than the equality with itself of each of these momentary forms— cannot be required by the law of Identity. That the reality of the one moment should be the same as that of the other, could not be more properly demanded as a consequence of this law than could the exact opposite of its meaning; namely that everything should be simply identical with everything else.'

44. If the view just stated were the true meaning of the theories which maintained the sole reality of Becoming, their fundamental thought would not be exactly expressed either by this conception of Becoming or by that of Change. It would not be expressed by the former, because when in connexion with such speculations we oppose Becoming to Being we do not commonly associate with it in thought any such continuity as has been described; a continuity according to which every later phase in the becoming, instead of merely coming into being *after* the earlier, issues *out of* it. It would not be expressed by the conception of change, because in it the later does in fact arise out of the complete extinction of the earlier; because *b* is consequently another than *a* and, apart from that constancy of connexion, there is no thought of a permanent residuum of *a* which would have undergone a change in adopting *b* as its state.

We may go on to remark that, however much of the interpretation given we may take to be of use, it is at once apparent that the theory is insufficient to explain everything which we believe to be presented to us in experience. It would be convincingly applicable only to the case of a development which, without any disturbance from without, gradually exhibited the phases *b*, *c*, *d*, lying in the direction of the moving *a*. In reality, however, we find no unmistakeable instance of such development. None but an artificial view, which we must notice later, has attempted to explain away what seems to be an obvious fact—the mutual influence of several such developments on each other, or the change that proceeds from the reciprocal action of different things. The next point for our consideration will therefore be, what we have to think in order to apprehend this mutual influence, taking it for the present to be matter of indifference how we judge of the metaphysical nature of the Things between which the influence is exchanged.

45. In the first instance we only find occasion for assuming the exercise of an influence by one element *a* over another *b* in a change to β which occurs in *b* when *a* having been constantly present incurs a change into *a*. It is not merely supposed that the contents of *a* and β, as they exist for thought, stand to each other once for all in the relation of reason and consequence; but that *a* sometimes is, sometimes is not, and that in accordance with this changeable major premiss the change from *b* into β sometimes will ensue, sometimes will not.

Now we know that it might be ordained by a law external to *a* and *b* that *b* should direct its course according to these different circumstances: but it would only obey this ordinance if it were superfluous and if its own nature moved it to carry out what the ordinance contains. In order to the possibility of this that difference of conditions, consisting in the fact that at one time *a* is, at another is not, must make a difference for *b* itself, not merely for an observer reflecting on the two. *b* must be in a different state, must be otherwise affected, must experience something different in itself, when *a* is and when *a* is not: or, to put it in a short and general form; if Things are to take a different course according to different conditions, they must *take note* whether those conditions exist or no. Two thoughts thus unite here. In order that *a* may be followed by β not by β¹ or β², *a* and β must stand in the relation of principle or *ratio sufficiens* and consequence. But in order that β may actually come into being and not remain the for ever vainly postulated consequence of *a*, the *ratio sufficiens* must become *causa efficiens*, the foundation in reason must become a productive agency: for the general descriptive conception of the agency of one thing on another consists in this that the actual states of one essence draw after them actual states of another, which previously did not exist. Now how it can come about that an occurrence happening to the one thing *a* can be the occasion of a new occurrence in the thing *b*, is just what constitutes the mystery of this interference or 'transeunt' action, with which we shall shortly be further occupied. We introduce it here, to begin with, only as a demand, which there must in some way be a possibility of satisfying, if an order of events dependent on conditions is to be possible between individual things.

46. Supposing us however to assume that this unintelligible act has taken place, from the impression which *b* has experienced as its own inner state we look for after effects within itself; a continuation of its Being or of its Becoming different from what it would have been without that excitement. To determine in outline the form of this continuation is a task which we leave to the sequel. As regards the question of its origin, we are apt to look on our difficulties as got rid of when this point is reached. This immanent operation, which develops state out of state within one and the same essential Being, we treat as a matter of fact, which calls for no further effort of thought. That this operation in turn remains completely incomprehensible in respect of the manner in which it comes about, we are meanwhile very well aware. For how a state a^1 of a thing a begins to bring about a consequent state, a^2, in the same thing, we do not understand at all better than how the same a^1 sets about producing the consequence β^1 in another being b. It is only that the unity of the essence, in which the unintelligible process in this case goes on, makes it seem superfluous to us to enquire after conditions of its possibility. We acquiesce therefore in the notion of immanent operation, not as though we had any insight into its genesis, but because we feel no hindrance to recognising it without question as a given fact. Conditions of the same subject, we fancy, must necessarily have influence on each other: and in fact if we refused to be guided by this fundamental thought, there would be no hope left of finding means of explanation for any occurrence whatever.

47. Towards these notions the two theories as to the essence of things, which we have hitherto pursued, stand in different relations. On the preliminary question how it comes about that the inwardly moving a attains an influence over the equally passing b the doctrine of Becoming must like every other admit ignorance for the present. But supposing this to have come about, it will look for the operation of this influence only in an altered form of Becoming, which a strives to impress on b. The next-following phase of b will consequently not be β, but a resultant compounded of β and the tendency imparted from without. Henceforth this new form would determine the progressive Becoming of that original b, if it continued to be left to itself: but every new influence of a c would alter its direction anew. If each of these succeeding phases is called a Thing, on the ground that it is certainly capable of receiving influences from without and of exerting them on its likes, then Thing will follow Thing and in its turn pass away, but it will be impossible to speak of the unity of a Thing which maintains itself under change. It is possible that the residuary effects of an original b in all members of the series may far outweigh the influence of action from without. In that case they would all, like different members of a single pedigree, bear a common family characteristic in spite of the admixture of foreign blood, but they would be no more one than are such members. It is another possible

case that b without disturbance from without should develope itself into its series b, β^1, β^2. Its members would then be comparable to the successive generations of an unmixed people, but again would form a real unity as little as do these. Even if b reproduced itself without change, each member of the series $b\,b\,b$ would indeed be as like the preceding one as one day is like another, but would as little *be* the preceding one as to-day is yesterday.

This lack of unity will afford matter of censure and complaint to the theory which treats the Thing as persistent; but it is time to notice that this theory has itself no unquestionable claim to the possession of such unity. Those who profess the theory rightly reject the notion which would represent the vanishing reality of one thing as simply followed by the incipient reality of the other without connecting the two by any inward tie ; but they think scorn of recognising this continuity in an actual, though unintelligible, becoming of the one *out of* the other and hope to make it intelligible by the interpolation of the persistent Essence. But this implies that they are in fact reduced simply to the impossibility, on which we have already touched, of attaining the manifold of change by a merely outward tie to the unchangeable stock of the Thing. This is merely disguised from them by the power of a word, the use of which we have found it impossible to avoid but are here called upon to rectify. When we called a^1, a^2, a^3 *states* of a, we could reckon only too well on the prospect that this expression would remain unchallenged and would be thought to contain the fulfilment of a demand, for which it merely supplies a name. Quite of itself this expression gives rise incidentally to the representation of an essence which is of a kind to sustain these states, to cherish them as *its own* and thus to maintain itself as against them. But what does this mean, and how can that be, which —under the impression that we are saying something that explains itself—we call the state of an essence ? And in what does that relation consist—a relation at once of inseparableness and difference— which we indicate by the innocent-seeming possessive pronoun ? So long as we maintain the position that a as in the state a^1 is something other than what it is as in the state a^2; so long again as we forego the assumption that there is present an identical residuum of a in a^1 and a^2, on which both alike might have a merely external dependence ; so long as we thus represent a as passing in complete integrity into both states—while this is so, the expressions referred to convey merely the wish or demand, that there should be something which would admit of being adequately expressed by them, or which

would satisfy this longing after identity in difference, after permanence in change. They do not convey the conception of anything
which would be in condition to satisfy this demand.

In saying this I must not be understood to take it as settled that
this Postulate cannot be fulfilled, only as unproven that it can be.
Reality is richer than Thought, nor can Thought make Reality after
it. The fact of Becoming was enough to convince us that there is
such a thing as a union of Being and not-Being, which we even when
it lies before us are not able to reconstruct in thought, much less
could have guessed at if it had not been presented to us. It is
possible that we may one day find a form of reality which may teach
us by its act how those unreconcilable demands are fulfilled, and prove,
in doing so, that in their nature they are capable of fulfilment, and that
the relation, seemingly so clear, between Thing and state is other than
an empty combination of words, to which nothing in reality corresponds. It will not be till a very late stage in these enquiries that we
shall have opportunity of raising this question again. For the present
we take the real permanent unity of the Thing under change of states
to be a doubtful notion, which is of no value for the immediate objects
of our consideration.

48. If a or a is to act on b, b must in all cases be differently
affected by the existence of a and by its non-existence. The 'transeunt' action of a on b would thus lead back to an operation 'immanent' in b. The proximate condition which brings about the change
of b, must have lain in b itself. We usually distinguish it as an impression from the reaction—a usage of speech on which we may have to
dwell below. For the present we satisfy ourselves with the reflection
that anything which b is to experience through the action of a must
result from the conflux of two principles of motion; from that which
a ordains or strives to bring about and from that which b, either in
self-maintenance or in self-transformation, would seek to produce, if
a were not. Two principles are thus present in b, of which in general
the one conditions something else than what the other conditions.
Neither of these two commands therefore could realise itself, if each
of them were absolute. For neither the one nor the other of
them would have any prerogative, both being, to revert to the old
phrase, states of the same essence, b. A determinate result is
only possible on supposition that not only a third general form of
consequence is thinkable, into which both impulses may be blended,
but that also the two principles have comparable quantitative values.
In the investigations of natural science it is not doubted that the deter-

mination of a result from various coincident conditions always presupposes, over and above the assignment of that which each condition
demands, the measure of the vivacity with which it demands it. It is not
merely in nature, however, but in all reality that something goes on
which has no place in the syllogistic system formed by the combination of our thoughts.   In the latter, of two opposite judgments only
one can be valid.   In reality different or opposite premisses confront
each other with equal claim to validity and both ask to be satisfied on
the ground of a common right.   I am therefore only filling a gap
which has hitherto been left unfilled in Metaphysic, when I seek to
bring out the necessity of this mathematical element in all our judgments of reality, leaving its further examination to the sequel.

49. 'Quo plus realitatis aut esse unaquaeque res habet, eo plura
attributa ei competunt.'   So says Spinoza[1]; and nothing seems to
forbid the converse proposition, that a greater or less measure of
Being or of reality belongs to things according to the degree of their
perfection.   I cannot share the disapproval which this notion of there
being various degrees of strength of Being has often incurred.   It is
no doubt quite correct to say that the general conception of Being,
identical with itself, is applicable in the same sense wherever it is
applicable at all, and that a large thing has no more Being in being
of large size than a little thing in being of small size.   I do not find
any reason, however, for emphasizing in Metaphysic this logical equality
of the conception of Being with itself, since Metaphysic is concerned
with this conception not as it is by itself but in its application to
its content—to the things that are.   But in this application it should
not, as it seems to me, be looked upon as if the 'Position' which it
expresses remained completely unaffected by the quantity of that
on which the 'Position' falls.   In the same way motions, the slowest
as well as the quickest, all enjoy the same reality.   We cannot say
that they *are*, but they all *take place*, one as much as another.   Neither
in their case does this reality admit of increase or diminution for any
single one of them.   The motion with the velocity C cannot, while
retaining this velocity, be taking place either more or less.   But for
all that the velocity is not matter of indifference in relation to the
motion.   When it is reduced to nothing the motion ceases; and conversely no motion passes out of reality into unreality otherwise than
by the gradual reduction of velocity.

Now that which we admit in the case of the extreme limit—the
connexion of Being, or in this case of taking place, with that which is

[1] [Eth. i. Prop. ix.]

or happens—why should we not allow to hold good within that interval, in which this quantity still has a real value? Why should we look on the velocity as a secondary property, only accidentally attaching to that character of the motion which consists in its being something that occurs, when after all it is just so far as this property vanishes that the motion continuously approximates to the rest in which nothing occurs? The fact is that the velocity is just the degree of intensity with which the motion corresponds to its own Idea, and the occurrence of the quicker motion is the more intensive occurrence. If now we apply the term ' Being,' as is proper in Metaphysic, not to the empty ' Position' which *might* fall upon a certain content, but to the filled and perfectly determinate reality as already including that on which the ' Position' has actually fallen, I should in that case have no scruple about speaking of different quantities or intensities of the Being of Things, according to the measure of the power with which each thing actively exerts itself in the course of change and resists other impulses. Nor in this argument am I by any means merely interested in rescuing a form of expression that has been assailed. I should seriously prefer this expression for the reason that it helps to keep more clearly in mind what I take to be the correct view; viz. that Being is really a continuous energy, an activity or function of things, not a doom thrust upon them of passive ' position'[1]. The constant reminder of this would be a more effectual security against shallow attempts to deduce the Real from the coincidence of a still unreal essence with a ' Position' supposed to be foreign to this content and the same for all Things indifferently.

---

[1] [' Passivischer Gesetzheit.']

CHAPTER V.

Of the Nature of Physical Action.

Our concern so far has been to give to the conception of Becoming a form in which it admits of being applied to the Real. In seeking to do so we were led to think that the connexion between a cause and its effects must be more than a *conditioning* of the one by the other; that it must consist in an *action* on the part of the cause, or require such an action for its completion. Only thus could it become intelligible that effects, which in a world of ideas are consequences that follow eternally from their premisses—premisses no less eternally thinkable, should in the world of reality sometimes occur, sometimes not. Many and various have been the views, as the history of Philosophy shows, which have been successively called forth by the need of supplying this complement to the idea of cause and by the difficulty of doing so without contradiction. Many of them, however, are for us already excluded, now that it becomes our turn to make the same attempt, by the preceding considerations.

50. In the first place we meet at times with a disposition—no longer indeed admitted among men of science but still prevalent in the untutored thoughts of mankind—to ascribe the nature and reality of a consequent wholly and exclusively to some one being, which is supposed to be the cause, the single cause, of the newly appearing event. The unreasonableness of this view is easily evinced. It condenses all productive activity into a single element of reality, while at the same time it deems it necessary that the results of the activity should be exhibited in certain other elements, which stand to the exclusively causal element in the relation of empty receptacles for effects with the form and amount of which they have nothing to do. As we have already seen, everything which we can properly call a receptivity consists, not in an absence of any nature of a thing's own, but in the active presence of determinate properties, which alone make it possible for the receptive element to take up into itself the

impressions tendered to it and to convert them into states of its own. Deprived of these qualities or condemned to a constant inability of asserting them, the elements in which the ordinance of the active cause is supposed to fulfil itself, would contribute no more to its realisation by their existence than by their non-existence. Instead of something being *wrought* by the cause, it would rather be created by it in that peculiar sense in which, according to a common but singular usage, we talk of a creation out of nothing. I call it a singular usage because we should properly speak simply of creation, to which we might add, merely in the way of negation, that the creation does not take place out of anything in particular. Trained by experience, however, to look upon new states merely as changes of what is already in existence, our imagination in this case gives an affirmative meaning even to the 'nothing' as the given material out of which something previously unreal is fashioned.

The same extraordinary process is repeated in that manner of conceiving the action of a cause of which I have just spoken. The supposition is allowed to stand of things which the active cause requires in order to fulfil its active impulse in them : but as these according to the conception in question contribute nothing to the nature of the new event, they are in fact merely empty images which serve to meet the requirements of our mental vision. They represent imaginary scenes upon which an act, wholly unconditioned by these scenes of its exhibition, originates, out of nothing and in nothing, some new reality. I reserve the question whether this conception of creation admits any application at all and, if so, in what case. It is certainly inapplicable in studying the course of the already existing universe ; inapplicable when the fact that requires explanation is this, that individual things in their changing states determine each other's behaviour. Were it possible for one of these finite elements, A or B, to realise its will, a or β, in other elements after this creative manner, without furtherance or hindrance from the co-operation of any nature which these other elements have of their own, there would be nothing to decide upon the conflicting claims which any one of these omnipotent beings might make on any other. The ordinances, a or β or γ, would be realised, with equal independence of all conditions, in all beings C, D, E. This notion, if it were possible to carry it out in thought, would at any rate not lead to the image of an ordered course of the universe, in which under definite conditions different elements are liable to different incidents, while other incidents remain impossible to them. Any assumption that A or B can only give reality to its command upon C or D, not upon E or F, would force us back upon the conception that C or D are not only different from E and F, but that in virtue of their own nature they are joint conditions of the character and reality of the new occurrence, which we previously regarded as due to a manifestation of power on one side only, to a single active cause.

51. Natural science, so long as it maintains its scientific character, is constrained by experience to recognise this state of the case. It has reduced it to the formula that every natural action is a reciprocal action between a plurality of elements. It was apt to be thought, however, that the proposition in this form expressed a peculiarity of natural processes, and it was a service rendered by Herbart to point out its universal validity as a principle of Metaphysics in his doctrine that every action is due to several causes. Though these things are ultimately self-evident, the mere establishment of a more exact phraseology calls for some enquiry. In the first place Reasons[1] and Causes[2] will have to be distinguished more precisely than is done in ordinary speech. By 'causes,' consistently with the etymology of the German term 'Ursache,' we understood all those real things of which the connexion with each other—a connexion that remains to be brought about—leads to the occurrence of facts that were not previously present. The complex of these new facts we call the effect, in German 'Wirkung'—an ambiguous term which we shall employ to indicate not the productive process but only the result produced. Wherever it shall appear necessary and admissible to take notice of this distinction, we shall reserve the infinitive 'Wirken' to express the former meaning. The 'Reason' on the other hand is neither a thing nor a single fact[3], but the complex of all relations obtaining between things and their natures; relations from which the character of the supervening effect is deducible as a logically necessary consequence.

Now just because we do not think of the new event as issuing from a creative activity independent of conditions, the explanation of any effect would require us, besides assigning the causes (Ursachen) to show the reason (Grund) which entitles the causes to be causes of just this effect and no other. Further, just because several constituents of this reason (Grund) are not merely given as possible in thought, but are embodied or realised in the form of real properties of real things and of actually subsisting relations between them, the consequence does not merely remain one logically necessary which we should be entitled to postulate, but becomes a postulate *fulfilled*, an actual effect

---

[1] ['Gründe.']   [2] ['Ursache.']   [3] ['Nicht Ding noch Sache.']

instead of an unreal necessity of thought. Finally, observation convinces us that things, without changing their nature, yet sometimes do,
sometimes do not, exercise their influence on each other. It appears
therefore that it is not the relations of similarity [1] or contrast between
the things—relations which upon comparison of their natures would
always be found the same—that qualify them to display their productive activity, but that, as a condition of this activity, there must
besides supervene a variable relation, *C*. I reserve the question
whether we are right in thinking of this relation as other than one of
those included in what we meant to be understood by the complete
Reason (Grund) of the effect. A doubt being possible on this point,
which will demand its own special investigation, we will provisionally
conform to the ordinary way of looking at the matter and speak of *C*
as the condition of the actual production of the effect—a condition
which is something over and above the Reason (Grund) that determines the form of the ensuing effect.

52. According to this usage of terms the causes (Ursachen) of a
gunpowder-explosion are two things or facts, viz. the powder *A* and
the heated body which forms the spark *B*. The condition, *C*, of their
action upon each other is presented to us in this case as their
approximation or contact in space. The reason (Grund) of the effect
lies in this, that the heightened temperature and the expansiveness of
the gaseous elements condensed in the powder are the two premisses
from which there arises for these elements a necessity of increase in
their volume as effect. The final question, how in this case the
efficient act takes place, we do not profess to be able to answer. Of
whatever conjecture as to the nature of heat we may avail ourselves
for the purpose, we find it impossible in the last resort to state how it
is that the heightened temperature operates in bringing about in the
expansive materials the movement of dilatation which they actually
undergo. It is only the effect, the result brought about, which in this
case is not a motionless state but itself a movement, that is open to
our observation.

In one respect this instance is unsatisfactory. In the case supposed
we have no experience as to what becomes of the spark which was
supposed to form one of the *two* causes of the total event. If on the
other hand we throw a red-hot body, *B*, into some water, *A*, we
notice, over and above the sudden conversion of water into steam,
which in this instance corresponds to the explosion of the powder in
the other, the change which *B* has undergone. Lowered in its tem-

[1] ['Aehnlichkeit.']

perature, perhaps with its structure shattered, or itself dissolved in what
is left of the water, there remains what was previously the heated body.
Thus even the effect in this case consists of several different changes
which are shared by the different concrete causes (Ursachen) that have
been brought into contact.   Finally, since the evaporating water dis-
sipates itself in the air, leaving behind it the cooled motionless body,
that contact between the two which previously formed the condition
of their effect upon each other, has changed into a new relation in
space between the altered bodies.   Combining all these circumstances,
we may say that, where a definite relation, C, gives occasion to an
exercise of reciprocal action between the things A and B, A passes
into a, B into β, and C into γ.

53. The particular forms and values which these transitions A—a,
B—β, C—γ, take in individual cases, can only be determined by so
many special investigations, and these would be beyond the province
of Metaphysics.   Even the task of merely showing that all kinds of
causation adjust themselves in general to the formula just given would
be one of inordinate length, and must be left to be completed by the
attentive reader.   The only point which I would bring into relief is
this, that alike the contributions which the several 'causes' (Ursachen)
make to the form of the effect, and the changes which they themselves
undergo through the process of producing it, admit of variation in a
very high degree.   In view of this variety the usage of speech has
created many expressions for states of the case, of which the distinction is well-founded and valuable for the collective estimate of the
importance of what takes place but which do not exhibit any distinctions that are fundamental in an ontological sense.   If elastic bodies,
meeting, exchange their motions with each other wholly or in part,
we have no doubt about the necessity of regarding both as metaphysically equivalent causes of this result.   They both contribute
alike, though in different measure, to determine the form of the result,
and the effect produced visibly divides itself between the two.

It is otherwise in the instance of the exploding powder.   Here
everything that conditions the form of the result appears to lie on one
side, viz. in the powder, in the capability of expansion possessed by
the elements condensed in it.   The spark contributes nothing but an
ultimate complementary condition—the high temperature, namely,
which is the occasion of an actual outburst on the part of the previously existing impulse to expansion, but which would not be qualified
to supply the absence of that impulse.   For this reason we look upon
these two causes of the effect in different lights.   It is not indeed as

if, in accordance with the reason given, we assigned the designation
'cause' *par excellence* to the powder. On the contrary this designation
is assigned by ordinary usage rather to the spark, which alone presents itself to our sensuous apprehension as the actively supervening
element in contrast with the expectant attitude of the powder. But
this usage at least we are ready to modify when we enter upon a more
scientific consideration of the case ; we then treat the spark as merely
an *occasional cause* which helps an occurrence, for which the prelimi-
naries were otherwise prepared, actually to happen. Though it is
undoubtedly important, however, to note that peculiarity of the case
which is indicated by the expression 'occasional cause,' yet from the
ontological point of view the spark, even in its character as occasional
cause, falls completely under the same conception of cause under
which we subordinate the powder. For whatever tendency to expansion we may ascribe to the elements united in the powder, taken by
itself this merely suffices to maintain the present state of things. It
is only the introduction of a heightened temperature that produces
the necessity of explosion. The 'occasional cause' therefore brings
about this result, not in the sense of giving to an event, for which the
reason (Grund) was completely constituted, but which still delayed to
happen, the impulse which projected it into reality, but in the sense
of being the last step in the completion of that 'reason' of the event
which was incompletely constituted before. Similar reflections will
have to be made in all those cases where one 'cause' seems only to
remove a hindrance which impedes the other causes in actually bringing about an effect for which the preliminary conditions are completely
provided by them. The setting aside of an obstruction can only be
understood as the positive completion of that which the obstruction
served to cancel in the complete 'Reason.'

Phenomena such as occur in the processes of life call for still
further distinctions of this kind. The same occasional causes, Light,
Warmth, and Moisture, excite the seeds of different plants to quite
different developments. In whatever amounts we combine these external forces, though we may easily succeed in destroying the power of
germination in any given seeds, we never succeed in eliciting different
kinds of plants from them. The same remark applies to the behaviour
of living things at a later stage, when fully formed. The form of
action which they exhibit, upon occasion being given from without, is
completely determined by their own organization, and we look upon
the occasional causes in this case as mere *stimuli*, necessary and fitted
to excite or check reactions of which the prior conditions are present
within the organism, but with no further influence on the form which

the reactions take. I do not pause to correct any inexactness that may be found in this last expression, nor do I repeat remarks which I have previously made and which would be applicable here. It is enough to say that, in a natural history of the various forms which the process of causation may assume, all those that have been just referred to, as well as many others, fully deserve to be distinguished by designations of their own and to have their peculiarity exhibited in full relief. It is the office of ontology, on the contrary, to hold fast the general outline of the relation of reciprocal action, in respect of which none of these forms contain any essential difference. In the view of ontology all causes of an effect are just as necessary to its production the one as the other. However great or small the share may be which each of them has in determining the form of the effect, no one of them will be wholly without such a share. Each of them is a contribution without which the complete 'reason' (Grund) of the actual effect cannot be constituted. No one of them serves as a mere means of converting into fact a possibility already, without it, completely determined in kind and quantity. It is exclusively with this ontological equivalence of the manifold causes of a fact that we are here concerned. It will only be at a later stage that it will become necessary to refer to those other characteristics of the causal relation of which the existence might even at this stage easily be established by the farther consideration of the instances already given. Such would be the fact that the effect produced does not attach itself exclusively to any one of the co-operative causes but rather distributes itself among them all, and, finally, the change, after the resulting action has been exerted, of the relation which served to initiate it.

54. After all these remarks, however, the proper object of enquiry has still been left untouched. How is this relation *C*, of which the establishment was necessary to elicit the effect, to be understood metaphysically? The need in which this question stands of special consideration is most readily apprehended if we transfer ourselves to the ontological position of Herbart. His theory started expressly from the supposition of a complete mutual independence on the part of the real Beings, of their being unconcerned with any Relation. If it allows the possibility of their falling into relations with each other, the readiness to make this admission rests simply on the supposition that they remain unaffected by so doing. At the same time this metaphysical theory recognises a relation, under the name of the coexistence [1] of the real Beings, which does away with their complete

---

[1] ['Zusammen,' lit. 'together.']

indifference towards each other, and compels them to acts of mutual disturbance and of self-maintenance.

In what, however, does this 'coexistence,' so pregnant with consequences, consist? So long as we confine ourselves to purely ontological considerations, we can find in this expression merely that indication of a postulate, not the indication of that by which this postulate is fulfilled. The 'coexistence' is so far nothing but that relation, as yet completely unknown, of two real Beings, upon the entry of which their simple qualities can no longer remain unaffected by each other but are compelled to assert an active reciprocal influence. Thus understood, let us call the 'coexistence' Γ. The term 'coexistence,' however, with its spatial associations, having once been chosen for this *Quaesitum*, appears to have been the only source of Herbart's cosmological conviction that, as a self-evident truth, the only form in which the ontological 'coexistence' Γ, the condition of efficient causation, can occur in the world, is that of coincidence in space. At least I do not find any further proof of the title to hold that the abstract metaphysical postulate Γ admits of realisation in this and in no other imaginable[1] form. I shall have occasion below to express an opinion against the material truth of this assumption; against the importance thus attached to contact in space as a condition of the exertion of physical action. Here we may very well concede the point to the common opinion, if appeal is made to the many instances in which, as a matter of fact, the approximation of bodies to each other presents itself to us as a necessary preliminary to their action upon each other. Assuming, then, that contact can be shown universally to be an indispensable preliminary condition of physical action, even then we should only have discovered or conjectured the empirical form C under which as a matter of fact that metaphysical Γ, the true ground of all physical action, presents itself in the world. The question would remain as to the law which entitles this connexion in space to make that possible and necessary which would not occur without it.

We are all at times liable to the temptation of taking that in the last resort to explain itself, of which continued observation has presented us with frequent instances. It cannot, therefore, be matter of surprise to me if younger and consequently keener intellects undertake to teach me that in this case I do not understand myself. Whatever my error may be, I cannot get rid of it. I must repeat that, so far as I can see, there is no such inner connexion between the conception

---

[1] ['Anschaulich.']

of contact in space and that of mutual action as to make it self-evident that one involves the other.   Granted that two Beings, *A* and *B*, are so independent of each other, so far removed from any mutual relation that each could maintain its complete existence without regard to the other, as it were in a world of its own ; then, though it may be easy to picture the two as ' coexisting ' in the same point of a space, it seems to me impossible to show that for this reason alone the indifference to each other must disappear.   The external union of their situations which we present to our mind's eye must remain for them as unessential as previously every other relation was.   Inwardly their several natures continue alien to each other, unless it can be shown that this ' coexistence ' in space, *C*, is more than a ' coexistence ' in space, that it includes precisely that metaphysical coexistence, which renders the Beings that would otherwise be self-sufficing, susceptible and receptive towards each other.   Not believing myself in the correctness, as a matter of fact, of this theory of contact, I have naturally no reason to attempt such a proof, which, moreover, would carry us prematurely beyond the province of ontology.   As a question of ontology, it only remains to ask what the Γ is, i. e. what is the condition which we must suppose fulfilled, if in any relation *C*, whether it be out of contact in space or of some wholly different form, we suppose things previously indifferent to each other to become subject to the necessity of having respect to each other and of each ordering its states according to the states of the other.   This question is the starting-point of the various views that have been held on the problem, how one thing comes to act on another.   None of them could avoid enquiring for a mode of transition of some sort or other from the state which is not one of coexistence to one that is so. It is according as they claim to have discovered the mode of transition or to be entitled to deny that there is any such transition, that they have resulted in notably divergent conceptions of the course of the universe.

55. The transfer of an *influence*, *E*, is the process by which according to the common view it is sought to explain the excitement of Things, previously unaffected by each other, to the exercise of their active force : and the process is generally conceived in a one-sided way as an emanation proceeding from an active Being only, and directed upon a passive Being.   That this representation only serves to indicate the fact of which an explanation is sought, becomes at once apparent if we attempt to define the proper meaning and nature of that to which, under the figurative name of influence, we

ascribe that transition from the one Being to the other. Only one
supposition would make the matter perfectly clear; the supposition,
namely, that this E which makes the transition is a Thing, capable of
independent reality, which detaches itself from its former connexion
with A and enters into a similar or different connexion with something else B. But precisely in this case unless something further
supervened, there would be no implication of that action of one thing on
another, which it is sought to render intelligible. If a moist body A,
becoming dry itself, makes a dry body B, moist, it is the palpable water
E which here effects this transition. If, however, what we understood by moisture was merely the presence of this water, at the end
of the transition neither A nor B would have undergone a change of
its own nature, such a change as it was our object to bring under the
conception of an effect attained by an active cause. The transition
itself is all that has taken place.

True, the withdrawal of the water alters the drying body, its accession alters the body that becomes moist. The connexion between
the minutest particles changes as the liquid forces its way among
them. As they are forced asunder, they form a larger volume and
the connexion between them becomes tougher, while the drying
body becomes more brittle as it shrinks in extent. These are effects
of the kind which we wish to understand, but the supposed transition
of the water does not suffice for their explanation. After the water
has reached its new position in the second body B, the question
arises completely anew what the influence is which, so placed, it is
able to exercise—an influence such that the constituents of B are
compelled to alter their relative positions. In like manner the question would arise how the removal of the water from A could become
for this body a reason for the reversal of its properties. This illustration will be found universally applicable. Wherever an element E,
capable of independent motion, passes from A to B—thus in all
cases where we observe what can properly be called a ' causa transiens'
—there universally this transition is only preliminary to the action [1] of
one body on another. This action follows the transition, beginning
in a manner wholly unexplained only when the transition is completed. Nor would it be of the slightest help if, following a common
tendency of the imagination, we tried to sublimate the transeunt element into something more subtle than a ' thing.' Whatever spiritual
entity we might suppose to radiate from A to B, at the end of its
journey it would indeed be in B, but the question how, being there,

---

[1] [' Wirkung.']

it might begin to exert its action upon constituents different from it, would recur wholly unanswered.

56. This difficulty suggests the next transformation of the common view. Instead of the causative thing (Ursache), we suppose a force, an action, or a state, E, to pass from A to B. We may suppose these various expressions, which are to some extent ambiguous, to have so far a clear notion attached to them that they denote something else than a thing. They thus avoid the question how the thing acts on other things after its transition has been effected. But in that case they are liable to the objection, familiar to the old Metaphysic : 'attributa non separantur a substantiis.' No state, E, can so far detach itself from the Thing A, of which it was a state, as to subsist even for an infinitesimal moment between A and B, as a state of neither, and then to unite itself with B in order to become its state.

The same remark would apply if that which passed from A to B were supposed, by a change of expression, to be an action, and thus not a state but an event. No event can detach itself from the A, in a change of which it consists, and leave this A unchanged behind it in order to make its way independently to B. According to this conception of it, so far as it is a possible conception at all, the action thus supposed to transfer itself would simply be the whole process of efficient causation which it is the problem to explain, not a condition, in itself intelligible, which would account for the result being brought about.

And after all these inadmissible representations would not even bring the advantage they were meant to bring. As in regard to the transition of independent causative things, so in regard to the transition of the state or event E from A to B the old question would recur. Granting that E could separate itself from A, what gave it its direction at the particular moment to B, rather than to C? If we assume that A has given it this direction, we presuppose the same process of causative action as taking place between A and E for which we have not yet found an intelligible account as taking place between A and B. Nor is this all. Since it will not be merely on B and C, but presumably on many other Beings that A will put forth its activity, we shall have to ask the further question what it is that at a given moment determines A to impart to E the direction towards B and not towards C, or towards C and not towards B. An answer to this question could only be found in the assumption that already at this moment A is subject to some action of B, and not at the same time to any action of C, and that there thus arises in it the counter-action, in the exercise of which it now enjoins upon E the transition to B and not to C. Thus for the second time we should have to presuppose an action which we do not understand before we could present to ourselves so much as the possibility of that condition which is no more than the preliminary to a determinate action.

Finally it is important to realise how completely impossible is the innocent assumption that the transferred E will all of a sudden become a state of B, when once it has completed its journey to B. Had this homeless state once arrived at the metaphysical place which B occupies, it would indeed be there, but what would follow from that? Not even that it would remain there. It might continue its mysterious journey to infinity and, as it was once a no-man's state, so remain. For the mere purpose of keeping it in its course, we must make the yet further supposition of an arresting action of B upon it. And given this singular notion, it would still be a long way to the consequence that E, being an independent state, not belonging to anything in particular, should not only somehow attach itself to the equally independent being B, but should become a state of this B itself, an affection or change of B. These accumulated difficulties make it clear that the coming to pass of a causative action can never be explained by the transfer of any influence, but that what we call such a transfer is nothing but a designation of that which has taken place in the still unexplained process of causation or which may be regarded as its result.

57. Apart from its being wholly unfruitful, the view of which we have been speaking has become positively mischievous through prejudices which very naturally attach themselves to it. It treats the transmitted effect E as one ready-made, and merely notices the change on the part of the things of which incidentally it becomes a state. No doubt there is a tacit expectation that, upon its being carried over to B, many further incidents will there follow in its train of which no more explicit account is taken. But in order that the view may have any sort of clearness, it must in any case assume that B will afford to E on its arrival the same possibility of reception and of existence in it which was offered it by A. There thus arise jointly the notions that the effect must be the precise counterpart of its cause or at least resemble[1] it, and that all beings, between which a reciprocal action is to be possible, must be qualified for it by homogeneity of nature.

[1] ['Gleich oder doch ähnlich sein müsse.' Cp. note on 'Gleichheit,' § 19 supra. Sect. 59 makes it clear that the term 'gleich' does not merely refer to the alleged *equality* of cause and effect.]

Our previous considerations compel us to contradict these views at every point. No thing is passive or receptive in the sense of its being possible for it to take to itself any ready-made state from without as an accession to its nature. For everything which is supposed to arise in it as a state, there is some essential and indispensable co-operating condition in its own nature. It is only jointly with this condition that an external impact can form the sufficient reason which determines the kind and form of the resulting change. So long as there is speaking generally a certain justification, owing to that peculiarity of the cases contemplated which we mentioned above, for treating one thing *A par excellence* as the cause, a second *B* as the sustainer of the effect or as the scene of its manifestation, in such cases we shall even find that the form of the effect produced by *A* depends in quite a preponderating degree on the nature of the *B*, which suffers it. It is only to forms of occurrence which are possible and appropriate to this its nature that *B* allows itself to be constrained by external influences. It is little more than the determination of the degrees in which these occurrences are to present themselves that is dependent on corresponding varieties in the external exciting causes. This is the case not only with living beings, but with inanimate bodies. Upon one and the same blow one changes its form yieldingly, another splits into fragments, a third falls into continuous vibrations, some explode. What each does is the consequence of its completely determinate structure and constitution upon occasion of the outward excitement.

This being so, if it is improper to speak of a transmission of a ready-made effect, it is still more so to speak of a universal identity in kind and degree[1] of cause with effect. It would in itself be an inexactness, to begin with, to try to establish an equation between the 'cause' (Ursache), which is a Thing, and the effect which is a state or an occurrence. All that could be attempted would be to maintain that what takes place in the one 'cause' considered as active is identical with that which will take place in the other considered as passive; or, to put the proposition more correctly, considering the number of objects which are equally entitled to be causes, each will produce in the other the same state in which it was itself. Expressed in this form, we might easily be misled into looking upon it as in fact a universal truth. The science of mechanics, at least, in the distribution of motions from one body to another, puts a number of instances at command which would admit of being reduced to this

[1] ['Gleichheit,' v. note on p. 104.]

point of view and which might awaken the conjecture that other occurrences of a different kind would upon investigation be found explicable in the same way. Against this delusion I must recall the previous expression of my conviction; that even in cases where as a matter of fact a perfectly identical reciprocal action, Z, is exercised between A and B, it yet cannot arise in the way of a transmission of a ready-made state, Z; that what takes place in A and B is even in these cases always the production anew of a Z, conformably with the necessity with which Z under the action of B arises out of the nature of A, and under the action of A arises out of the nature of B; that, while it is a possible case, which our theory by no means excludes, that these two actions should be the same; their equality is not a universal condition which we are to consider in the abstract as essential to the occurrence of any reciprocal action.

58. The fatal error, on which we have been dwelling, is not one to be lightly passed over. The conviction must be established that of the alleged identity between cause and effect nothing is left but the more general truth with which we are familiar. This truth is that the natures of the Things which act on each other, the inner states in which for the moment they happen to be and the exact relation which prevails between them—that all this forms the complete 'reason' from which the resulting effect as a whole issues. Even that this consequence is *contained* in its reason is more than we should be entitled to say, unless we at least conceive as immediately involved in the nature of the things and already in living operation those highest grounds of determination, according to which it is decided what consequence shall follow from what reason in the actual world. And this tacit completion of our thought would emphatically not lead back to the view which we are here combating. For of what is contained in those highest conditions which determine what shall emanate from what, in the actual world, as consequent from cause or reason, we have not in fact the knowledge which we might here be inclined to claim. There is nothing to warrant the assurance that it is exclusively by general laws, the same in innumerable instances of their application, that to each state of facts, as it may at any time stand, the new state, which is to be its consequence, is adjusted. It is an assurance in which the wish is father to the thought. It naturally arises out of our craving for knowledge, for it is doubtless only upon this supposition that any consequence can be derived analytically from its 'reason' or be understood as an instance of a general characteristic.

But what is there to exclude '*in limine*' the other possibility; that some one plan, which in the complex of reality only once completes itself and nowhere hovers as a universal law over an indefinite number of instances, should assign to each state of facts that consequence which belongs to it as a further step in the realisation of this one history—so belongs to it, however, but once at this definite point of the whole, never again at any other point? On that supposition indeed our knowledge would no longer confront reality with the proud feeling that it can easily assign its place to everything that occurs in it, as a known instance of general laws, and can predetermine analytically the consequence which must attach to it. The series of events would unfold itself for us synthetically; an object of wondering contemplation and experience, but not an object of actual understanding till we should have apprehended the meaning of the whole, as distinguished from that which repeats itself within the whole as a general mode of connexion between its several members.

59. We will not, however, pursue these ultimate thoughts. I merely hint at them here in order to dislodge certain widely-spread prejudices from their resting-place, but cannot now work them out. We will take it for granted that every effect in the world admits of being apprehended in accordance with the requirements of knowledge as the conclusion of a syllogism, in which the collective data of a special case serve as minor premiss to a major premiss formed by a general law. Even on this supposition it would still be an unwarrantable undertaking to seek to limit the content of that general law itself and that relation between its constituent members which is supposed to serve as a model for the connexion between the facts given in the minor premiss. Supposing this content of the law to be symbolised by $a + \beta = f$, we are not to go on for ever attempting to deduce the title of $a + \beta$ to be accepted as the reason of f from higher and more general laws. Each of these higher laws which we might have reached would repeat the same form $a_1 + \beta_1 = f_1$ and would compel us at last to the confession that while undoubtedly a conception of the individual admits of being derived analytically from the general, the most general laws are given synthetic relations of reason and consequent, which we have simply to recognise without in turn making their recognition dependent on the fulfilment of any conditions whatever. No doubt, in the plan of the world as a whole these given relations are not isolated, unconnected, data. Any one who was able to apprehend and express this highest idea would find them bound together, not indeed necessarily by a logical connexion; but by an

æsthetic necessity and justice. From finite knowledge this actual system of reality is hidden. It has no standard at command for deciding with what combination $a+\beta$ this system associates a consequence f, to what other combination $a_1+\beta_1$ it forbids every consequence. In judging of particular phenomena the natural sciences conform to this sound principle. It is to experience alone that they look for enlightenment as to all those simplest and most primary modes of action of bodies upon each other, to which by way of explanation they reduce the individual characteristics of the given cases.

This makes us wonder the more at the general inclination to venture recklessly, just at this most decisive point, upon an *a priori* proposition of a kind from which science would shrink if it were a question of the primary laws of matter and motion, and to make the possibility of any reciprocal action depend on identity of kind and degree[1], comparability or likeness on the part of the agents between which it is to take place. Where this identity really exists, it does not help to explain anything—neither the nature of the effect nor the manner in which it is brought about. For our minds, no doubt, *a* and *a* upon coming together form the sum $2a$, but how they would behave in reality—whether one would add itself to the other, whether they would fuse with each other, would cancel, or in some way alter each other—is what no one' can conjecture on the ground of this precise likeness between them. As little can we conjecture why they should act upon each other at all and not remain completely indifferent. In spite of this likeness they were, on the supposition, two mutually independent things before they came together. Why their likeness[1] should compel them to become susceptible to each other's influence is far less immediately intelligible than it would be that difference and opposition should have this effect. These at least imply a demand for an adjustment to be effected by a new event, whereas from an existing likeness the absence of any reciprocal action would seem the thing to be naturally looked for. Such considerations however simply settle nothing. · All that we can be certain of is the complete groundlessness of every proposition which connects the possibility of reciprocal action, between things with any other homogeneity on the part of the things than that which is guaranteed by the fact of this reciprocal action. To connect the reciprocal action with *this* homogeneity is an identical proposition. *If* the things act upon, and are affected by, each other, they have just this in common that they fall under the conception of substance, of which the essence

[1] ['Gleichheit,' v. note on § 57. 'Equality' would not suit the argument here.]

is determined merely by these two predicates. But there is no other obligation to any further uniformity on their part in order to their admitting of subsumption under this conception of substance.

60. There have been two directions in which the mischievous influence of the prejudices we have been combating has chiefly asserted itself. One of its natural consequences was the effort to reduce whatever happens to a single common denomination, to discover perhaps in spatial motion, at present, for instance, in the favourite form of vibration, not one kind of event, but that in which all events, as such, consist; the primary process, variations of which— none of them being more than variations in quantity—had not only to afford to all other events, differing in kind and form, the occasions for their occurrence, but to produce them as far as possible entirely out of themselves, as an accession to their own being, though indeed an unintelligible one. This impoverishment of the universe, by re-duction of its whole many-coloured course to a mere distribution of a process of occurrence which is always identical, was in fact scarcely avoidable if every effect in respect of all that it contained was to be the analytical consequence of its presuppositions. It is enough here to have raised this preliminary protest against the ontological prin-ciples on which this reduction is founded. There will be occasions later for enlarging further upon the objections to it.

The other equally natural consequence of the prejudice in question was the offence taken at the manifold variety in the natures of things. This has been at the bottom of views now prevalent on many ques-tions, and especially on that of the reciprocal action between soul and body. On this point ancient philosophy was already under the in-fluence of the misleading view. That[1] like can only be known by like was an established superstition to which utterance had been given before the relation of causality and reciprocal action became an object of enquiry in its more general aspect. What truth there may be in this ancient view is one of the questions that must be deferred for special investigation; but I can scarcely pass it over at once, for do I not already hear the appeal, 'If the eye were not of the nature of the sun, how could it behold the light[2]?' But the finest verses do not settle any metaphysical question, and this greatly misapplied utterance of Goethe's is not an exception. To the logical analyst, in search for clearness, it conveys another impression than to the sensi-

[1] ['Gleich.']
[2]   ['Wär nicht das Auge sonnenhaft
         Wie könnte es das Licht erblicken!'—Zahme Xenien IV.]

bility that demands to be excited. It is not the eye at all that sees the sun : the soul sees it. Nor is it the sun that shines, but the seen image [1], present only in the soul, that yields to the soul the beautiful impression of illumination. Light in that sense in which it really issues from the sun—the systematic vibratory motions of the ether— we do not see at all, but there supervenes upon it owing to the nature of our soul the new phenomenon, wholly incomparable with it, of luminous clearness. What confirmation then could there be in Goethe's inspired lines for the assumption that like can only be known by like, kin by kin? To the poet it is no reproach that he should have seized and expressed a general truth of great interest in a beautiful form, though the persuasive force of that form of expression lies less in its exactness than in the seductive presentation to the mind's eye of a fascinating image. Perhaps this poet's privilege has been somewhat too freely used in these charming verses, of which the matter is false in every single fibre ; but we must candidly confess what we all feel, that at all events they express forcibly and convincingly the pregnant thought of a universal mutual relativity which connects all things in the world, and among them the knowing spirit with the object of its knowledge, and which is neither less real nor less important if it is not present in the limited and one-sided form of a homogeneity of essence. The truth on the contrary is that there is no limit to the possible number and variety of the ties constituted by this relativity, by the mutual susceptibility and reciprocal action of things. The metaphysician, who stands up for this wealth of variety against every levelling prejudice which would attenuate it without reason, is certainly in deeper sympathy with the spirit of the great poet than are those who use this utterance, itself open to some objection, as a witness in favour of a wholly objectionable scientific mistake.

61. So much by way of digression. Let us return to the object before us. It was impossible, we found, in the case of two causes operating on each other, to represent anything as passing from each to the other which would explain their reciprocal influence. Yet it appeared to be only under this condition that the conception of causal action was applicable. The only alternative left, therefore, is to render the course of the universe explicable, without presupposing this impossible action.

The first attempt in this direction is the doctrine of Occasionalism—

---

[1] [I know of no other word than 'image' by which 'Bild' can here be rendered, but it must be understood that no meaning of *likeness* attaches to the word in this connexion.]

the doctrine which would treat a relation C arising between A and B only as the occasion upon which in A and B, without any mutual influence of the two upon each other, those changes take place into a and β, which we commonly ascribe to reciprocal action between them. In this simple form there would be little in the doctrine to excite our attention. It is easy to see that an occasion which cannot be used is no occasion. But in order to be used, it must be observable by those who are to make use of it. If A and B, upon an occasion C, are to behave otherwise than they would have done upon an occasion γ, they must already in case C be otherwise affected than they would have been in case γ. That this should be so is only thinkable on supposition that some action, wherever it may have come from, has already taken effect upon them. The occasion, accordingly, which was to make it possible for the active process to be dispensed with, presupposes it on the contrary as having already taken place. Otherwise the occasion could not serve as an occasion for a further reaction. Occasionalism therefore cannot be accepted as a *metaphysical* theory. The notion that it can is one that has only been ascribed to me by a misinterpretation which I wish expressly to guard against. As I remarked above, I can only regard 'Occasionalism' as a precept of Methodology, which for the purpose of definite enquiries excludes an insoluble question—one at any rate which does not press for a solution—in order to concentrate effort upon the only attainable, or only desirable, end. If it is a question of the reciprocal action between soul and body, it is of importance to investigate the particular spiritual processes that are in fact so associated with particular bodily ones according to general rules that the manifold and complex occurrences, presented to us by our inner experience, become reducible to simple fundamental relations, and thus an approximate forecast of the future becomes possible. On the other hand, it is for this purpose a matter of indifference to know what are the ultimate means by which the connexion between the two series of events is brought about. Thus for this question as to body and soul—and it was this that, as a matter of history, the doctrine of Occasionalism was framed to meet—it may be as serviceable as for Physics, which itself is content to enquire in the first instance into the different modes of connexion between different things, not into the way in which the connexion is brought about. Metaphysics, however, having this latter problem for its express object, cannot be satisfied with passing it over, but must seek its solution.

62. Meanwhile I may mention a special expression of this view,

which is not without some plausibility. 'Why,' it will be asked, 'if it is once allowed that the relation C between A and B is the complete reason of a definite consequent F, do we go on to seek for something further by which the sequence of this consequent is to be conditioned? What power in the world could there be which would be able to hinder the fulfilment of a universal law of nature, if all conditions are fulfilled to the realisation of which the law itself attaches the realisation of its consequent?' Such is the argument that will be used, and it may be supplemented by a previous admission of our own, that whenever there is an appearance as if the occurrence of a consequent, of which all the conditions are present, were yet delayed, pending a final impulse of realisation, it will always be found on closer observation that in fact the sum of conditions was not completed and that it was for its completion, not for the mere realisation of something of which the cause was already completely given, that the missing detail required to be added [1].

This argument, however, is only a new form of an old error, and our rejoinder can do no more than repeat what is familiar. The assertion that there obtains a general law, which not only connects necessary truths with each other but reality with reality, is simply an expression of the recollection, observation, and expectation that in all cases where the condition forming the hypothesis of the law has been, is, or will be realised, the event forming its conclusion has occurred, is occurring, or will occur. We are therefore not entitled to treat the validity of the law as an independently thinkable fact, to which its supervening fulfilment attaches itself as a necessary consequence. Rather it is simply the observed or expected fulfilment itself, and we should have to fall back on the barren proposition that wherever the law fulfils itself it does fulfil itself, while the question how this result comes about would remain wholly unanswered. Or, to express the same error in another way; were we really to conceive the law to be valid merely as a law, it would follow that it was only hypothetically valid, and was not in a state of constant fulfilment: for in the latter case it would be no law, but an eternal fact. Even on this supposition it will only fulfil itself when the conditions involved in its antecedent, which form the sole legitimation of its conclusion, have been actually realised. If then the force compelling the realisation proceeded from the law, this must be incited to the manifestation of its force by the given case of its application, which implies that it must itself be otherwise affected in that case than in the case where it is not

[1] [Cp. § 53.]

applicable. We should thus be clearly presupposing an action exercised upon the law itself in order, by help of the power of the law, to dispense with the action of the things upon each other.

If, then, we decide to give up these peculiar views in which the law is treated as a thing that can act and suffer; if we allow that, whatever be the ordinance of the law, it must always be the things that take upon themselves to execute it, then A and B, at the moment when they find themselves in the relation C, must be in some way aware of this fact and must be affected by it otherwise than they would be by any other relation γ, not at present obtaining. The upshot of these considerations is that neither the validity of a general law nor the mere subsistence of a relation between two things is enough to explain the new result thereupon arising without the mediation of some action. On the contrary, what we call in this connexion the action supervening in consequence of the relation, is in fact only the reaction upon another action that precedes it and to which the things had already been subject from each other. It was our mistake to look upon this as a relation merely subsisting but not yet operative, a relation merely introducing and conditioning the causative action. The recognition of this truth is of fundamental importance. We shall be often occupied in the sequel with its further exposition. This preliminary statement of it may serve to throw light on the complete untenableness of Occasionalism even in this refined form and to show that it can as little dispense as can any other theory with the problematical process of causative action, by help of which alone it can explain how it is that a law is alternately fulfilled and not fulfilled according as its conditions are fulfilled or no.

63: Another series of kindred attempts may be grouped under the name given by Leibnitz to the most elaborate of them, that of the 'Pre-established Harmony.' In laying down the principle that 'the Monads are without windows,' Leibnitz starts from the supposition of a relation of complete mutual exclusion between the simple essences on which he builds his universe. The expression is one that I cannot admire, because I can find no reason for it, while it summarily excludes a possibility as to which at any rate a question still remained to be asked. That Monads, the powers of which the world consists, are not empty spaces which become penetrated by ready-made states through openings that are left in them, was a truth that did not need explanation, but this proved nothing against the possibility of a less palpable commerce between them, to which the name 'reciprocal action' might have been fitly applied. It would not therefore have caused me any surprise if Leibnitz had employed the same figure in an exactly opposite way and had taught that the Monads had windows, through which their inner states were communicated to each other. There would not have been less reason, perhaps there would have been more, for this assertion than for that which he preferred. To let that pass, however, when once reciprocal action had been rejected, there was nothing left for explanation of the *de facto* correspondence which takes place between the states of things but an appeal to a higher all-encompassing bond, to the deity which had designed their developments. Before the understanding of God there hover innumerable images of possible worlds: each of them so ordered in the multitude of its details as is required with consistent necessity by certain eternal laws of truth, binding for God himself and not alterable at his pleasure. In this inner arrangement of each world God can alter nothing. If in the various worlds his wisdom finds various degrees of perfection, he yet cannot unite their scattered superiorities into one wholly perfect world. His will can only grant for that one which is relatively most perfect, just as it is, admission to reality.

The further elaboration of the doctrine might be looked for in either of two different directions. It might have been expected either to take the line of confining the original determination to the general laws governing the world that has been called into existence, as distinct from the sum of the cases in which these laws may be applied, or that of supposing these cases of their application also to have been once for all irrevocably determined. The first assumption would only have led back to the embarrassments of Occasionalism just noticed. Leibnitz decided unhesitatingly for the second. Just as in our first parents the whole series of descendants is contained, with all details of their individuality, with their acts and destinies, so is every natural occurrence, down to the direction which the falling rain-drop takes to-day in the storm, completely predetermined. But this is not to be understood as if the manifold constituent agents of the world by their co-operation at each moment brought about what is contained in the next moment of the world's existence. For each single constituent the series of all its states is established from the beginning, and the inner developments of all take place after the manner of a parallel independent course, without interference with each other. The correspondence which is nevertheless maintained between them is the unavoidable consequence of their first arrangement, if we consider the world as a creation of the divine design, or simply their *de facta* character, if we consider it merely as an unalterable object of the divine intellect.

64. This notable theory impresses us in different ways, according as one or other of its features is put in clearer relief. The doctrine of a thorough mutual relation between all elements of the universe, and the other doctrine of the independence of those elements, are in it alike carried to a degree of exaggeration at which both conceptions seem to approach the unintelligible. The whole content of the Universe and of its history is supposed to be present to the divine understanding at one and the same time as a system of elements mutually and unalterably conditioned in manifold ways, so that what appears in time as following an antecedent is not less the condition of that antecedent than is any antecedent the condition of that which it precedes. Thus Leibnitz could say that not merely do wind and waves impel the ship but the motion of the ship is the condition of the motion of wind and waves. The immediate consequence of thus substituting the connexion of a system of consistent ideas for a connexion in the way of active causation is to take away all intelligible meaning from the Reality which God is supposed to have vouchsafed to this world, while he denied it to the other imaginary worlds which were present to his intellect as consistent articulations of what was contained in other ideas. The development in time adds nothing to the eternally predetermined order. It merely presents it as a succession. What new relation then is constituted for God or the world by this reality, so that it should count for something more and better than the previous presentation of the idea of a world to the mind of God? It is of no avail to say that then the world was merely thought of, whereas now it *is*. It is not open to us consistently with the system of Leibnitz, as it might be elsewhere, simply to recognise this antithesis as one that is given, however hard to define. When the supposition is that of a wise will, which had the alternative of allowing reality to an idea or of refusing it, the question, what new Good could arise merely by the realisation of what previously was present to Thought, must be plainly answered.

If the artist is not satisfied with the completed image of the work, which hovers before his mind's eye, but wishes to see it in bodily form with the bodily eye; or if the hearer of a tale betrays his interest by enquiring whether it is true; what is the source of the craving for reality in these two cases, which we may compare with the case in question? In the first case, I think, it is simply this, that there is a tacit expectation of some growth in the content of the work of art arising from its realisation. To walk about in the building as actually built is something different from the range of imagination through the details of the plan. Not only the materials of the building, but the world outside it, among the influences of which—influences subject to incalculable change—the work, when realised, is placed, create a multitude of new impressions, which the inventive fancy might indeed hope for but without being able to create the impressions themselves. This advantage of realisation is one that Leibnitz could not have had in view since his theory of the Pre-establishment of all that is contained in the world had excluded the possibility of anything new as well as the reciprocal action from which alone anything new could have issued. The other wish—the wish that a story heard may be true or (in other cases) that it may not be true, arises from the interest which the heart feels in the depicted relations of the figures brought on the scene. It is not enough that every happy moment of spiritual life should merely be a thought of the Poet and an enjoyment imparted to the hearer, of which the exhibition of unreal forms is the medium. We wish these forms themselves to live, in order that it might be possible for them also to enjoy the good which delights us in the imaginary tale. In like manner we console ourselves with the unreality of what we hear or read, if we are distressed by the images presented to us of unhappiness or wrong.

This line of thought was not excluded by the conception with which Leibnitz began, but it could only be worked out on one supposition. To give reality to an idea of a world was only worth doing if the sum of the Good was increased by the sum of those who m'ght become independent centres of its enjoyment; if, instead of that which was the object of God's approval remaining simply His thought, the beings, of whom the image and conception were included in the approved plan of a world, were enabled themselves to think it and have experience of it in their lives. I reserve the question how far this view corresponds with Leibnitz' theory. Alien to him it was not. Something at least analogous to spiritual life was accepted by him, for whatever reason, as the concrete import of the being which his Monads possessed.

65. This line of thought, however, which alone seems to me to correspond to the notion of an admission to reality of a world otherwise only present in idea to God, is scarcely consistent with the complete pre-establishment of all events. When we turn to the implications of natural science, we find that it too, if it allows no limits to its principle of causality and denies the possibility of any new starting-point for events, cannot avoid the conclusion that every detail in the established course of the universe is a necessary consequence of the past, and ultimately, though this regress can never be completed, of some state of the universe which it decides to regard as the primary state.   But it does not take this doctrine to mean that the sum of all these consequences has been fixed in some primary providential computation.   The consequences are supposed really to come into being for the first time, and the validity of universal laws is taken to be sufficient to account for their realisation without any such pre-arrangement.   These laws are enough to provide for limitation to a definite direction in the development of the new out of the old.   In their ultimate consequences the two doctrines coincide so far as this, that they lead to the belief in an irrevocable arrangement of all events. Yet in the actual pursuit of physical investigations something else seems to me to be implied.   We shrink from surrendering ourselves to this last deduction from the causal nexus.   No natural law, as expressed by a universal hypothetical judgment, indicates by itself the cases in which it comes to be applied.   It waits for the requisite points of application to be supplied from some other quarter.

We know, of course, that upon supposition of the universal validity of the causal nexus neither accident nor freedom is admissible ; that accordingly what remains undetermined in our conception of the law cannot be really undetermined ; that thus every later point of application of a law is itself only a product of earlier applications. This is admitted without qualification in reference to every limited section of reality, since behind it one still uninvestigated may be conceived in the past, as to which silence may be kept.   But with every inclination to treat the spiritual life in its turn according to like principles, we shrink from pronouncing flatly that the whole of reality, including the history of spirits, is only the successive unfolding of consequences absolutely predetermined.   That in the real passage of events something should really come to pass, something new which previously was not ; that history should be something more than a translation into time of the eternally complete content of an ordered world ; this is a deep and irrepressible demand of our spirit, under the influence of which we all act in life.   Without its satisfaction the world would be, not indeed unthinkable and self-contradictory, but unmeaning and incredible.   When we admit the universal validity of laws, it is at bottom only in the tacit hope that, among the changing points of application which are presented to those laws in the course of events, there may turn out to be new ones introduced from which the consequences of the laws may take directions not previously determined. Natural sympathy, therefore, is what the Pre-established Harmony does not command. Even if it fulfilled its metaphysical purpose, this hypothesis of Leibnitz would have an artificiality which would prevent it from commending itself to our sense of probability. I admit that this repugnance rests more upon feeling than upon theoretical reasons; more at any rate than upon such reasons as fall within the proper domain of Metaphysics. It remains, therefore, for us to enquire how far this view serves the purpose of a *theoretical* explanation of the universe.

66. In each single Monad, according to Leibnitz, state follows upon state through an immanent action, which is accepted as a fact, unintelligible indeed but free from contradiction. It was only 'transeunt' action of which the assumption was to be avoided. If this exclusion of transeunt action is to accord with the facts, the two states a and β of the Monads A and B, which observation exhibits to us as apparent products of a reciprocal action, must occur in the separate courses of development of the two beings at the same moment. If we had a right to assume that a was separated from a previous state a of A by as many intervening phases as β from a state b corresponding to a, we should not need to ascribe anything but an equal velocity to the progress of the development of all Monads. But since a may be removed from a by a larger number of phases than β from b, we should be obliged to attribute to every single Monad its special velocity of development in order to understand the coincidence of the corresponding states. This assumption does not seem to me in contradiction with the fundamental view which governs the theory in question. As was above remarked, the thought of Leibnitz approximates to that interpretation of becoming which we conceived to be the pre-supposition of Heraclitus: once grant that the being of every Thing, if the name 'Thing' is to be accepted for a closed cycle of phases, consists in a constant effort to pass from one state to another, then it is natural that different things should be distinguished from each other not merely by the direction but also by the velocity of their becoming, i.e. by an intensity of their being or reality which, if it is to express itself subject to the form of time, will appear partly at least as velocity.

I cannot recall any explanation given by Leibnitz on this point. He might have refused any answer. He might have said that the hidden rationality, without which no image of a world would have been possible at all, had provided for this correspondence of all

occurrences that go together. Only in that case it would be difficult
to say how the whole doctrine was distinguished from the modest
explanation, that everything is from the beginning so arranged that
the universe must be exactly what it is. The feeling which Leibnitz
had of the necessity of accounting in some way for the correspondence is betrayed, I think, by his reference to the example, borrowed
from Geulinx, of the two clocks which keep the same time; for it
was scarcely required as a mere illustration of the meaning of his
assertion, which is simple enough. As an explanation, however, this
comparison is of no avail. Mutual influence, it is true, the two clocks
do not exercise. But in order that they should at every moment
point to the same time, it was not enough that the artificer ordered it
so to be. And on the other hand the mechanism, which he had to
impart to them with a view to this end, is according to its idea precisely not transferable to the Monads, shut up in themselves as they
are supposed to be. Each of the two clocks, *A* and *B*, is a system
of different, mutually connected parts. The materials of which they
are constructed, as well as the movements which may be imparted to
these, are subject to general mechanical laws, which apply to one as
much as to the other. From them it follows that with reference to a
time, which is measurable according to the same standard for the rate
of motion of *A* and *B*, different quantities of matter can be so
arranged that the entire systems, *A* and *B*, can pass at the same
moments into constantly corresponding positions, *a* and *b*, *a* and *β*.
But that which in this case carries out the corresponding transition is
nothing but the 'transeunt' action, which one element by communication of its force and motion exercises on the other. The independence of mutual influence on the part of the two clocks is compen-·
sated by the carefully pre-arranged influence which the elements of
each of them exercise upon each other. It is merely the place,
therefore, of the 'transeunt' action that is shifted by this comparison.
It is not shown that it can be dispensed with in accounting for the
correspondence of the events.

All this indeed is of little importance. For it must certainly be
admitted that in this case of the clocks, as much as in any other,
Leibnitz would deny the 'transeunt' action which appears to us to be
discoverable in it. It is not, he would say, that one wheel of the
clock acts motively on the other; it is of its own impulse that the
latter wheel puts itself in motion—the motion which according to our
ordinary apprehension is the effect of the former wheel. Upon this
it may be remarked that comparisons are usually employed in order

that some process which, as described generally, seems improbable
or cannot be brought before the mind's eye, may be illustrated by an
instance in which it is presented with a clearness that allows of no
contradiction.   The cases therefore which one selects for comparison
are not such as, before they can supply the desired demonstration,
require, like Leibnitz' clocks, to be rendered by an effort of thought
into instances of the process of which a sensible illustration is sought.
Granting all this, however, our enquiry will have shown no more
than what was well known without it, that Leibnitz was never very
happy in his comparisons.   The possibility in itself of what he maintains must nevertheless be allowed.

67. For the complete reconciliation of theory and experience one
thing more is needed.   That the connexion of occurrences according to general laws is intelligible, we may, at least with reference to
all natural events, regard as a fact.   It is a fact however which, like
any other, would demand its explanation—not indeed an explanation
of how it comes about, for that would be pre-established like everything else, but an explanation of the meaning which its pre-establishment would have in the Leibnitzian theory of the universe taken as a
whole.   Images of possible worlds, to which God might vouchsafe
reality, we found distinguished from impossible ones, which must
always remain without reality.   The advantage of consistency, which
distinguishes the former sort, we might suppose to lie in this, that
they not merely combine their manifold elements according to a
plan, but that at the same time the elements which, in so doing, they
bring together are such as are really connected with each other
according to general laws.   It is obvious, that is to say, that every
.imaginary world must appear as a whole, and its development in time
as the realisation of a preconceived plan, in which for all phases of
the internally moved Monads—for a^1, a^2, a^3 . . . and β^1, β^2, β^3, as for
the several pieces of a mosaic, their sequence and their coincidence
are prescribed.   But there was no necessity for any single one of
these phases to occur more than once in this whole.   It was accordingly no self-evident necessity that there should be *general* laws—
laws connecting the repetitions of a with repetitions of β.   Without
any such repetition, these series of events might still be constantly
carrying out a predetermined plan.   It is a somewhat arbitrary interpretation which I take leave to adopt, since Leibnitz himself gives us
no light on the matter, when I understand that rationality, which
distinguishes the realisable images of worlds from the unrealisable, to
imply not merely an agreement with logical truths of thought, but

this definite character of conformity to general laws, which in itself is no necessity of thought : in other words, the fact that the demands made by the realisation of the world-plan are met by help of a multiplicity of comparable elements, which fall under common generic conceptions, and by repetitions of comparable events, which fall under general laws.

But neither with this interpretation nor without it are we properly satisfied. If in the last resort it is the greatest perfection which determines the divine choice between different rational images of worlds, is it then self-evident that among the indispensable preconditions of the perfection is to be reckoned above all this conformity to universal law, and that anything which lacked it was not even open to choice ? For the coherence of our scientific efforts this conformity to law, which is the sole foundation for our knowledge of things, has indeed attained such overpowering importance, that its own independent value seems to us almost unquestionable. Yet, after all, is it certain that intrinsically a greater good is attained, if every a is always followed by the same β, than if it were followed sometimes by β, sometimes by γ, sometimes by δ, just as was at each moment required by the constantly changing residue of the plan still to be fulfilled ? Might there not be as good reason to find fault with those general laws as at bottom vexatious hindrances, cutting short a multitude of beautiful developments which but for their troublesome intervention might have made the system of the most perfect world still more perfect ? If we pursue this thought, it becomes clear what is for us the source of confidence in the necessary validity of universal laws. In a dream, which needs no fulfilment, we find a succession possible of the most beautiful events, connected only by the coherence of their import : and the case would be the same if a realisation of this dream could come about through the instantaneous spell of its admission as a whole to reality, without the requirement by each successive constituent of a labour of production on the part of the previous ones.

If, on the other hand, we follow our ordinary conception of the world which finds this labour necessary, the state of the case is different. Supposing that in the moment t an element a of the world happened to be in the state a, and supposing it to be indispensable that, in order to the completion of the plan of the world or to the restoration of its equilibrium or to consecutiveness in its development, at the same moment t, b also should pass into the state β, then the fact s of this necessity, i.e. the present state of the remaining elements, R, of the world together with the change of a into a, must exert an action upon b. But in order that only β and not any other consequence may arise in b, s and β—therefore also a and β—must merely in respect of their content, without reference to the phase of development of the universe as a whole, belong together as members that condition each other : and for that reason in every case of the repetition of a the same consequence β will occur, so far as it is not impeded by other relations that condition the state of the case for the moment. Upon this supposition, therefore, which is habitual to us, that the course of the world is a gradual becoming produced by active causation, its connexion according to general laws appears to us to be necessary. But this way of thinking is not reconcileable with the views of Leibnitz. He looks upon the whole sum of reality as predetermined in all the details of its course and as coming into being all at once through that mysterious admission to existence which he has unhappily done so little to define. No work is left to be gradually done within it. But if this supposition is granted him, the limitation of realisability to such projected worlds as have their elements connected according to general laws is an arbitrary assumption. Any combination whatever of manifold occurrences—any dream—might in this way have just as well obtained a footing in reality. We have here therefore an inconsistency in Leibnitz' doctrine. If the necessity of general laws was to be saved from disappearing, there were only, it would seem, two ways of doing it. He should either have exhibited them as a condition of that perfection of the world which renders it worthy of existence—and it is not improbable that he would have decided for this alternative—or we should have given up the attempt to substitute for the unintelligible action of one thing on another an even more unintelligible pre-establishment of all things.

CHAPTER VI.

The Unity of Things.

68. THERE is ohly one condition, as we have found, under which the conception of a 'transeunt' operation can be banished from our view of the world and replaced by that of a harmony between independent inner developments of Things. The condition is that we make up our minds to a thoroughly consistent Determinism, which regards all that the world contains as collectively predetermined to its minutest details. So long, however, as we shrink from this conclusion, and cling to the hope, for which we have in the meantime no justification but which is still insuppressible, that the course of Things in which we live admits of events being initiated, which are not the necessary consequence of previous development—so long as this is the case the assumption of 'transeunt' operation cannot be dispensed with by help either of the theory of a predetermined sympathetic connexion, or by that of an unconditioned validity of universal laws. Our final persuasion, therefore, might seem to depend on the choice we make between the two above-mentioned pre-suppositions (that of complete determinism, and that which allows of new beginnings)—a choice which theoretical reasons are no longer sufficient to decide. But if this were really the case—a point which I reserve for later investigation—the option left open to us would be a justification for developing, in the first place hypothetically, the further conceptions which we should have to form as to 'transeunt' operation if having adopted the second of the suppositions stated we maintained the necessity of assuming such operation. I cannot however apply myself to this task without once again repeating, in order to prevent misunderstandings, a warning that has already been often given.

My purpose cannot be to give such a description of the process by which every operation comes about as may enable the reader to present it to his mind's eye, and thus by demonstrating *how* it happens to give the most convincing proof that it *can* happen. The object in view is merely to get rid of the difficulties which make the conception of a ' transeunt ' operation obscure to us while, although in fact understanding just as little how an ' immanent ' operation comes about, we make no scruple about accepting it as a given fact. How in any case a condition, if realised, begins in turn to give reality to its effect, or how it sets about uprooting a present state of anything and planting another state in the real world — of that no account can be given. Every description that might be attempted would have to depict processes and modes of action which necessarily presuppose the very operation that has to be explained as already taking place many times over between the several elements which are summoned to perform it. Indeed the source of many of the obscurities attaching to our notion of operation lies in our persistent effort to explain it by images derived from complex applications of the notion itself, which for that reason lead necessarily to absurdity if supposed to have any bearing on its simplest sense. If we avoid these unprofitable attempts, and confine ourselves to stating that which operation actually consists in, we must state it simply thus: that the reality of one state is the condition of the realisation of another. This mysterious connexion we allow so long as its product is merely the development of one and the same Being within the unity of that Being's nature. What seems unthinkable is how it can be that something which occurs to one Being, *A*, can be the source of change in another, *B*.

69. After so many failures in the attempt to bridge a gulf of which we have no clear vision, in the precise mode demanded by imagination, we can only hope for a better result if we make the point clear in which the cause of our difficulty lies. In the course of our consideration of the world we were led, at the outset, to the notion of a plurality of Things. Their multiplicity seemed to offer the most convenient explanation for the equally great multiplicity of appearances. Then the impulse to become acquainted with the unconditioned Being which must lie at the foundation of this process of the conditioned was the occasion of our ascribing this unconditioned Being without suspicion to the very multiplicity of elements which we found to exist. If we stopped short of assigning to every reality a pure Being that could dispense with all relations to other Beings, yet even while allowing relations we did not give up the independence of Things as against each other which we assumed to begin with. It was as so many independent unities that we supposed them to enter into such peculiar relations to each other as compelled their self-sufficing natures to act and react upon each other. But it was im-

possible to state in what this transition from a state of isolation to metaphysical combination might consist, and it remained a standing contradiction that Things having no dependence on each other should yet enter into such a relation of dependence as each to concern itself with the other, and to conform itself in its own states to those of the other. This prejudice must be given up. There cannot be a multiplicity of independent Things, but all elements, if reciprocal action is to be possible between them, must be regarded as parts of a single and real Being. The Pluralism with which our view of the world began has to give place to a Monism, through which the 'transeunt' operation, always unintelligible, passes into an 'immanent' operation.

A first suggestion of the impossibility of that unlimited pluralism was, strictly speaking, afforded as soon as we felt the necessity of apprehending the events which form the course of the world, as Consequents that can be known from Antecedents. If no elements of the world admitted of comparison any more than do our feelings of sweet and red, it would be impossible that with the union of the two A and B in a certain relation C there should be connected a consequence F, to the exclusion of all other consequences. For in that case the relation of A to B, which alone could justify this connexion, would be the same—the two elements being completely incomparable and alien to each other—as that between any two other elements, A and M, B and N, M and N. There would accordingly be no legitimate ground for connecting the consequence with one rather than another pair of related elements, or indeed for any definite connexion whatever. Hence it appears that the independent elements of the world, the many real essences which we supposed that there were, could by no means have had unlimited licence of being what they liked as soon as each single one by simplicity of its quality had satisfied the conditions under which its 'position' was possible. Between their qualities there would have had to be throughout a commensurability of some kind which rendered them, not indeed members of a single series, but members of a system in which various series are in some way related to each other. All however that this primary unity necessarily implied on the part of the elements of the world was simply this commensurability. Their origin from a single root, or their permanent immanence in one Being, it only rendered probable. It is not till we come to the consideration of cause and effect that we find any necessity to adopt this further view—to hold that Things can only exist as parts of a single Being, separate relatively to our apprehension, but not actually independent.

70. This conclusion of our considerations requires so much to be added in the way of justification and defence that to begin with my only concern is to explain it. Let M be the single truly existing substance, A, B, and R the single Things into which, relatively to our faculties of presentation and observation, the unity of M somehow resolves itself—A and B being those upon the destinies of which our attention has to be employed, R the sum of all the other things to which has to be applied, by help of analogy, all that we lay down about A and B. Then by the formula $M = \phi (A B R)$ we express the thought that a certain definite connexion of $A B$ and R, indicated by ϕ, exhibits the whole nature of M.

If we allow ourselves further to assume that one of the individual elements has undergone a transition from A into a—however the excitement to this transition may have arisen—then the former equation between $\phi (a B R)$ and M will no longer hold. It would only be re-established by a corresponding change on the part of the other members of the group, and $\phi (a b R^1) = M$ would anew express the whole nature of M. Let us now admit the supposition that the susceptibility, which we had to recognise in every finite Being—a susceptibility in virtue of which it does not experience changes without maintaining itself against them by reaction—that this belongs also to the one, the truly existing M; then the production of the new states b and R^1 in B and R will be the necessary consequence of the change to a that has occurred in A. But this change a was throughout not merely a change of the one element A, for such a change would have needed some medium to extend its consequences to B and R. It was at the same time, without having to wait to become so, a change of M, in which alone, in respect of Being and content, A has its reality and subsistence. In like manner this change of M does not need to travel, in order as by transition into a domain not its own, to make its sign in B and R. It too, without having to become so by such means, is already a change of B and R, which in respect of what they contain and are, equally have reality and subsistence only in M. Or— if we prefer another expression, in which we start from the apparent independence of $A B$ and R—the only mediation which causes the changes of B and R to follow on those of A consists in the identity of M with itself, and in its susceptibility which does not admit a change a without again restoring the same nature M by production of the compensatory change b and R^1. To our observation a presents itself as an event which takes place in the isolated element A; b as a second event which befalls the equally isolated B. In accordance with this appearance we call that a 'transeunt' operation of A upon B, which in truth is only an immanent operation of M upon M. A process thus seems to us to be requisite to bring the elements A and B, originally indifferent towards each other, into a relation of mutual sympathy. In truth they always stand in that relation, for at every moment the reality which they simultaneously possess has its connexion in the import of M, and A or a is the complement to B and R, or to b and R^1 (as the case may be), required by M in order to the maintenance of its equality with itself, just as B or b is the complement required to A and R or to a and R^1.

Our earlier idea, therefore, of manifold original essences, unconditionally existing and of independent content, which only came afterwards to fall together into variable actions and reactions upon each other, passes into a different idea, that of manifold elements, of which the existence and content is throughout conditioned by the nature and reality of the one existence of which they are organic members; whose maintenance of itself places them all in a constant relation of dependence on each other as on it; according to whose command, without possibility of offering resistance or of rendering any help which should be due to their own independent reality, they so order themselves at every moment that the sum of Things presents a new identical expression of the same meaning, a harmony not pre-established, but which at each moment reproduces itself through the power of the one existence.

71. Before passing to details, let me remark that I would not have these statements regarded as meant to describe a process which needed to be hit upon by conjecture, and did not naturally follow from the metaphysical demand which it was its purpose to satisfy. Or, to use another expression, I do not imagine myself to have stated what we have to think in order to render reciprocal action intelligible, but what we in fact do think as soon as we explain to ourselves what we mean by it. If we suppose a certain Being A to conform itself to the state b of another Being B and to fall into the state a, this thought directly implies the other, that the change b which at first seemed only to befall B is also a change for the other Being, A. There may be required investigation of the mode in which b is a change also for A, but there can be no doubt that it has to be brought under the same formal conception of a state of A which we at first only applied to a. But the idea that the states of a Being B are at the same time states of another Being A, involves the direct negation of the proposition that A and B are two separate and independent

Beings: for a unity of the exclusive kind by which each would set a barrier between itself and the other, if it is to be more than verbally maintained—if it is to be measured according to what may be called its practical value—can only consist in complete impenetrability on the part of the one against all conditions of the other.

Thus it was not necessary that the unity of all individual Beings should be conjectured or discovered as an hypothesis enabling us to set aside certain difficulties that are in our way. It is, as it seems to me, a thought which by mere analysis can be shown to be involved in the conception of reciprocal action. If we fancy it possible to maintain that Things are to begin with separate and mutually independent Unities, but that there afterwards arises between them a relation of Union in operation, we are describing, not an actual state of Things or a real process, but merely the movement of thought which begins with a false supposition and afterwards, under the pressure of problems which it has itself raised, seeks in imperfect fashion to restore the correct view which it should have had to start with.

72. Moreover, in the logical requisites of a theory, this view of the original unity of all Things in M is by no means inferior to the other view of their changeable combinations. It might be urged indeed that our view represents all Things too indiscriminately as comprehended once for all in the unity of M, and thus has no place for the gradations that exist in the intimacy of their relations to each other; that the opposite view, by recognising on the one hand the progress from a complete absence of relation to an ever greater closeness of relation, and on the other the relaxation of relations that previously existed, alone admits of due adjustment to experience, which testifies in one case to a lively action and reaction of Things upon each other, in another to their mutual indifference. In truth the reverse seems to me to be the case. So far we regard M as expressing only the formal thought of the one all comprehensive Being. As to the concrete content of that which is to occupy this supreme position of M we know nothing, and therefore can settle nothing as to the form ϕ, in which according to its nature it at each moment comprehends the sum of finite realities. There is nothing, however, against our assuming the possibility of the various equations; $M = \phi\,(A\,B\,R)$, $M = \phi\,(A\,B\,r\,\rho)$, $M = \phi\,(A\,\beta\,R^1)$, $M = \phi\,(a\,\beta\,R)$. Of these equations the second would express the possibility of a change in the sum of the members R into r—a change which is balanced by a second ρ, and therefore does not require a compensatory change on the part of A and B. This being so, the two latter would appear unaffected by the alteration of the rest of the world in which they are included. Of the third equation the meaning would be that another change of R, viz. into R^1, only requires a change β in B, to which A would appear indifferent; while the fourth would represent a reciprocal action which exhausts itself between A and B, leaving the rest of the world unaffected.

It thus appears that our view is not irreconcileable with any of the gradations which the mutual excitability of the world's elements in fact exhibits. There would be nothing to prevent us even from ascribing to the unity, in which they are all comprehended, at various moments various degrees of closeness down to the extreme cases in which two elements, having no effect whatever on each other, have all the appearance of being two independent entities; or in which, on the other hand, limited to mutual operation, they detach themselves from all other constituents of the world as a pair of which each belongs to the other. But the source of these gradations would not be that elements originally independent were drawn together by variable relations ranging in intensity from nought to any degree we like to imagine. Their source would be that the plan of that unity which holds things permanently together, obliges them at every moment either to new reciprocal action of definite kind and degree or to the maintenance of their previous state, which involves the appearance of deficient reciprocal action. Thus the reason why things take the appearance of independence as against each other is not that the Unity M, in which they are always comprehended, is sometimes more, sometimes less, real, or even altogether ceases to be, but that the offices which M imposes on them vary: so that every degree of relative independence which things exhibit as against each other is itself the consequence of their entire want of independence as against M, which never leaves them outside its unity. That relations, on the other hand, which did not previously subsist between independent things, can never begin to subsist, I have already pointed out, nor is it necessary to revert to this impossible notion.

73. The next question to be expected is, not indeed what M consists in but how, even as a mere matter of logical relation, the connexion assumed between it, the One, and the multiplicity of elements dependent on it is to be thought of. We have contented ourselves with describing these elements as parts of the infinite M. We should find no lack of other designations if we cared to notice all the theories which the history of philosophy records as having on various grounds arrived at a similar Monism. We might read of

modifications of the infinite substance, of its developments and dif-
ferentiations, of emanations and radiations from it.   Much discussion
and enthusiasm has gathered round these terms.   Their variety serves
in some measure to illustrate the variety of the needs by which men
were led to the same persuasion.   Stripped of their figurative clothing
—a clothing merely intended to serve the unattainable purpose of
presenting to the mind's eye the process by which the assumed relation between the one and the multitude of finite beings is brought
about—all that they collectively contain in regard to the import of
this relation amounts merely to a negation.   They all deny the independent reality of finite things, but they cannot determine positively
the nature of the bond which unites them.

This inability by itself would not to my mind form any ground of
objection to the view stated.   The exact determination of a postulate,
whether effected by means of affirmations or by means of negations,
may claim to be a philosophic result even when it is impossible to
present anything to the mind's eye by which the postulate is fulfilled.
An intuition, however—a presentation to the mind's eye—of that
which according to its very idea is the source of all possibility of
intuition—is what we shall not look for.   Neither the One, before its
production of the manifold capable of arrangement in various outlines, nor the metaphysical process, so to speak, by which that production is brought about, can be described by help of any figure, for
the possibility of presentation as a figure depends on the previous existence of the manifold, and the origin of the manifold world in the
case before us is just the point at issue.   But it does not follow that
there is no meaning in the conception of that relation of dependence
of the many upon the one.   Though unable to state what constitutes
the persistent force of the bond which connects individual things in
reality, we can yet seek out the complex modes in which its un-
imaginable activity conditions the form of their connexion : and the
general ideas, which I have already indicated on the subject, in their
application to our given experience, warrant the hope, on this side,
of an unlimited growth of our knowledge.

74. In saying this however I do not overcome the objection which
our view excites.   It will readily be allowed that the relation of the
One being to the many does not admit of being exhibited in any
positive way.   It will be urged however that it ought not to involve
a contradiction if it is to be admitted even as a postulate ; yet how is
it to be conceived that what is one should not only cause a manifold
to issue out of itself, but should continue to be this manifold?   This ·

question has at all times formed one of the difficulties of philosophy for the reason that in fact, whatever may have been the point of departure, a thousand ways lead back to it. I need not go further back than the latest past of German philosophy. For the idealistic systems, which ended in Hegel, not merely the relativity of everything finite, but also the inner vitality of the infinite which projects the fullness of the manifold out of its unity, was a primary certainty which forced itself on the spirit with an æsthetic necessity and determined every other conviction accordingly. It must be allowed that this prerogative of the so-called reason in the treatment of things, as against the claims made by the understanding on behalf of an adherence to its law of identity, has been rather vigorously asserted than clearly defended against the attacks made on it in the interest of this law. In the bold paradox, that it is just in contradiction that there rests the deepest truth, that which had originally been conceived as the mystery of things came to be transferred in a very questionable way to our methods of thought. There ensued in the philosophy of Herbart a vigorous self-defence on the part of formal logic against this attack—a defence which no doubt had its use as restoring the forms of investigation that had disappeared during the rush and hurry of 'dialectical development,' but which in the last resort, as it seems to me, can only succeed by presupposing at the decisive points the actual existence, in some remote distance, of that unity of the one and the many, which in its metaphysic it was so shy of admitting. On this whole question, unless I am mistaken, there is not much else to be said than what is objected by the young Socrates in the 'Parmenides' to the assertions of Zeno. 'Is there not one idea of likeness and another of unlikeness? And are we not called like or unlike according as we partake in one or the other? Now if something partook in each of the opposed ideas, and then had to be called like and unlike at the same time, what would there be to surprise us in that? No doubt if a man tried to make out likeness as such to be equivalent to unlikeness as such, that would be incredible. But that something should partake in both ideas and in consequence should be both like and unlike, that I deem as little absurd as it is to call everything one on account of its participation in the idea of unity and at the same time many on account of its equal participation in the idea of multiplicity. The only thing that we may not do is to take unity for multiplicity, or multiplicity for unity.'

75. It may seem at first sight as if Socrates had only pushed the

difficulty a step further back. The possibility, it may be said, of simultaneous participation in those two ideas is just what the laws of thought forbid to every subject. With this objection I cannot agree. I have previously pointed out the merely formal significance of the principle of identity. All that it says is that $A = A$; that one is one and that many are many; that the real is real and the impossible impossible; in short, that every predicate is equivalent to itself, and every subject no less so. By itself it says nothing as to the possibility of attaching several predicates simultaneously, or even only one, to a single subject. For that which we properly mean by connecting two thinkable contents S and P, as subject and predicate—the metaphysical copula subsisting between S and P which justifies this mode of logical expression—is what cannot itself be expressed or constructed by means of any logical form. The only logical obligation is when once the connexion has been supposed or recognised, to be consistent with ourselves in regard to it. Therefore the law of excluded middle in its unambiguous form asserts this, and only this; that of two judgments which severally affirm and deny of the same subject S the same predicate P only one can be true. For even that metaphysical copula, which unites S and P, whatever it may consist in, must be equivalent to itself. If it is V, it cannot be non-V; if non-V, it cannot be V. Thus the propositions, S is P, and S is not P, are irreconcileable with each other; but the propositions, S is P, and S is non-P, are reconcileable until it is established as a matter of fact that there is no non-$P = Q$ which can be connected with S by a copula, W, that is reconcileable with V. No one therefore disputes the simultaneous validity of the propositions, ' the body S is extended P,' and ' S has weight Q.' Logic finds them compatible. It could not however state the reason of their compatibility, for the metaphysical copula, V, between S and P—i. e. the real behaviour on the part of the body which constitutes its extension, or the mode in which extension attaches to its essence—is as unknown as the copula W—the behaviour which makes it heavy. Still less could we show positively how it is possible for V and W to subsist undisturbed along with each other. That is and remains a mystery on the part *of the thing*.

Let us now apply these considerations to the matter in hand. If M is one, then it is untrue that it is *not* this unity, P. If it is many, then it is impossible that it should not be this multiplicity, Q. If it is at once unity and multiplicity, then it is impossible that either should be untrue of it. But from the truth of one determination there is no inference to the untruth of the other. This would only be the case if it could be shown that the concrete nature of M is incapable of uniting the two modes of behaviour in virtue of which severally it would be unity and multiplicity. On the contrary, it might be held that their reconcileability is logically shown by pointing out that the apparently conflicting predicates are not applicable to the same subject, since it was not the one M that we took to be equivalent to many M, but the one unconditioned M that we took to be equivalent to the many conditioned m. But, although this is correct, yet the material content of our proposition is inconsistent with this logical justification. For M was supposed to be neither outside the many m nor to represent their sum. It was supposed to possess the same essential being, that of a real existence, which belongs to every m. Not even the activity which renders it one would, upon our view, be other than that which renders it many. On the contrary, by the very same act by which it constitutes the multiplicity, it opposes itself to this as unity, and by the same act by which it constitutes the unity it opposes itself to this as multiplicity. Thus here, if anywhere, we expressly presuppose the essential unity of the subject to which we ascribe at once unity and multiplicity.

At the same time that other consideration must be insisted on ; that it is quite unallowable to leave out of sight the peculiar significance of the whole procedure which our theory ascribes to M, and to generate a contradiction by thinking of unity and multiplicity as united with M in that meaningless way which the logical schemata of judgment express by the bald copula, *is*. If this word is to have an unambiguous logical meaning of its own, it can only be the meaning of an identity between the content of two ideas as such. The various meanings of the metaphysical copula, on the contrary, it never expresses—that copula which, as subsisting between one content and another, justifies us in connecting them, by no means always in the same sense, but in very various senses, as subject and predicate. While it cannot be denied, then, that the one *is* the many, if we must needs so express ourselves, still in this colourless expression it is impossible to recognise what we mean to convey. The one is by no means the many in the same neutral sense in which we might say that it is the one. It is the many rather in the active sense of bringing it forth and being present in it. This definite concrete import of our proposition—the assertion that such procedure is really possible—is what should have been disputed. There is no meaning whatever in objections derived from the treatment of unity and multiplicity, *in abstracto,* apart from their actual points of relation, as opposite conceptions. That they are, and cannot but be so opposed, is self-evident. Every one allows it the moment he speaks of a unity of the manifold. For there would be no meaning in what he says if he did not satisfy the principle of identity by continuing to understand unity merely as unity, multiplicity merely as multiplicity. Neither this principle, then, nor that of excluded middle, is violated by our doctrine. On the other hand, they are alike quite insufficient to decide the possibility of a relation, of which the full meaning cannot be brought under these abstract formulæ. In applying them we fall into an error already noticed. From the laws which our thought has to observe in connecting its ideas as to the nature of things, we deem ourselves able immediately to infer limitations upon what is possible in this nature of things.

76. I must dwell for a moment longer on this point, which I previously touched upon. Reality is infinitely richer than thought. It is not merely the case that the complex material with which reality is thronged can only be presented by perception, not produced by thought. Even the universal relations between the manifold do not admit of being constructed out of the logical connexions of our ideas. The principle of identity inexorably bids us think of every A as $= A$. If we followed this principle alone and looked upon it as an ultimate limit of that which the nature of reality can yield, we should never arrive at the thought of there being something which we call Becoming. Having recognised, however, the reality of becoming, we persuade ourselves that it at every moment satisfies the principle of Identity, though in a manner which outrages it in the total result, and that its proper nature can be comprehended by no connexion, which Logic allows, of elements identical or not-identical. For certainly if a passes through the stages a^1 a^2 a^3 into b, it is true that at each moment $a=a$, $a^1=a^1$, $a^2=a^2$, $a^3=a^3$, $b=b$, and the principle of Identity is satisfied; but, for all that, it remains the fact that the same a which was real is now unreal, and the b which was unreal is real. How this comes about—how it is that the reality detaches itself from one thing, to which it did belong, and attaches itself to another from which it was absent—this remains for ever inexplicable by thought, and even the appeal to the lapse of time does not make the riddle clearer. It is true that between the extremities, a and b, of that chain, our perception traverses the intermediate links, a^1, a^2, and so on. But each of these passes in an indivisible moment into its successor. If we thought of a^2 as broken up into the new chain a_1 a_2 a_3,

each of these links in turn would be identical with itself, so long as it remained in existence, and even if the immediately sequent a_4 were separated by an interval of empty time from a_3, still the transition of a_3 from being into not-being would have to be thought of as taking place in one and the same moment, and could not be expanded into a new series of transitions.

Undoubtedly therefore, if we want to *think of* Becoming, we have to face the requirement of looking upon being and not-being as fused with each other. This, however, does not imply that the import of either idea is apprehended otherwise than as identical with itself and different from the other. How the fusion is to be effected we know not. Even the intuition of Time only presents us with the *de facto* solution of the problem without informing us how it is solved. But we know that in fact the nature of reality yields a result to us un-thinkable. It teaches us that being and not-being are not, as we could not help thinking them to be, contradictory predicates of every subject, but that there is an alternative between them, arising out of a union of the two which we cannot construct in thought. This ex-plains how the extravagant utterance could be ventured upon, that it is just contradiction which constitutes the truth of the real. Those who used it regarded that as contradictory which was in fact superior to logical laws—which does not indeed abrogate them in their legitimate application, but as to which no sort of positive conjecture could possibly be formed as a result of such application.

77. The like over-estimate of logical principles, the habit of re-garding them as limitations of what is really possible, would oblige us to treat as inadmissible the most important assumptions on which our conception of the world is founded. All ideas of conditioning, of cause and effect, of activity, require us to presuppose connexions of things, which no thought can succeed in constructing. For thought occupies itself with the eternally subsisting relations of that which forms the content of the knowable, not with real existence and with that which renders this existence for ever something more than the world of thoughts. In regard, however, to all the rest of these assumptions the imaginings of 'speculation' have been busied, though in our eyes ineffectually, in banishing them from our theory of the world. It was only Becoming itself that it could not deny, even after reducing professedly every activity to a relation of cause and effect, and every such relation to a mere succession of phenomena. Even if in the outer world it substituted for the actual succession of events a mere appearance of such succession, it could not but

recognise a real Becoming and succession of events at least in those beings in and for which the supposed appearance unfolded itself. It is to this one instance, therefore, of Becoming, that we confine ourselves in order to convey the impression of how much may exist in reality without possibility of being reproduced by a logical connexion of our thoughts. One admission indeed must be made. Of the fact of Becoming at any rate immediate perception convinced us. It cannot similarly convince us that the connexion which we assumed between the one unconditioned real and the multiplicity of its conditioned forms, is more than a postulate of our reflection, that it is a problem eternally solved in a fashion as mysterious as is Becoming itself.

This makes it of the more interest to see how this requirement of the unity of the manifold, in one form or another, is always pressing itself upon us anew. Even the metaphysic of Herbart, though so unfavourably disposed to it, has to admit it among those ' accidental ' ways of looking at things, by which it sought to make the perfectly simple qualities, a and b, of real beings, so far comparable with each other as to explain the possibility of a reciprocal action taking place between them. If the simple a was taken to $= p + x$, the no less simple b to $= q - x$, these substitutions were to be called ' accidental ' only for the reason that the preference of these to others depended on the use to which it was intended to put them, not on the nature of the things. If the object had been the explanation of another process, a might just as well have been taken to $= r + y$ in order to be rendered comparable with (say) $c = s - y$. However unaffected, therefore, by these ' accidental ' modes of treatment the essence of things might be held to be, their application always involves the presupposition that the perfect simplicity of quality, from which any sort of composition is held to be excluded, may in respect of its content be treated as absolutely equivalent not merely to some one but to a great number of connected multiplicities.

The ease with which, in mathematics, a complex expression can be shown to be equivalent to a simple one, has made the application of this view to the essence of things seem less questionable than it is. For that which is indicated by those simple mathematical expressions makes no sort of claim to an indissoluble metaphysical unity of content as do the real essences. On the contrary, the possibility of innumerable equivalents being substituted for a rests in this case on the admitted infinite divisibility of a, which allows of its being broken up, and the fragments recompounded, in any number of forms; or else, in geometry, on the fact that *a* is included in a system of relations of position, which implies the possibility in any given case of bringing into view those external relations of *a* to other elements of space by which it may contribute to the solution of a problem proposed without there being any necessity for an alteration in the conception of the content of *a* itself. The essence of things cannot be thought of in either of these ways. The introduction of mathematical analogies could only serve to illustrate, not to justify, this metaphysical use of 'accidental' points of view. Whoever counts it admissible maintains, in so doing, the new and independent proposition that the unity of the uncompounded quality, by which one real essence is distinguished from another, is identical with many mutually connected multiplicities.

78. A further step must be taken. The 'accidental views' are not merely complex expressions, by which our thought according to a way of its own contrives to present to itself one and the same simple essence; not merely our different ways of arriving at the same end. The course of events itself corresponds to them. In the presentation of *a* as $=p+x$ and of *b* as $=q-x$ there was more than a mere view of ours. In the opposition that we assumed to take place between $+x$ and $-x$, which would destroy each other if they could, lay the active determining cause of an effort of self-maintenance on the part of each being, which was not elicited by the mutually indifferent elements, *p* and *q*. Now whether we do or do not share Herbart's views as to the real or apparent happening of what happens and as to the meaning of self-maintenance, this in any case amounts to an admission that not merely the content of the simple qualities is at once unity and multiplicity, but also that the things, so far as they are things, in their doing and suffering are at once one and many. It is only with that element *x* of its essence that *a* asserts itself and becomes operative, which finds an opposite element in *b*. But for all that *x* remains no less in indissoluble connexion with *p*, which for the present has no occasion for activity, and which would come into play if in another being *d* it met with a tendency, $-p$, opposed to it.

For reasons to be mentioned presently I cannot adopt this way of thinking. I have only pursued it so far in order to show that it asserts the unity of the manifold, and that in regard to the real, though in a different place from that in which it seemed to me necessary. That which in it is taken to be true of every real essence is what in our theory is required of the one Real; except that with

Herbart that abrupt isolation of individual beings continues in which we find a standing hindrance to the real explanation of the course of the world. Herbart was undoubtedly right in holding that an unconditioned was implied in the changes of the conditioned. But there was no necessity to seek this unconditioned straightway in the manifold of the elements which no doubt have to be presupposed as proximate principles of explanation for the course of events. The experiment is not made of admitting this multiplicity, but only as a multiplicity that is conditioned and comprehended in the unity of a single truly real Being. Yet it is only avoided at the cost of admitting in the individual real a multiplicity so conditioning itself as to become one, of the very same kind as that which is ostensibly denied to the Real as a whole.

79. I return once more to Leibnitz. He too conceives manifold mutually-independent Monads as the elements of the world, in antithesis, however, to the unity of God, by whose understanding, according to Leibnitz, is determined the content of what takes place in the world, even as its reality is determined by his will. If we can make up our minds to abstain from at once dismissing the supports drawn from a philosophy of religion, which Leibnitz has given to his theory, there is nothing to prevent us from going back still further to an eternally mobile Phantasy on the part of God, the creative source of those images of worlds which hover before His understanding. Those of the images which by the rationality of their connexion justify themselves to this understanding are the possible worlds the best among which His will renders real. Now so long as we think of a world-image, *A*, as exposed to this testing inspection on the part of the divine Being, so long we can understand what is meant by that truth, rationality or consistency, on which the possibility of its realisation is held to depend. It is the state of living satisfaction on the part of God, which arises out of the felt frictionless harmony between this image as unfolding itself in God's consciousness and the eternal habits of his thought. In this active divine intelligence which thinks and enjoys every feature of the world image in its connexions with other features—in it which knows how to hold everything together—the several lines of the image are combined and form not a scattered multiplicity but the active totality of a world which is possible because it forms such a complete whole. I have previously noticed the difficulty of assigning any further determination which accrues to this world, already thought of as possible, if it is not merely thought but by God's will called into reality. Howsoever this may be, it could only enjoy this further something which reality yielded under one of two conditions. It must either continue within the inner life of God as an eternal activity of his Being, or enter on an existence of its own, as a product which detaches itself from him, in an independence scarcely to be defined.

The first of these suppositions—that of the world's Immanence in God—we do not further pursue. It will lead directly back to our view that every single thing and event can only be thought as an activity, constant or transitory, of the one Existence, its reality and substance as the mode of being and substance of this one Existence, its nature and form as a consistent phase in the unfolding of the same.

If, on the other hand, we follow Leibnitz in preferring the other supposition that the real world is constituted by a sum of developments of isolated Monads—developments merely parallel and not interfering with each other, in what precise form has this world preserved the very property on which rested its claim to be called into reality? I mean that truth, consistency, or rationality, which rendered it superior to the unrealisable dreams of the divine Phantasy? What would be gained by saying that in this world, while none of its members condition each other, everything goes on as if they all did so; that accordingly, while it does not .really form a whole, yet to an intelligence directed to it, it will have the appearance of doing so; that, in one word, its reality consists in a hollow and delusive imitation of that inner consistency which was pronounced to be, as such, the ultimate reason why its realisation was possible? I can anticipate an objection that will here be made; doubtless, it will be said, between the elements of this world there exist reciprocal conditions, though it may not follow that the elements actually operate on each other in accordance with these conditions; they exist in the form of a sum of actually present relations of all elements to all, but the presence of these relations does not imply an Intelligence that comprehends them; like any truth, they continue to hold though no one thinks of them.

The substance of what I have to say against the admissibility of such views I postpone for a moment. Here I would only remind the reader that all this might equally be said of the unrealised world-image *A* as supposed to be still hovering before the divine understanding. At the same time something more might be said of it. For in this living thought of God it was not merely the case that a part *a* of this image stood to another part *b* in a certain relation,

which might have been discovered by the attention of a mind directed
to it. For in fact this consciousness actually was constantly directed
to it, and in this consciousness, in its relating activity, these relations
had their being. The presentation of *a* was in fact in such an instance the efficient cause which brought the presentation of *b* into
the divine consciousness, or—if this is held to be the office of the
Phantasy—which at any rate retained it in consciousness and recognised it as the consistent complement to *a*. The active conditioning of *b* by *a* is absent from the elements of reality and is expressly
replaced, according to the theory in question, by the mere coexistence, without any active operation of one on the other, of things the
same in content with the presentations of the divine consciousness.
Thus, to say the least, the realised world, so far from being richer,
is poorer in consequence of its supposed independent existence as
detached from the Divine Being—in consequence of its course resulting no longer from the living presence of God but only from an
order of relations established by him. The requirement that God
and the world should not be so blended as to leave no opposition
between them is in itself perfectly justified. But the right way to
satisfy it would have been not by this unintelligible second act of
constitution, by the realisation of what was previously an image of a
merely possible world, but by the recognition that what in this theory
is presented as a mere possibility and preliminary suggestion (to the
mind of God) is in fact the full reality, but that nevertheless the one
remains different from all the manifold, which only exists in and
through the one.

80. I now return to the thesis, of which I just now postponed
the statement for an instant. It at once forms the conclusion of a
course of thought previously entered on and has a decisive bearing
on all that I have to say in the sequel. At the outset of this discussion we came to the conclusion that the proposition, 'things
exist,' has no intelligible meaning except that they stand in relations
to each other. But these relations we left for the present without a
name, and contented ourselves, by way of a first interpretation of our
thought, with reference to various relations in the way of space, time,
and of cause and effect, of which the subsistence between things
constituted for our every-day apprehension that which we call the
real existence of the world. But between the constituents of the
world of ideas—constituents merely thinkable as opposed to real—
we found a complex of relations no less rich. Nay, our mobile
thought, it seemed, had merely to will it, and the number of these

relations might be indefinitely increased by transitions in the way of comparison between points selected at pleasure. This consideration could not but elicit the demand that the relations on which the being of things rests should be sought only among those which obtain objectively between them, not among such as our subjective process of thinking can by arbitrary comparisons establish between them.

This distinction however is untenable. I repeat in regard to it what I have already in my Logic[1] had opportunity of explaining in detail. In the passage referred to I started with considering how a *representation* of relations between two matters of consciousness, *a* and *b*, is possible. The condition of its possibility I could not find either in the mere succession or in the simultaneity of the two several presentations, *a* and *b*, in consciousness, but only in a relating activity, which directs itself from one to the other, holding the two together. ' He who finds red and yellow to a certain extent different yet akin, becomes conscious, no doubt, of these two relations only by help of the changes which he, as a subject of ideas, experiences in the transition from the idea of red to that of yellow ;' but, I added, he will not in this transition entertain any apprehension lest the relation of red to yellow may in itself be something different from that of the affections which they severally occasion in him ; lest in itself red should be like yellow and only appear different from it to us, or lest in reality there should be a greater difference between them than we know, which only appears to us to involve nevertheless a certain affinity. Doubts like these might be entertained as to the external causes, to us still unknown, of our feelings. But so long as it is not these causes but only our own ideas, after they have been excited in us, that form the object of our comparison, we do not doubt that the likenesses[2], differences, and relations which these exhibit on the part of our presentative susceptibility indicate at the same time a real relation on the part of what is represented to us. Yet how exactly is this possible ? How can the propositions, *a* is the same as *a*, and, *a* is different from *b*, express an objective relation, which, as objective, would subsist independently of our thought and only be discovered or recognised by it ? Some one may perhaps still suppose himself to know what he means by a self-existent identity[3] of *a* with *a*; but what will he make of a self-existent distinction between *a* and *b* ? and what objective relation will correspond to this ' *between*,' to which we only attach a meaning, so long as it suggests to us the distance in

[1] [§§ 337, 338.]　　　　　　　　　[2] ['Gleichheiten.']
[3] ['Gleichheit. gleich,' v. note on § 19.]

space which *we*, in comparing *a* and *b*, metaphorically interpolated
for the purpose of holding the two apart, and at the same time as a
connecting path on which our mind's eye might be able to travel from
one to the other ?   Or—to put the case otherwise—since difference,
like any other relation, is neither a predicate of *a* taken by itself nor of
b taken by itself, of what is it a predicate ?   And if it only has a
meaning when *a* and *b* have been brought into relation to each other,
what objective connexion exists between *a* and *b* in the supposed case
where the relating activity, by which we connected the two in consciousness, is not being exercised ?

The only possible answer to these questions we found to be the
following.   If *a* and *b*, as we have so far taken to be the case, are not
things belonging to a reality outside and independent of our thought,
but simply contents of possible ideas like red and yellow, straight and
curved, then a relation between them exists only so far as we think it
and by the act of our thinking it.   But our soul is so constituted, and
we suppose every other soul which inwardly resembles our own to be
so constituted, that the same *a* and *b*, how often and by whomsoever
they may be thought, will always produce in thought the same relation—a relation that has its being only in thought and by means of
thought.   Therefore this relation is independent of the individual
thinking subject, and independent of the several phases of that subject's thought.   This is all that we mean when we regard it as having
an existence in itself between *a* and *b* and believe it to be discoverable
by our thought as an object which has a permanence of its own.   It
really has this permanence, but only in the sense of being an occurrence which will always repeat itself in our thinking in the same way
under the same conditions. So long therefore as the question concerns
an *a* and *b*, of which the content is given merely by impressions and
ideas, the distinction of objective relations obtaining between them,
from subjective relations established between them by our thought, is
wholly unmeaning.   All relations which can be discovered between
the two are predicable of them on exactly the same footing ; all, that
is to say, as inferences which their own constant nature allows to our
thought and enjoins upon it ; none as something which had an existence of its own between them prior to this inferential activity on our
part.   The relation[1] of *a* to *b* in such cases means, conformably to the
etymological form of the term, our act of reference[2].

81. We now pass to the other case, which concerns us here as

---

[1] ['Beziehung.']
[2] ['Unsere Handlung des Beziehens.']

dealing no longer with logic but with metaphysics. Let *a* and *b* indicate expressly Realities, Entities, or Things. The groups, *a* and *b*, of sensible or imaginable qualities, by which these things are distinguished from each other, we can still submit with the same result as before to our arbitrary acts of comparison, and every relation which by so doing we find between the qualities will have a significance for the two things *a* and *b* equally essential or unessential, objective or non-objective. No relation between them could be discovered if it were not founded on the nature of each, but none is found before it is sought.

But it is not these relations that we have in view if, in order to render intelligible a connexion of the things *a* and *b* which experience forces on our notice, we appeal to a relation *C*, which sometimes does, sometimes does not, obtain between *a* and *b*; which is thus not one that belongs to the constant natures, *a* and *b*, of the two things, but a relation into which the things, as already constituted independently of it, do or do not enter. In this case the conclusion is unavoidable that this objective relation *C*, to which we appeal, cannot be anything that takes place *between* *a* and *b*, and that just for that reason it is not a relation in the ordinary sense of the term, but more than this. For it is only in our thought, while it passes from the mental image or presentation of *a* to that of *b*, that there arises, as a perception immediately intelligible to thought, that which we here call a *between*. It would be quite futile to try, on the contrary, to assign to this *between*, at once connecting and separating *a* and *b*, which is a mere memorial of an act of thought achieved solely by means of the unity of our consciousness, a real validity in the sense of its having an independent existence of its own apart from the consciousness which thinks it. We are all, it is true, accustomed to think of things in their multiplicity as scattered over a space, through the void of which stretch the threads of their connecting relations; whether we insist on this way of thinking and consider the existence of things to be only possible in the space which we see around us, or whether we are disposed with more or less clearness, as against the notion of a sensible space, to prefer that of an intelligible space which would afford the web composed of those threads of relation equal convenience of expansion. But even if we cannot rid ourselves of these figures, we must at least allow that that part of the thread of relation which lies in the void between *a* and *b*, can contribute nothing to the union of the two immediately but only through its attachment to *a* and *b* respectively. Nor does its mere contact with *a* and *b* suffice to yield this result. It must communicate to both a definite tension, prevalent throughout its own length, so that they are in a different condition from that in which they would be if this tension were of a different degree or took a different direction.

It is on these modifications of their inner state, which *a* and *b* sustain from each other—on these alone—that the result of the relation between them depends; and these are obviously independent of the length and of the existence of the imagined thread of relation. The termini *a* and *b* can produce *immediately* in each other these reciprocal modifications, which they in the last resort must produce even on supposition that they communicated their tension to each other by means of the thread of relation; since no one would so far misuse the figure as to make the thread, which was ostensibly only an adaptation to sense of the relation between the termini, into a new real material, capable of causing a tension, that has arisen in itself from the reciprocal action of its own elements, to act on inert things, *a* and *b*, attached to it. Let us discard, then, this easy, but useless and confusing, figure. Let us admit that there is no such thing as this interval between things, in which, as its various possible modifications, we sought a place for those relations, *C*, that we supposed to form the ground of the changing action of things upon each other. That which we sought under this name of an objective relation between things can only subsist if it is more than mere relation, and if it subsists not between things but immediately in them as the mutual action which they exercise on each other and the mutual effects which they sustain from each other. It is not till we direct our thought in the way of comparison to the various forms of this action that we come to form this abstract conception of a *mere* relation, not yet amounting to action but preceding the action which really takes place as its ground or condition.

CHAPTER VII.

Conclusion.

82. We may now attempt by way of summary to determine how many of the ontological questions, so far proposed, admit of a final answer. In the first place, to stand in relations appeared to us at the beginning of our discussion to be the only intelligible import of the being of things. These relations are nothing else than the immediate internal reciprocal actions themselves which the things unremittingly exchange. Beside the things and that which goes on in them there is nothing in reality. Everything which we regard as mere relation—all those relations which seem to extend through the complete void of a ' *between-things*,' so that the real might *enter into* them—subsist solely as images which our presentative faculty on its own account makes for itself. They originate in it and for it, as in its restless activity it compares the likeness, difference, and sequence of the impressions which the operation of A, B, C upon us brings into being—this operation at each moment corresponding to the changeable inner states a, b, c, which A, B, C experience through their action on each other. To pursue this Thesis further is the problem of Cosmology, which deals with things and events as resting or passing in the seemingly pre-existent forms of space and time, and which will have to show how all relations of space and time, which we are accustomed to regard as prior conditions of an operation yet to ensue, are only expressions and consequences of one already taking place.

We find an answer further to the enquiry as to that metaphysical C, that relation which it seems necessary should supervene, in order that things, which without it would have remained indifferent to each other, might be placed under the necessity, and become capable, of operation on each other. The question is answered to the effect that such a thing as a non-C, a separation which would have left the things indifferent to each other, is not to be met with in reality and that therefore the question as to the transition from this state into that of combination is a question concerning nothing. The unity of M is this eternally present condition of an interchange of action, unremitting but varying to the highest degree of complexity. For neither does this unity ever really exist in the general form indicated by this conception and name of unity and by this sign M. It really exists at each moment only as a case, having a definite value, of the equation for which I gave the formula[1], and in such form it is at the same time the efficient cause of the actuality of the state next-ensuing as well as the conditioning ground of what this state contains. Thus the stream of this self-contained operation propagates itself out of itself from phase to phase. If a sensible image is needed to help us to apprehend it, we should not think of a wide-spread net of relations, in the meshes of which things lie scattered, so that tightening of the threads, now at this point, now at that, may draw them together and force them to share each other's states. We should rather recall the many simultaneous 'Parts' of a piece of polyphonic music, which without being in place are external to each other in so far as they are distinguished by their pitch and tone, and of which first one and then another, rising or falling, swelling or dying away, compels all the rest to vary correspondingly in harmony with itself and one another, forming a series of movements that result in the unity of a melody which is consistent and complete in itself.

83. Our last considerations started from the supposition that in a certain element A of M a new state a has somehow been introduced. It is natural that now a further question should be raised as to the possibility of this primary change, from the real occurrence of which follows the course of reactions depicted. This question as to the beginning of motion has been a recognised one since the time of Aristotle, but it has been gradually discovered that the answer to it cannot be derived from the unmoved, which seemed to Aristotle the ultimate thing in the world. The most various beliefs as to the nature and structure of reality agree upon this, that out of a condition of perfect rest a beginning of motion can never arise. Not merely a multiplicity of originally given real elements, but also given motions between them, are presupposed in all the theories in which professors of the natural sciences, no less than others, strive to explain the origin of the actual course of the world out of its simplest principles. To us, with that hunger for explanation which characterises our thought, it looks like an act of despair to deny the derivability from anything else of some general fact, when in regard to its

[1] [Cp. § 70.]

individual forms one is accustomed to enquire for the conditions of their real existence. We experience this feeling of despair if we find ourselves compelled to trace back the multiplicity of changeable bodies to a number of unchangeable elements. Yet the question, why it is just these elements and no others that enjoy the prerogative of original reality, does not force itself upon us. Our fancy does not avail, beyond the elements given by experience, to produce images of others, which might have existed but were in some unintelligible way cheated of their equal claim to reality. Of the motions, on the contrary, of which these elements, once given, are capable, we see first one and then another take place in reality according as their changing conditions bring them about. None of them appears to us so superior to the rest that it exclusively, and without depending in its turn on similar conditions, should claim to be regarded as the first actual motion of the real.

These considerations lead on the one side to an endless regress in time. It is not necessary however at this point to complicate our enquiry by reference to the difficulties connected with occurrence in time. Our effort will be to exclude them for the present. But, no matter whether we believe ourselves to reach a really first beginning or whether we prolong the chain of occurrence in endless retrogression, the established course of the world is anyhow a single reality in contrast with the innumerable possibilities, which would have been realised if either the primary motion had been different, as it might have been, or if, which is equally thinkable, the endless progression, as a whole, had taken a different direction. For whether in reality it be finite or infinite, in either case its internal arrangement admits of permutations which, as it is, are not real.

All these doubts, however, are only different off-shoots of a general confusion in our way of thinking and a complete misunderstanding of the problems which a metaphysical enquiry has to solve. The world once for all is, and we are in it. It is constituted in a particular way, and in us for that reason there lives a Thought, which is able to distinguish different cases of a universal. Now that all this is so, there may arise in us the images and conceptions of possibilities which in reality are not; and then we imagine that we, with this Thought of ours, are there *before* all reality and have the business of deciding what reality should arise out of these empty possibilities, which are yet all alike only thinkable because there is a reality from which this Thought springs. When once, in this Thought, affirmation and denial of the same content have become possible, we

can propose all those perverted questions against which we have so
often protested—Why there is a world at all, when it is thinkable that
there should be none? Why, as there is a world, its content is M
and not some other drawn from the far-reaching domain of the
non-M? Given the real world as M, why is it not in rest but in
motion? Given motion, why is it motion in the direction X and not
in the equally thinkable direction Z? To all these questions there is
only one answer. It is not the business of the metaphysician to
make reality but to recognise it; to investigate the inward order of
what is given, not to deduce the given from what is not given. In
order to fulfil this office, he has to guard against the mistake of
regarding abstractions, by means of which he fixes single determinations of the real for his use, as constructive and independent elements
which he can employ, by help of his own resources, to build up the
real.

In this mistake we have often seen metaphysicians entangled.
They have formed the idea of a pure being and given to this a
significance apart from all relations, in the affirmation of which and
not otherwise it indicates reality. They have petrified that reality
which can only attach to something completely determined, into a
real-in-itself destitute of all properties. They have spoken of laws
as a controlling power between or beyond the things and events in
which such laws had their only real validity. In like manner we
are inclined to think at the outset of the truly existing M, the complex
of all things, as a motionless object of our contemplation; and we are
right in doing so as long as in conceiving it we think merely of the
function, constantly identical with itself, which it signifies to us. From
this function, it is true, simply as conceived, no motion follows. But
we forget meantime that it is not this conception of this function that
is the real, but that which at each moment the function executes, and
of which the concrete nature may *contain* a kind of fulfilment of the
function, which does not *follow* from that conception of it. In what
way that one all embracing M solves its problem—whether by maintaining a constant equality of content, or by a succession of innumerable different instances, of which each satisfies the general equation prescribed by its plan—that is its own affair. Between these two
thinkable possibilities it is not for us to choose as we will. Our
business is to recognise whichever of them is given as reality. Now
what is given to us is the fact of Becoming. No denial of ours can
banish it from the world. It is not therefore as a stationary identity
with itself but only as an eternally self-sustained motion that we have

to recognise the given being of that which truly is. And as given with it we have also to recognise the direction which its motion takes.

84. I have referred to the theories which agree with my own in being Monistic. In all of them motion is at the same time regarded as an eternal attribute of the supposed ultimate ground of the world. This motion, however, was generally represented as a ceaseless activity, on the opposition of which, as living and animating, to the unintelligible conception of a stark and dead reality the writers referred to loved to dwell. Such language shows that the metaphysical reasons for believing in the Unity of Being have been reinforced by æsthetic inclinations which have yielded a certain prejudice as to the nature of the Being that is to be counted supreme. It was not the mere characteristic of life and activity but their worth and the happiness found in the enjoyment of them which it was felt must belong in some supreme measure to that in which all things have their cause and reason. Such a proposition is more than at this stage of our enquiry we are entitled to maintain. Life and Activity only carry the special meaning thus associated with them on supposition of the spirituality of the Being of which they are predicated. The only necessary inference, however, from the reasoning which has so far guided us is to an immanent operation, through which each new state of what Is becomes the productive occasion of a second sequent upon it, but which for anything we have yet seen to the contrary may be a blind operation. I would not indeed conceal my conviction that there is justification, notwithstanding, for a belief in the Life of that which is the ground of the world, but it is a justification of which I must postpone the statement. I would only ask, subject to this proviso, to be allowed the use of expressions, for the sake of brevity, of which the full meaning is indeed only intelligible upon a supposition, as we have seen, still to be made good, but which will give a more vivid meaning to the propositions we have yet to advance than the constant repetition of more abstract terms could do.

85. So long as all we know of M is the function which it is required to fulfil—that, namely, of being the Unity which renders all that the world contains what it is—so long we can derive nothing from this thought but a series of general and abstract deductions. Every single being which exists, exists in virtue not of any being of its own but of the commission given it, so to speak, by the one M; and it exists just so long as its particular being is required for the fulfilment of the equation $M = M$. Again, it is what it is not abso-

lutely and in immemorial independence of anything else; it is that which the one M charges it to be. One thing, finally, operates on another not by means of any force of its own, but in virtue of the One present in it, and the mode and amount of its operation at each moment is that prescribed it by M for the re-establishment of the equation just spoken of.

To the further interpretation of these propositions in detail I return presently. That which is implied in all of them is a denial of any knowledge antecedent to all experience—a denial which goes much deeper, and indeed bears quite another meaning than is understood by those who are so fond of insisting on this renunciation of *a priori* knowledge. It is not in philosophy merely, but in the propositions on which scientific men venture that we trace the influence of the prejudice that, independently of the content realised in this world, $M = M$, there are certain universal modes of procedure, certain rights and duties, which self-evidently belong to all elements, as such, that are to be united in any possible world, and which would be just as valid for a wholly different world, $N = N$, as for that in which we actually live. There has thus arisen in philosophy a series of propositions which purport to set forth the properties and prerogatives of substances as such independently of that course of the world in which they are inwoven. They obviously rest on the impression that every other order of a universe, whatever it might be, that could ever come into Being, would have to respect these properties and prerogatives and could exact no function from Things other than what, in virtue of a nature belonging to them antecedently to the existence of a world, they were fitted and necessitated to render. And no less in the procedure of the physical sciences, however many laws they may treat as obtaining merely in the way of matter of fact, there is yet implied the notion of there being a certain more limited number of mechanical principles, to which every possible nature, however heterogeneous from nature as it is, would nevertheless have to conform. The philosophers, it is true, have imagined that the knowledge of the prerogatives of Substance was to be attained by pure thinking, while the men of science maintain that the knowledge of ultimate laws is only to be arrived at by experience. But as to the metaphysical value of that which they suppose to be discovered in these different ways they are both at one. They take it as the sum of pre-mundane truth, which different worlds, $M = M$ and $N = N$, do but exhibit in different cases of its application.

This is the notion which I seek to controvert. Prior to the world, or prior to the first thing that was real, there was no pre-mundane or pre-real reality, in which it would have been possible to make out what would be the rights which, in the event of there coming to be a reality, each element to be employed in its construction could urge for its protection against anything incompatible with its right as a substance, or to which every force might appeal as a justification for refusing functions not imposed on it by the terms of its original charter. There is really neither primary being nor primary law, but the original reality, M or N. Given M or N, there follows from the one M for its world, $M = M$, the series of laws and truths, which hold good for this world. If not M but N were the original reality, then for the world $N = N$ there would follow the other series of regulated processes which would hold good for this other world. There is nothing which could oppose to these ordinances M or N any claim of its own to preservation or respect.

86. Here the objector will interpose: ' Granting this, are you not liable to the charge of having here in your turn given utterance to one of those pre-mundane truths, of which you refuse to admit the validity? Have you not of your own accord expressly alleged the case of two worlds, M and N, which you suppose would both be obliged to conform to the general rule stated?' Now I have purposely chosen these expressions in order to make my view, which certainly stands in need of justification against the above objection, perfectly clear. In the first place, as regards the world N, which I placed in opposition to the real world M, I have to repeat what I have already more than once pointed out. The world M is, and we, thinking spirits, are in it, holding a position which M in virtue of its nature as M could not but assign to us. To this position are adjusted those general processes of our Thought, by which we are to arrive at what we call a know-ledge of the rest of the world. Among these is that very important one, no doubt corresponding to the plan on which the world M is ordered, which enables us not only to form general ideas as such, but to subsume any given manifold under any one of its marks, of which a general idea has been formed, as a species or instance thereof. This intellectual capability, once given, does not subject itself to any limits in its exercise. Even that which, when we consider it meta-physically, we recognise as in reality the all-containing and uncon-ditioned, we may as a matter of logic take for one of the various instances admitting of subsumption under the general idea of the un-conditioned. Hence, while it is only of particular things that we assert multiplicity as a matter of reality, we attempt on the other

hand to form a plural of the conception 'Universe,' and oppose the real M to many other possible Universes.

But the capacity of doing this we owe not to the knowledge of a law to which M and N alike are subject, but only to that which actually takes place in M, and to a certain tendency transferred from it to us as constituents of M: the tendency to think of everything real as an instance of a kind, of which the conception is derived by abstraction from that thing, and thus at last to think even of the primary all-embracing Real, M itself, as an instance representing the idea we form of it, and so to dream of other instances existing along with it. Thus arises the notion of that world N, a perfectly empty fiction of thought to which we ascribe no manner of reality, and of no value, except, like other imaginary formulæ, to illustrate the other conception M, which is not imaginary. And I employed N exclusively for this purpose. Further, when we said that, if N existed, the laws valid for N would flow from the equation $N = N$ in just the same way as those valid for M flow from the equation $M = M$, this was not a conclusion drawn from knowledge of an obligation binding on both of them. On the contrary, it was an analogy in which what was true of the real M was transferred to the imaginary N. In reality we have no title to make this transfer, for—to put it simply—who can tell what would be and would happen if everything were other than it is? But if we do oppose this imaginary case to the real one in order to explain the latter, we must treat it after the type of the real. Otherwise, as wholly disparate, it would not even serve the purpose of illustrating the real by contrast with it—the only purpose for which it is introduced.

87. Yet a third objection remains to be noticed. The statement that from M follows the series of laws that hold good for this world M, obviously does not mean merely that these laws proceed anyhow from M; it means that they are the proper consequences of its nature. But what is meant by a 'proper consequence' when it can no longer be distinguished from an improper consequence as corresponding to some rule to which the improper consequence does not correspond? Have we not after all to presuppose some law of the necessity or possibility of thought, absolutely prior to the world and reality, which determines, in regard to every reality that may come to be, what development of its particular nature can follow consistently from the nature of the primary real, M or N, in distinction from such a development as would be inconsistent?

This variation of the old error can only be met by a variation of the old answer. At first sight it seems a pleonasm to demand that actual consequences should not be inconsequent. Still the expression has a certain meaning. Hitherto we have taken the idea of reason and consequent to be merely this, that from a determinate something there flows another determinate something. The question, what determinate something admits of being connected with what other, by coherence of this sort, has been left aside. The idea of reason and consequent, as above stated, would be satisfied, if with the various reasons $g^1 g^2 g^3$ the completely determinate consequences $p\, q\, r$ were as a matter of fact associated, without there being any affinity between $p\, q$ and r corresponding to that between $g^1 g^2 g^3$. We shall find that our knowledge of reality is in fact ultimately arrested by such pairs of cohering occurrences. For instance, between the external stimuli on which the sensations of sight and hearing depend, we are able to point out affinities which make it possible to present those several modes of stimulation as kinds, g^1 and g^2, of one process of vibration, g. But between sounds and colours we are quite unable to discover the same affinity, or to prove that, if sensations of sound follow upon g^1, sensations of colour must in consistency present themselves on occasion of g^2.

This example illustrates the meaning of that consistency of consequence which, in our view as stated above, can within certain limits be actually discovered and demonstrated in the real world, but beyond those limits is assumed to obtain universally in some form or other. The Unity of Being, without which there would be no possibility of the reciprocal action within a world of the seemingly though not really separate elements of that world, excludes the notion of a multiplicity of isolated and fatalistic ordinances, which without reference to each other should bind together so many single pairs of events. There must be some rule or other according to which the connexion of the members of each single pair, g^1 and f, with each other determines that of all the other pairs, g^m and f^m. It is only in reference to the comparison of various cases with each other, which thus becomes possible, that there is any meaning in speaking as we did of 'consistency.' The expression has no meaning in relation to any single pair, g and f, which we might have made the point of departure for our preliminary consideration of the rest. The coherence between two members would at the outset be an independent fact of which nothing could be known but simply that it was the fact. For

---

[1] ['g' and 'f' stand for 'Grund' and 'Folge' here, as on p. 83. Cp. also p. 96 where 'Grund' (Reason) is distinguished from 'Ursache' (Cause).]

supposing we chose to think of their adjustment to each other as connected with the fulfilment of a supreme condition Z requiring consistency, they would still only *correspond* to this condition. The actual concrete mode in which they satisfied it, the content in virtue of which they subordinated themselves to it, would be something which it would be impossible to suppose determined by Z itself; the more so in proportion as Z was more expressly taken to be an ordinance that would have to be fulfilled indifferently in innumerable cases, nay even in the most various worlds. Supposing Z to be neither the determining ground of the content of g and f, nor the productive cause of their real existence, the proposition that a connexion between the two ensues in accordance with Z, cannot be a statement of a real metaphysical order of supremacy and subordination : but is just the reverse of the real order. The primary independent fact of the connexion between g^1 and f^1 is of such a character that the comparison of it with g^2 and f^2, g^3 and f^3, enables us first to apprehend a universal mode of procedure on the part of the various connexions of events in the world—a concrete procedure, peculiar to this world M—and then, upon continued abstraction, to generate the conception of a condition Z, which would hold good for the organization of any world, N, so long as the mental image of N was formed after the pattern of the given reality, M.

88. At the present day few will understand the reasons for the persistency with which I dwell on these considerations and so often return to them. We live quickly, and have forgotten, without settling, a controversy which forty years ago was still a matter of the liveliest interest among the philosophers of Germany. The difficulties involved in Hegel's system of thought were then beginning to make themselves felt even by those who looked with favour on his enterprise—of repeating in thought by a constructive process the actual development of the world from the ground of the absolute. It was not after Hegel's mind to begin by determining the subjective forms of thought, under which alone we can apprehend the concrete nature of this ground of the Universe—a nature perhaps to us inaccessible. From the outset he looked on the motion of our thought in its effort to gain a clear idea of this still obscure goal of our aspiration as the proper inward development of the absolute itself, which only needed to be pursued consistently, in order gradually to bring into consciousness all that the universe contains.

Thus the most abstract of objects came to be thought of as the root of the most concrete—a way of thinking which it was soon found impossible to carry out. Even in dealing with the phenomena of nature, though they were forced into categories and classifications without sufficient knowledge, it had to be supposed that the process of development, once begun, was carried on with a superabundance in the multiplication of forms for which no explanation was to be found in the generalities which preceded the theory of nature. All that these could do was to make us anticipate some such *saltus*; for the transition of one determination into its opposite, or at any rate into an 'otherness,' had been one of the supposed characteristics of the motion which was held to generate the world. The same difficulty might have been felt when the turn came for the construction of the spiritual and historical world, into which nature was supposed to pass over. There are many reasons, however, even in actual life, for not being content with the derivation of our ideas of the beautiful and the good from the living feeling which in fact alone completely apprehends their value, but for giving them greater precision by requiring them to satisfy certain general formal determinations. It is true that they too undergo a sensible degradation if they are looked on merely as instances of abstract relations of thought, but this was taken almost less notice of than the same fact in regard to the phenomena of nature, for owing to the latter being objects of perception, it could not be ignored how much more they were than the abstract problems which according to the Hegelian philosophy they had to fulfil.

Hegel himself was quite aware of the error involved in this way of representing the world's course of development. He repeatedly insists that what appears in it as the third and last member of the dialectical movement described is in truth rather the first. And assuredly this remark is not to be looked upon as an after-thought of which no further application is made, but expresses the true intention of this bold Monism, which undertook far more than human powers can achieve, but of which the leading idea by no means loses its value through the great defects in its execution. From the errors noticed Schelling thought to save us. It was time, he told us, that the higher, the only proper, antithesis should be brought into view—the antithesis between freedom and necessity, in apprehending which, and not otherwise, we reach the inmost centre of philosophy. I will not dwell on the manner in which he himself worked out this view in its application to the philosophy of religion. It was Weisse who first sought to develope it systematically. That which Hegel had taken for true Being, he looked upon merely as the sum of prior conditions without which such Being would be unthinkable and could not be, but which

themselves have not being. Thus understood, they formed in his
view the object of a certain part of philosophy, and that comparatively
speaking a negative part, namely Metaphysic. It was for experience
on the other hand—the experience of the senses and that of the moral
and religious consciousness—as a positive revelation to give us knowledge of the reality built on that abstract foundation.

Such expressions might easily be explained in a sense with
which we could agree. It would be a different sense, however,
from that which they were intended to convey. According to that
original sense the general thoughts, which it was the business of
Metaphysic to unfold, were more than those forms of apprehending
true Being without which we cannot think. They were understood
indeed to be this, but also something more. In their sum they were
held to constitute an absolutely necessary matter for which it was
impossible either not to be or to be other than it is, but which, notwithstanding this necessity, notwithstanding this unconditional being,
was after all a nothing, without essence and without reality; while
over against it stood the true Being, for which according to this
theory, it is possible not to be or to be other than it is, thus being
constituted not by necessity but by freedom. I shall not spend time
in discussing this usage of the terms, freedom and necessity. I would
merely point out that the latter term, if not confined to a necessity of
thought on our part, but extended to that which is expressly held to
be the unconditioned condition of all that is conditioned, would have
simply no assignable meaning and would have to be replaced by the
notion of a *de facto* universal validity. The adoption of the term
' Freedom' to indicate the other sort of reality expressly recognised as
merely *de facto*—the reality of that which might just as well not be—
is to be explained by the influence of ideas derived from another
sphere of philosophy—the philosophy of religion—which cannot be
further noticed here. Taken as a whole, the theory is the explicit
and systematic expression of that Dualism which I find wholly unthinkable, and against which my discussions have so far been directed.
In this form at any rate it cannot be true. It is impossible that there
should first be an absolute *Prius* consisting in a system of forms that
carry necessity with them and constitute a sort of unaccountable Fate,
and that then there should come to be a world, however created,
which should submit itself to the constraint of these laws for the
realisation of just so much as these limits will allow. The real alone
is and it is the real which by its Being brings about the appearance
of there being a necessity antecedent to it, just as it is the living body

that forms within itself the skeleton around which it has the appearance of having grown.

89. We have not the least knowledge how it is that the seemingly homogeneous content of a germ-vesicle deposits those fixed elements of form, around which the vital movements are carried on. Still less shall we succeed in deducing from the simple original character, M, of a world, the organization of the necessity which prevails in it. There are two general ways, however, of understanding the matter, alike admissible consistently with our assumption of the unity of the world, which remain to be noticed here. I will indicate them symbolically by means of our previous formulæ, $M = \phi\,[A\,B\,R]$, and the converse $\phi\,[A\,B\,R] = M$. By the former I mean to convey that M is to be considered the form-giving Prius, of which the activity, whether in the way of self-maintenance or development, at every moment conditions the state of the world's elements and the form of their combination, both being variable between the limits which their harmony with M fixes for them. In the second formula M is presented as the variable resulting form, which the world at each moment assumes through the reciprocal effects of its elements—this form again being confined within limits which the necessity, persistently and equally prevalent in these effects, imposes. I might at once designate these views as severally Idealism and Realism, were it not that the familiar but at the same time somewhat indefinite meaning of these terms makes a closer investigation necessary.

90. Availing ourselves once again, for explanatory purposes, of the opposition between two worlds, M and N, we might designate the form in which, according to the sense of the former view, we should conceive the different characters of the two worlds to be alike comprehended, so that of an Idea[1] or, Germanicè, as that of a Thought[2]. It is thus that in Æsthetic criticism we are accustomed to speak of the Idea or Thought of a work of Art, in the sense of the principle which determines its form in opposition to the particular outlines in which indeed the principle is manifested but to which it is not so absolutely tied that other kindred means, even means wholly different, might not be combined to express it. So again in active life we speak of a project as an Idea or Thought, when we mean to censure it for including no selection between the manifold points capable of being related by the combination of which it might be carried out. If now we drop the imaginary world N, we cannot thereupon suppose that the real world M lacks that concrete character

[1] ['Idee.']   [2] ['Gedanke.']

by which we distinguished it from N, although that character would
no longer be needed for the purpose of distinguishing it from something else now that it is understood that there is nothing external
to it. It would therefore be incorrect to call the Idea, simply as the
Idea, the supreme principle of the world. Even the absolute idea,
although, in opposition to the partial ideas which it itself conditions
as constituents of its meaning, it might fitly be called unlimited, would
not on that account be free from a definitely concrete content, with
which it fills the general form of the Idea.

In other cases it is more easy to avoid this logical error of putting
an abstract designation of essence, as conceived by us, in place of the
subject to which the essence belongs. We are more liable to it in
the present case, where the reality, being absolutely single, can only
be compared with imaginary instances of the same conception. We
are then apt to think that every determinate quality which we might
leave to this reality would rest on a denial of the other determinate
qualities which we excluded from it, and which, in order to the
possibility of such exclusion, must at the same time be classed with
that which excludes them as coordinate instances of a still higher
reality. This reality can then only be reached by an extinction of
all content whatever. Thus the tendency, which so often recurs in
the history of philosophy, spins out its thread—the tendency to look
on the supreme creative principle of the world not merely as undefinable by any predicates within our reach but as in itself empty
and indefinite. These ways of thinking are only justifiable so far as
they imply a refusal to ascribe to the supreme M, as a sort of presupposition of its being, a multitude of ready-made predicates, from
which as from a given store it was to collect its proper nature. It is
no such doctrine that we mean to convey in asserting that the supreme
principle of reality is to be found in a definitely concrete Idea, M,
and not in the Idea merely as an Idea. The truth is rather this.
M being in existence, or in consequence of its existence, it becomes
possible for our Thought, as included in it, to apprehend that which
M is in the form of a *summum genus* to which M admits of being
subordinated and as a negation of the non-M. It is not every determination that rests on negation. On the contrary, there is an original
Position without which it would be impossible for us to apprehend
the content of that Position as a determination and to explain it by
the negation of something else.

91. The mode of development, accordingly, which is imposed on
the world by the Idea of which it is the expression, would depend on

the content of the Idea itself, and could only be set forth by one who had previously made himself master of this content. So to make himself master of it must be the main business of the Idealist as much as of any one else.* The only preliminary enlightenment which he would have to seek would relate to that characteristic of the cosmic order in the way of mere form which is implied in the fact that, according to him, it *is* in the form of a governing Idea that the content just spoken of, whatever it may be, constitutes the basis of this order. For him M means simply a persistent Thought, of which the import remains the same, whatever and how great soever in each instance of its realisation may be the collection of elements combined to this end. The world therefore would not be bound by M either to the constant maintenance of the same elements or to the maintenance of an identical form in their connexion. Not only would ABR admit of replacement by abr and $a\beta\rho$, but also their mode of connexion ϕ by χ or ψ, if it was only in these new forms that those altered elements admitted of being combined into identity with M. It would be idle to seek universally binding conditions which in each single form of M's realisation the coherent elements would have to satisfy simply in order to be coherent. What each requires on the part of the other in these special cases is not ascertainable from any source whatever either by computation or by syllogism. We have no other analogy to guide us in judging of this connexion than that—often noticed above—of æsthetic fitness which, when once we have become acquainted with the fact of a combination between manifold elements, convinces us that there is a perfect compatibility, a deep-seated mutual understanding, between them, without enabling us to perceive any general rule in consequence of which this result might have come about. The relation, however, of the Idea M to the various forms, thus constituted, of its expression—$\phi[ABR]$, $\chi[abr]$, $\psi[a\beta\rho]$—is not that of a genus to its species. It passes from one into the other—not indifferently from any one into any other, but in definite series from ϕ through χ into ψ. No Idealism at any rate has yet failed to insist on the supposition—a supposition which experience bears out—that it is not merely in any section of the world which might be made at any given moment, but also in the succession of its phases, that the unity of the Idea will assert itself.

The question may indeed be repeated, What are the conditions which ϕ and χ have to satisfy in order to the possibility of sequence upon each other, while it is impossible for ψ to arise directly out of ϕ? Of all theories Idealism is most completely debarred from an

appeal to a supra-mundane mechanism, which makes the one succession necessary, the other impossible. In consistency it must
place the maintenance of this order as unconditionally as the formation of its successive members in the hands of the Idea itself which
is directed by nothing but its own nature. On this nature will depend the adoption of one or other of certain courses ; or rather it
will consist in one or other of them. It will require either a perfectly unchanged self-maintenance, or the preservation, along with
more or less considerable variations, of the same idea and outline
in the totality of phenomena ; either a progress to constantly new
forms which never returns upon itself or a repetition of the same
periods. It is only the first of these modes of procedure which
observation contradicts in the case of the given world. Of the others
we find instances in detail; but if we were called to say which of
them bears the stamp of reality as a whole, our collective experience would afford no guide to an answer. All that we know is
that the several phases of the cosmic order, whatever the nature of the
coherent chain formed by their series as a whole, are made up of
combinations of comparable elements, that is, as we are in the habit
of supposing, of states and changes of persistent things. This is
the justification of our way of employing the equivalent letters of
different alphabets to indicate the constituents which in different
sections of the cosmic order seem to replace each other. If we
allow ourselves then to pursue this mode of representation and concede to Idealism that the Idea M determines the series of its forms
without being in any way conditioned by anything alien to itself,
still by this very act of determination it makes each preceding phase,
with its content, the condition of the realisation of that which follows.
It is no detached existence, however, that we can ascribe to the
Idea, as if it were an as yet unformed M apart from all the several
forms of its possible realisation. We may not present it to ourselves as constantly dipping afresh into such a repertory of forms,
with a definite series in view, for the purpose, after discarding the
prior phase, of clothing itself in the new one which might be next in
the series. At each moment the Idea is real only in one of these
forms. It is only as having at this particular time arrived at this particular expression of its meaning, that it can be the determining ground
for the surrender of this momentary form and for the realisation of
the next succeeding one. The æsthetic or, if that term is preferred,
the dialectic connexion between such phases of reality as stand in a
definite order of succession, which was implied in their being re-

garded as an expression of one Idea, must pass over into a causal connexion, in which the content and organization of the world at each moment is dependent on its content and organization at the previous moment.

92. The difficulties involved in this doctrine have been too much ignored by Idealism, in the forms which it has so far taken. In seeking to throw light on them, I propose to confine myself to the succession of two phases of the simple form $\phi\,[ABR]$ and $\phi\,[a\,b\,R]$, which were treated in § 72 as possible cases. This determinate succession can never become thinkable, if each of these phases is represented as an inert combination of inert elements: for in that case each is an equivalent expression for M and the transition from each into each of the innumerable other expressions or phases of M is equally possible and equally unnecessary. Either the included elements must be considered to be in a definitely directed process of becoming, or the common form of combination, ϕ, must be considered a motion which distributes itself upon them in various definite quantities. This assumption is not inconsistent either with the principles previously laid down, according to which a stationary being of things could not be held to be anything but a self-maintenance of that which is in constant process of becoming, or with the spirit of Idealism ; for Idealism includes in its conception of every form of being the dialectical negativity, which drives the being out of one given form of its reality into another. For these two unmoving members therefore we should have at once to substitute the one independent fact of a process by which A passes into a and B into b, while R remains the same. Now this fact is an equivalent expression of that form of becoming which at this moment constitutes the reality of M. A-a and B-b, accordingly, are two occurrences of which, in the expression of the idea which constitutes M, one cannot take place without the other. Taken by themselves, indeed, they would have no such mutual connexion. The connexion does not represent any supra-mundane law, holding good for the world N as well as for the real M. It is only in this real M—which means for us in fact *unconditionally*—that they belong together as each the condition of the other, so long as there is no change on the part of the remaining member R to affect the pure operation of the two on each other.

Supposing it, now, to come about in the course of this world M, that certain preceding phases once again gave rise to the occurrence A-a and along with it to an unchanged R or an R changed only in respect of internal modifications without external effect, then we should infer that in this case of repetition of A-a, the occurrence B-b must also reappear as its consequence required by the nature of M. If, however, the preceding phases necessitated along with A-a a transition of R to r, then the tendency of the former occurrence to produce B-b, while continuing, would not be able to realise itself purely. What would really take place would be a resulting occurrence, the issue of those two impulses, determined by a relation of mutual implication in M just in the same way as, in the case of the indifference of R, B-b is determined by A-a. Or—to express the same generally—the transition of the one phase ϕ into the other χ is brought about by the combination of the reciprocal effects, which the several movements contained in ϕ once for all exercise in virtue of their nature, independently of the phase in which they happen to be combined or of the point in the world's course at which they from time to time appear.

We thus come to believe in the necessity of a mechanical system, according to which each momentary realisation of the Idea is that which the preceding states of fact according to certain laws of their operation had the power to bring about. Nor is it, in any fatalistic way, as an alien necessity imposing itself on the Idea, that this mechanism is thought of, but as an analytical consequence of our conception of the Idea—of the supposition that it enjoins upon itself a certain order in its manifold possible modes of manifestation and by so doing makes the one an antecedent condition of that which follows. So long, however, as Idealism continues to regard the import of the Idea as the metaphysical Prius which determines the succession of events, so long there lies a difficulty in this twofold demand—the demand that what is conditioned by the Idea *a fronte* should be always identical with that to which this mechanism of its realisation impels *a tergo*. At a later stage of my enquiry I shall have occasion to return to this question. It will be at the point, to which the reader will have been long looking forward, where the appearance within nature of living beings brings home to us with special cogency the thought of relation to an end as governing the course of things, or of an ideal whole preceding the real parts and their combination. The question can then be discussed on more definite premisses. In the region of generality to which I at present confine myself Idealism could scarcely answer otherwise than by the mere assertion; 'Such is the fact: such is the nature of the concrete Idea, and such the manner of its realisation at every moment, that everything which it ordains in virtue of its own import must issue as a necessary result in ordered succession from the blind co-operation of all the several movements into which it distributes itself, and according to the general laws which it has imposed on itself.'

93. It is not every problem that admits of a solution, nor every goal, however necessarily we present it to ourselves, that can be reached. We shall never be able to state the full import of that Idea *M*, which we take to be the animating soul of the Cosmos. Not the fragmentary observation, which is alone at our command, but only that complete view of the whole which is denied, could teach us what that full import is. Nay, not even an unlimited extension of observation would serve the purpose. To know it, we must live it with all the organs of our soul. And even if by some kind of communication we had been put in possession of it, all forms of thought would be lacking to us, by which the simple fulness of what was given to us in vision could be unfolded into a doctrine, scientifically articulated and connected. The renunciation of such hopes has been prescribed to us by the conclusion to which we were brought in treating of Pure Logic. It remains, as we had there to admit[1], an unrealisable ideal of thought to follow the process by which the supreme Idea draws from no other source but itself those minor Premisses by means of which its import, while for ever the same, is led up to the development of a reality that consists in a manifold change. Here, however, as there we can maintain the conviction that in reality that is possible which our thoughts are inadequate to reproduce[2]. It is not any construction of the world out of the idea of which the possibility is thus implied, but merely a regressive interpretation, which attempts to trace back the connexion of what is given us in experience, as we gradually become acquainted with it, to its ineffable source.

To this actual limitation upon our possibilities of knowledge the second of the views above[3] distinguished—'Realism'—adjusts itself better than Idealism, though it has not at bottom any other or more satisfactory answer to give to the questions just raised. Realism does not enquire how the course of the world came to be determined as it is. It contents itself with treating the collective structure of the world at any moment as the inevitable product of the forces of the past operating according to general laws. On one point, however, I think the ordinary notion entertained by those who hold this view has already been corrected. They commonly start from the assumption of an indefinite

[1] Logic, § 151.   [2] Logic, *loc. cit.*   [3] [§ 89.]

number of mutually independent elements, which are only brought even into combination by the force of laws. That this is impossible and that for this Pluralism there must be substituted a Monism is what I have tried to show and need not repeat. It is not thus, from the nature of objects[1], but from the nature of the one object[2], that we must, even in Realism, derive the course of things. In fact, the distinction between the two views would reduce itself to this, that while the Idealist conceives his one principle as a restlessly active Idea, the Realist conceives his as something objective[3], which merely *suffers* the consequences of an original disintegration into a multitude of elements that have to be combined according to law—a disintegration which belongs to the *de facto* constitution of its nature, as given before knowledge begins. The mode of their combinations may become known to us through the elaboration of experience : and this knowledge gives us as much power of anticipating the future as satisfies the requirements of active life. An understanding of the universe is not what this method will help us to attain. The general laws, to which the reciprocal operations of things conform—in the first instance special to each group of phenomena—are presented as limitations coeval with knowledge, imposed by Reality on itself and within which it is, as a matter of fact, compelled to restrain the multiplicity of its products. The overpowering impression, however, which is made by the irrefragability of these limits, is not justified by any value which in respect of their content they possess for our understanding.

They would thus only satisfy him who could content himself with the mere recognition of a state of things as unconditional matter of fact. But even within the range of realistic views the invincible spiritual assurance asserts itself that the world not merely is but has a meaning. To succeed in giving to the laws, that are found as a matter of fact to obtain, such an expression as makes the reason in them, the *ratio legis*, matter of direct apprehension, is everywhere reckoned one of the finest achievements of science. Nor can the realistic method of enquiry resist the admission that the ends to which events contribute cannot always be credibly explained as mere products of aimless operation. It is not merely organic structures to which this remark applies. Even the planetary system exhibits forms of self-maintenance in its periodic changes, which have the appearance of being particular cases especially selected out of innumerable equally possible, or more easily possible, results of such operations. It is true that our observation is unable to settle the question whether

---

[1] ['Sachen.']   [2] ['Sache.']   [3] ['Sache.']

these cases of adaptation to ends are to be thought of as single islands floating in a boundless sea of aimless becoming, or whether we should ascribe a like order in its changes to the collective universe. Realism can find an explanation of these special forms only in the assumption of an arrangement of all operative elements, which for all that depends on the general laws might just as well have been another, but which, being what it is and not another, necessarily leads in accordance with those laws to the given ends. It thus appeals on its part to the co-operation, as a matter of fact, of two principles independent of each other which it knows not how to unite; on the one hand the general laws, on the other hand the given special arrangement of their points of application. In this respect Realism can claim no superiority over Idealism. At the same time it is only enquiries conducted in the spirit of Realism that will satisfy the wishes of Idealism. They will indeed never unveil the full meaning of the Idea. But there is nothing but recognition of the *de facto* relations of things that can make our thoughts at least converge towards this centre of the universe.

94. The conception of a Thing which we adopt has been exposed to many transformations, hitherto without decisive issue. Doubts have at last been raised whether the union of oneness of essential being with multiplicity of so-called states has any meaning at all and is anything better than an empty juxtaposition of words. In approaching our conclusion on this point we must take a roundabout road. The misgiving just expressed reaches further. In all the arguments which we ultimately adduced, and in which we passed naif judgments on the innermost essence of the real, on what is possible and impossible for it, according to principles unavoidable for our thought, what warranted the assurance that the nature of things must correspond to our subjective necessities of thought? Can such reasonings amount to more than a human view of things, bearing perhaps no sort of likeness to that which it is credited with representing?

This general doubt I meet with an equally general confession, which it may be well to make as against too aspiring an estimate of what Philosophy can undertake. I readily admit that I take Philosophy to be throughout merely an inner movement of the human spirit. In the history of that spirit alone has Philosophy its history. It is an effort, within the presupposed limits, even to ourselves absolutely unknown, which our earthly existence imposes on us, to gain a consistent view of the world—an effort which carries us to something beyond the satisfaction of the wants of life, teaching us to set before ourselves and to attain worthy objects in living. An absolute truth, such as the archangels in heaven would have to accept, is not its object, nor does the failure to realise such an object make our efforts bootless. We admit therefore the completely human subjectivity of all our knowledge with the less ambiguity, because we see clearly moreover that it is unavoidable and that, although we may forego the claim to all knowledge whatever, we could put no other knowledge in the place of that on which doubt is thrown, that would not be open to the same reproach. For in whatever mind anything may present itself which may be brought under the idea of knowledge, it will always be self-evident that this mind can never gain a view of the objects of its knowledge as they would seem if it did not see them, but only as they seem if it sees them, and in relation to it the seeing mind. It is quite superfluous to make this simple truth still more plain by a delineation of all the several steps in our knowledge, each monotonously followed by a proof that we everywhere remain within the limits of our subjectivity and that every judgment, in the way of recognition or correction, which we pass from one of the higher of these steps upon one of the lower, is still no more than a necessity of thought for us. At most it is worth the trouble to add that—still, of course, according to our way of thinking—this is no specially preju- dicial lot of the human spirit, but must recur in every being which stands in relation to anything beyond it.

Just for this reason this universal character of subjectivity, belong- ing to all knowledge, can settle nothing as to its truth or untruth. In putting trust in one component of ostensible knowledge while we take another to be erroneous we can be justified only by a con- sideration of the import of the two components. We have to reject and alter all the notions, which we began by forming but which cannot be maintained without contradiction when our thoughts are systematized, while they can without contradiction be replaced by others. As regards the ultimate principles, however, which we follow in this criticism of our thoughts, it is quite true that we are left with nothing but the confidence of Reason in itself, or the certainty of belief in the general truth that there is a meaning in the world, and that the nature of that reality, which includes us in itself, has given our spirit only such necessities of thought as harmonise with it.

95. Of the various forms in which the scepticism in question reappears the last is that of a doubt not as to the general capacity for truth on the part of our cognition, but as to the truth of one of its utterances—a determinate though very comprehensive one. It relates to that whole world of things which so far, in conformity with the

usual way of thinking, we have taken for granted. After the admirable exposition which Fichte has given us of the subject in his 'Vocation of Man,' I need not show over again how everything which informs us as to the existence of a world without us, consists in the last resort merely in affections of our own *ego*, or—to use language more free from assumption—in forms which hover before our consciousness, and from the manifold variations and combinations of which there arises the idea—and always as our idea—of something present without us, of a world of things.   Now we have a right to enquire what validity this idea, irrespectively of its proximate origin, may claim in the whole of our thoughts; but it would have been a simple fallacy merely on account of the subjectivity of all the elements out of which it has been formed, to deny its truth and to pronounce the outer world to be merely a creation of our imagination.   For the state of the case could be no other, were there things without us or no.   Our know-ledge in the one case, our imagination in the other, could alike only consist in states or activities of our own being—in what we call im-pressions made on our nature, supposing these to be things, but on no supposition in anything other than a subjective property of ours.

As is well known, Fichte did not draw the primary inference which —offensive as it is—would be logically involved in the error noticed, the inference, namely, that the single subject, adopting such a philo-sophy, would have to consider itself the sole reality, which in its own inner world generated the appearance of a companion Universe.   In regard to Spirits he followed the conviction which I just now stated. It is only by means of subjective effects produced upon him, like those which mislead him into believing in things, that any one can know of the existence of other Spirits; but just because this must equally be the case if there really are Spirits, this fact proved nothing against their existence.   If therefore Fichte allowed the exist-ence of a world of Spirits, while he inexorably denied that of a world of Things, the ground of his decision would only lie in the judgment which he passed on the several conceptions in respect simply of their content—in the fact that he found the conception of Spirit not only admissible but indispensable in the entirety of his view of the world, that of the Thing on the contrary as inadmissible as superfluous.   To this conviction he was constant.   To have no longer an eye for mere things was in his eyes a requirement to be made of every true philosophy.

98. I proceed to connect this brief historical retrospect with the difficulties which, as we saw, have still to be dealt with.   We found

it impossible for that to be unchangeable which we treated as a thing, *a*. It did not even admit of being determined by varying persistencies on the part of different qualities[1]. We were forced to think of it as in continuous becoming, either unfolding itself into the one series, a^1, a^2, a^3, or maintaining itself, in the other, *a, a, a*, by constantly new production. Each of these momentary phases, however, we saw must be exactly like itself, but $a^1 = a^1$ is different from every other. Even the exactly similar members of the latter series, though exactly similar, were not one and the same. For all that we asserted that in this change the Unity of a thing maintained itself. We could not but assert this if we were to conceive the mutual succession of the several forms, which could not arise out of nothing but only out of each other. We were not in a condition, however, to say what it was that remained identical with itself in this process of becoming. We took advantage of the term 'states'[2], which we applied to the changing forms, but we came to the conclusion that in so doing we were only expressing our mental demand without satisfying it. We saw that an immediate perception was needed to show us this relation of a subject to its states as actually under our hands and thereby convince us of its possibility.

Perhaps the reader then cherished the hope that there would be no difficulty in adducing many such instances in case of need. Now, on returning to this question, we only find one being, from the special nature of which the possibility of that relation seems inseparable. This is the spiritual subject, which exercises the wonderful function not merely of distinguishing sensations, ideas, feelings from itself but at the same time of knowing them as its own, as its states, and which by means of its own unity connects the series of successive events in the compass of memory. I should be misunderstood if this statement were interpreted to mean that the Spirit understands how to bring itself and its inner life in the way of logical subsumption under the relation of a subject to its states or to recognise itself as an instance of this subordination. It experiences the fact of there being this relation at the very moment when it lives through the process of its own action. It is only its later reflection on itself which thereupon generates for it in its thinking capacity the general conception of this relation—a relation in which it stands quite alone without possibility of another homogeneous instance being found. It is only in the sensitive act, which at once repels the matter of sense from us as something that exists for itself and reveals it to us as our own, that

[1] [§ 24 ff.]   [2] [§ 47.]

we become aware what is meant by the apprehension of a certain *a* as a state of a subject *A*. It is only through the fact that our attention, bringing events into relation, comprehends past and present in memory, while at the same time there arises the idea of the persistent Ego to which both past and present belong, that we become aware what is meant by Unity of Being throughout a change of manifold states, and that such unity is possible. In short it is through our ability to appear to ourselves as such unities that we are unities. Thus the proximate conclusion to which we are forced would be this. If there are to be things with the properties we demand of things, they must be more than things. Only by sharing this character of the spiritual nature can they fulfil the general requirements which must be fulfilled in order to constitute a Thing. They can only be distinct from their states if they distinguish themselves from their states. They can only be unities if they oppose themselves, as such, to the multiplicity of their states.

97. The notion that things have souls has always been a favourite one with many and there has been some extravagance in the imaginative expression of it. The reasoning which has here led us up to it does not warrant us in demanding anything more than that there should belong to things in some form or other that existence as an object for itself which distinguishes all spiritual life from what is only an object for something else. The mere capacity of feeling pain or pleasure, without any higher range of spiritual activity, would suffice to fulfil this requirement. There is the less reason to expect that this psychical life of things will ever force itself on our observation with the clearness of a fact. The assumption of its existence will always be looked on as an imagination, which can be allowed no influence in the decision of particular questions, and which we can only indulge when it is a question, in which no practical consequences are involved, of making the most general theories apprehensible.

It is therefore natural to enquire whether after all it is necessary to retain in any form that idea of an existence of Things which forced this assumption upon us. There are two points indeed which I should maintain as essential: one, the existence of spiritual beings like ourselves which, in feeling their states and opposing themselves to those states as the unity that feels, satisfy the idea of a permanent subject [1]: the other, the unity of that Being, in which these subjects in turn have the ground of their existence, the source of their peculiar nature, and which is the true activity at work in them. But why over

[1] ['Eines Wesens.']

and above this should there be a world of things, which themselves gain nothing by existing, but would only serve as a system of occasions or means for producing in spiritual subjects representations which after all would have no likeness to their productive causes? Could not the creative power dispense with this roundabout way and give rise directly in spirits to the phenomena which it was intended to present to them? Could it not present that form of a world which was to be seen without the intervention of an unseen world which could never be seen as it would be if unseen? And this power being in all spirits one and the same, why should there not in fact be a correspondence between the several activities which it exerts in those spirits of such a kind that while it would not be the same world-image that was presented to all spirits but different images to different spirits, the different presentations should yet fit into each other, so that all spirits should believe themselves planted at different positions of the same world and should be able to adjust themselves in it, each to each, in the way of harmonious action? As to the effects again which Things interchange with each other and which according to our habitual notions appear to be the strongest proof of their independent existence—why should we not substitute for them a reciprocal conditionedness on the part of innumerable actions, which cross and modify each other within the life of the one Being that truly is? If so, the changes which our world-image undergoes would at each moment issue directly from the collision of these activities which takes effect also in us, not from the presence of many independent sources of operation bringing these changes about externally to us.

In fact, if the question was merely one of rendering the world, as phenomenally given to us, intelligible, we could dispense with the conception of a real operative atom, which we regard only as a point of union for forces and resistances that proceed from it, standing in definite relations to other like atoms and only changing according to fixed laws through their effect upon it. We could everywhere substitute for this idea of the atom that of an elementary action on the part of the one Being—an action which in like manner would stand in definite relations to others like it, and would through them undergo a no less orderly change. The assumption of real things would have no advantage but such as consists in facility of expression. Even this we could secure if, while retaining the term 'things,' we simply established this definition of it; that 'things' may be accepted in the course of our enquiry as secondary fixed points, but for all that are not real

existences in the metaphysical sense, but elementary actions of the
one Being which forms the ground of the world, connected with each
other according to the same laws of reciprocal action which we commonly take to apply to the supposed independent things.

98. For the prosecution of our further enquiries it is of little importance to decide between the two views delineated.   But a third
remains to be noticed which denies the necessity of this alternative,
and undertakes to justify the common notion of a Thing without a
Self.   When we set about constructing a Being which in the change
of its states should remain one, it was the experience of spiritual life,
it will be said, which came to our aid, and by an unexpected actual
solution of the problem convinced us that it was soluble.   What
entitles us, however, to reckon this solution the only one? Why might
there not just as well be another, of which we can form no mental
picture only for the reason that we have had no experience of it as
our own mode of existence?   Why may not the 'thing' be a Being
of its own particular kind, defined for us only by the functions which
it fulfils, but not bound in the execution of these to maintain any such
resemblance to our Spirit as, with the easy presumption of an anthro-
pomorphic imagination, we force upon it?

This counter-view is one that I cannot accept. So long as what we
propose to ourselves is to give shape to that conception of the world
which is necessary to us, we allow ourselves to fill up the gaps in our
knowledge by an appeal to the unknown object, to which our thoughts
converge without being able to attain it ; but we may not assume an
unknown object of such a kind as would without reason conflict with
the inferences which we cannot avoid.   Now it seems to me that the
suggestions just noticed imply a resort to the unknown of this unwarrantable kind.   In the first place it is not easy to see why the
conception of the Thing, in the face of the duly justified objections to
it, needs to be maintained at the cost of an appeal to what is after all
a wholly unknown possibility of its being true.   Secondly, while
readily allowing that anything which really exists may have its own
mode of existence, and is not to be treated as if it followed the type
of an existence alien to it, we must point out that where such
peculiarity of existence is asserted the further predicates assigned to
it must correspond.   What manner of being, however, could we consistently predicate of that from which we had expressly excluded the
universal characteristics of animate existence, every active relation to
itself, every active distinction from anything else?   Of that which had
no consciousness of its own nature and qualities, no feeling of its

states, which in no way possessed itself as a Self? Of that of which the whole function consisted in serving as a medium to convey effects, from which it suffered nothing itself, to other things like itself, just as little affected by those effects, till at last by their propagation to animate Beings there should arise in these, and not before, a comprehensive image of the whole series of facts. If we maintain that in fact such a thing cannot be said to *be*, it is not that we suppose ourselves to be expressing an inference, which would still have to be made good as arising out of the notion of such a thing : it is that we find directly in the description of such a thing the definition of a mere operation, which, in taking place, presupposes a real Being from which it proceeds and another in which it ends, but *is* not, itself, as a third outside the two. That our imagination will nevertheless cling to the presentation of independent and blindly-operating individual things, we do not dispute nor do we seek to make it otherwise ; but in the effort to find a metaphysical truth in this mode of expression we cannot share. It is not enough to try to give a being to these things outside their immanence in the one Real, unless it is possible to show that in their nature there is that which can give a real meaning to the figure of speech conveyed in this '*outside.*'

As to the source of our efforts in this direction and their fruitlessness, I may be allowed in conclusion to repeat some remarks which in a previous work[1] I have made at greater length. We do not gain the least additional meaning for Things without self and without consciousness by ascribing to them a being outside the one Real. All the stability and energy which they ensure as conditioning and motive forces in the changes of the world we see, they possess in precisely the same definiteness and fulness when considered as mere activities of the Infinite. Nay it is only through their common immanence in the Infinite, as we have seen, that they have this capability of mutual influence, which would not belong to them as isolated beings detached from that substantial basis. Thus for the purpose of any being or function that we would ascribe to things as related to and connected with each other, we gain nothing by getting rid of their immanence. It is true however that things, so long as *they* are only states of the infinite, are nothing in relation to themselves : it is in order to make them something in this relation or on their own account that we insist on their existence outside the Infinite. But this genuine true reality, which consists in relation to self—whether in being *something* as related to self or in that relation simply as such—is not acquired by

[1] Mikrokosmus, iii. 530.

things through a detachment from the one Infinite, as though this 'Transcendence,' to which in the supposed case it would be impossible to assign any proper meaning, were the antecedent condition on which the required relation to self depended as a consequence. On the contrary, it is in so far as something is an object to itself, relates itself to itself, distinguishes itself from something else, that by this act of its own it detaches itself from the Infinite. In so doing, however, it does not acquire but possesses, in the only manner to which we give any meaning in our thoughts, that self-dependence of true Being, which by a very inappropriate metaphor from space we represent as arising from the impossible act of 'Transcendence.' It is not that the opposition between a being *in* the Infinite and a being *outside* it is obviously intelligible as explaining why self-dependence should belong to the one sort of being while it is permanently denied to another. It is the nature of the two sorts of being and the functions of which they are capable that make the one or the other of these figurative expressions applicable to them. Whatever is in condition to feel and assert itself as a Self, that is entitled to be described as detached from the universal all-comprehensive basis of being, as *outside* it : whatever has not this capability will always be included as 'immanent' within it, however much and for whatever reasons we may be inclined to make a separation and opposition between the two.

BOOK II.

OF THE COURSE OF NATURE (COSMOLOGY).

———•———

CHAPTER I.

Of the Subjectivity of our Perception of Space.

In the course of our ontological discussion it was impossible not to mention the forms of Space and Time; within which, and not otherwise, the multiplicity of finite things and the succession of their states are presented to perceptive cognition. But our treatment did not start from the first questions that induce enquiry, rather it presupposed the universal points of view which have already been revealed in the history of philosophy. We were able therefore to deal with abstract ontological ideas apart from these two forms which are the conditions of perception. Any further difficulties must look for a solution to the Cosmological discussions on which we are now entering. Among the subjects belonging to Cosmology it may seem that Time should come first in our treatment ; seeing that we substituted the idea of a continual Becoming for that of Being as unmoved 'position[1].' Accessory reasons however induce us to speak first of Space, which indeed is as directly connected with our second requirement, that we should be able in every moment of time to conceive the real world as a coherent unity of the manifold.

99. In proposing to speak of the metaphysical value of Space, I entirely exclude at present various questions which, with considerable interest of their own, have none for this immediate purpose. At present we only want to know what kind of reality we are to ascribe to space as we have to picture it, and with what relation to it we are to credit the real things which it appears to put in our way. No answer to this, nor materials for one, can be got from psychological discussions

[1] [v. Bk. I. § 38.]

as to the origin or no-origin of our spatial perception. To designate it as an *a priori* or innate possession of the mind is to say nothing decisive, and indeed, nothing more than a truism; of course it is innate, in the only sense the expression can bear[1], and in this sense colours and sounds are innate too. As surely as we *could* see no colours, unless the nature of our soul included a faculty which could be stimulated to that kind of sensation, so surely could we represent to ourselves no images in space without an equally inborn faculty for such combination of the manifold. But again, as surely as we *should* not see colours, if there were no stimulus independent of our own being to excite us to the manifestation of our innate faculty, so surely we should not have the perception of space without being induced to exert our faculty by conditions which do not belong to it.

On the other hand, one who should regard our spatial perception as an abstraction from facts of experience, could have nothing before him, as direct experience out of which to abstract, beyond the arrangement and the succession of the sense-images in his own mind. He might be able to show how, out of such images, either as an unexplained matter of fact, or by laws of association of ideas which he professed to know, there gradually arose the space-perception, as a perception in our minds. He might perhaps show too, how there originated in *us* the notion of a world of things outside our consciousness as the cause of these spatial appearances. We shall find this a hard enough problem, later on; but granting it completely solved, still the mere development-history of our ideas of space would be in no way decisive of their validity as representing the postulated world of things, nor of the admissibility of this postulate itself. As was said above, the way in which a mode of mental representation grows up can be decisive of its truth or untruth, only in cases where a prior knowledge of the object to which it should relate convinces us that its way of growth must necessarily lead whether to approximation or to divergence. Therefore, for this latter view, as well as for the former which maintains the *a priori* nature of the space-perception, there is only one sense in which the question of its objective validity is answerable : namely, whether such a perception as we in fact possess and cannot get rid of, however it arose, is consistent with our notions of what a reality apart from our consciousness must be ; or whether, directly or in its results, it is incompatible with them.

100. A further introductory remark is called for by recent investigations. We admitted that our ideas of Space are conditioned by the

¹ Logic, § 324.

stimuli which are furnished to our faculty for forming them. It is conceivable that these stimuli do not come to all minds with equal completeness, and that hence the space-perception of one mind need not include all that is contained in that of another. But this indefiniteness in the object of our question is easily removed. Modes of mental presentation which are susceptible of such differences of development may have their simplest phases still in agreement with the object to which they relate, while their consistent evolution evokes germs of contradiction latent before. Therefore when their truth is in question, we have only to consider their most highly evolved form; in which all possibility of further self-transformation is exhausted, and their relation to the entirety of their object is completed.

We all live, to begin with, under the impression of a finite extension, which is presented to our senses as surrounding us, though with undetermined or unregarded limits; it is our subsequent reflection that can find no ground in the nature of this extension for its ceasing at any point, and brings the picture to completion in the idea of infinite space. This then, the inevitable result of our mode of mental portrayal when once set in motion, is the matter whose truth and validity are in question. But scepticism has gone further. It is no longer held certain and self-evident that the final idea of a space uniform and homogeneous in all directions, at which men have in fact arrived, and which geometry had hitherto supported, is the only possible and consistent form of combination for simple perceptions of things beside one another. Some hold that other final forms are conceivable, though impossible for men; some credit even mankind with the capacity to amend their customary perception of space by a better guided habituation of their representative powers. This last hope we may simply neglect, till the moment when it shall be crowned with success; the former suggestion, in itself an object of lively interest, we are also justified in disregarding for the present: for all the other forms of space whose conceivability these speculations undertake to demonstrate, would share the properties on which our decision depends with the only form which we now presuppose; that, namely, whose nature the current geometry has unfolded.

101. The kind of reality which we ought to ascribe to the content of an idea must agree with what such a content claims to be; we could not ascribe the reality of an immutable existence to what we thought of as an occurrence; nor endow what seemed to be a property with the substantive persistence which would only suit its substratum. Therefore we first try to define what space as represented

in our minds claims to be; or, to find an acknowledged category of established existence under which if extended to it, it could fairly be said to fall.

Some difficulty will be found in the attempt. The only point which is clear and conceded is that we do not regard it as a *thing* but distinguish it from the *things* which are moveable in it; and that though many determinations which are possible in space are properties of things, space itself is never such a property. Further; the definitions actually attempted are untenable; space is not a limit of things, but every such limit is a figure in space; and space itself extends without interruption over any spot to which we remove the things. It is neither form, arrangement, nor relation of things, but the peculiar principle which is essential to the possibility of countless different forms, arrangements, and relations of things; and, as their absolutely unchangeable background, is unaffected by the alternation and transition of these determinations one into another. Even if we called it 'form' in another sense, like a vessel which enclosed things within it, we should only be explaining it by itself; for it is only in and by means of Space that there can be vessels which enclose their contents but are not identical with them. These unsuccessful attempts show that there is no known general concept to which we can subordinate space; it is *sui generis*, and the question of what kind its reality is, can only be decided according to the claims of this its distinctive position.

102. As the condition of possibility for countless forms, relations, and arrangements of things, though not itself any definite one of them, it might seem that Space should be on a level with every universal genus-concept, and as such, merit no further validity. Like it, a genus-concept wears none of the definite forms, which belong to its subordinate species; but contains the rule which governs the manifold groupings of marks in them, allows a choice between certain combinations as possible, and excludes others as impossible.

Just such is the position of Space. Although formless in comparison with every outline which may be sketched in it, yet it is no passive background which will let any chance thing be painted on it; but it contains between its points unchangeable relations, which determine the possibility of any drawing that we may wish to make in it. It is not essential to find an exhaustive expression for these relations at this moment; we may content ourselves, leaving much undetermined, with defining them thus far:—that any point may be placed with any other point in a connexion homogeneous with that in which any

third point may be placed with any fourth; that this connexion is
capable of measurable degrees of proximity and that its measure
between any two points is defined by their relations to others.   No
matter, as I said, what more accurate expression may be substituted
for that given, in as far as our perception of space contains such a
legislative rule we might regard every group of manifold elements,
which satisfied this rule, as subordinate to the universal concept of
Space.   But we should feel at once, that such a designation was
unsuitable; such a group might be called a combination of multiplicity
in space, but not an instance of space, in the sense in which we regard
every animal whose structure follows the laws of his genus as a species
or instance of that genus.   The peculiarities of what we indicated
above as the law of space in general [1] create other relations between
the different cases of its application, than obtain between the species
of natural Genera.   Each of the latter requires indeed that its rule of
the grouping of marks shall be observed in each of its species; but it
puts the different species which do this in no reciprocal connexion.
They are therefore subordinate to it; but when we call them, as
species of the same genus, co-ordinate with one another, we really
mean nothing by this co-ordination but the uniformity of their lot in
that subordination.   Supposing we unite birds, fishes, and other
creatures under the universal concept ‘ animal,' all we find is that the
common features of organization demanded by the concept occur in
all of them; this tells us nothing of the reciprocal attitude and behaviour of these classes; the most we can do is, conversely, to attempt
afterwards a closer systematic union, by the formation of narrower
genera, between those which we have ascertained from other sources
of experience to possess reciprocal connexions.

On the other hand, the character of Space in general [1], requiring
every point to be connected with others, forbids us to regard the
various particular figures which may satisfy its requirements as isolated
instances; it compels us to connect them with each other under the
same conditions under which points are connected with points within
the figures themselves.   If we conceive this demand satisfied, as far
as the addition of fresh elements brings a constantly recurring possibility and necessity of satisfying it, the result which we obtain is
‘*Space*’ [2]: the single and entire picture, that is not only present by the
uniformity of its nature in every limited part of extension, but at the
same time contains them all as *its* parts, though of course it is not,
as a whole, to be embraced in a single view: it is like an integral

---

[1] [‘ Räumlichkeit.’]   [2] [‘ der Raum.’]

obtained by extending the relation which connects two points, to the infinite number of possible points. The only parallel to this condition, is in our habit of representing to ourselves the countless multitudes of mankind not merely as instances of their genus, but as parts united with the whole of Humanity; in the case of animals the peculiar ethical reasons which bring this about are wanting, and we are not in the habit of speaking in the same sense of ' animality.'

103. Of course, in the above remarks, I owe to the guidance of Kant all that I have here said in agreement with his account in Sect. 2 of the Transcendental Aesthetic; as regards what I have not mentioned here, I avoid for the moment expressing assent or dissent, excepting on two points which lie in the track of my discussion. ' It is impossible,' Kant says [1], ' to represent to one's self that there is no space, though it is possible to conceive that no objects should be met with in space.' Unnecessary objections have been raised against the second part of this assertion, by requiring of the thought of empty space, which Kant considers possible, the vividness of an actual perception, or of an image in the memory recalling all the accessory conditions of the perception. Then, of course, it is quite right to pronounce that a complete vacuum could not be represented to the mind, without at least reserving a place in it for ourself; for whatever place, outside the vacuum which we were observing, we might attempt, as observer, to assign ourself, we should unavoidably connect that place in its turn, by spatial relations, with the imagined extension. We should have the same right to assert that we could not conceive space without colour and temperature; an absolutely invisible extension is obviously not perceptible or reproducible as an image in memory: it must be one which is recognised by the eye at least as darkness, and in which the observer would include the thought of himself with some state of skin-sensation, which, like colour, he transfers as a property to his surroundings. But the question is not in the least about such impossible attempts; the admitted mobility of things is by itself a sufficient proof that we imply the idea of completely empty space, as possible in its own nature, even while we are actually considering it as filled with something real. This is most simply self-evident for atomistic views; if the atoms move, every point of the space they move in must be successively empty and full; but motion would mean nothing and be impossible, unless the abandoned empty places retained the same reciprocal positions and distances which they had when occupied; the empty totality of space is there-

---

[1] [Trans. Aesth. 2. (2).]

fore unavoidably conceived as the independent background, for which the occupation by real matter is a not unvarying destiny.

To prefer the dynamical view of continuously filled space leads to the same result. Degrees of density could mean absolutely nothing, and would be impossible, unless the same volume could be continuously occupied by different quantities of real matter; but this too implies that the limits of the volume possess and preserve their geometrical relations independently of the actual thing of which they are the place; and they would continue to possess them, if we supposed the density to decrease without limit and to approach an absolute vacuum. Therefore it is certain that we cannot imagine objects in space without conceiving its empty extension as a background present to begin with; although no remembered image of a perception of it is possible without a remembrance of the objects which made it perceptible to sense.

104. With this interpretation we may also admit the first part of the Kantian assertion. It is true that we cannot *represent* to ourselves the non-existence of space as something that can be experienced, and re-experienced in memory. It is however not *inconceivable* to us absolutely; but only under the condition that an aggregate of actual existence, capable of combination, in short a real world, is to be given, and that the subjects which have to bring it before them are our minds. Now this real world is given us; metaphysic rests entirely on this fact, and only investigates its inner uniformity without indulging in contemplation of the unreal: it is enough then for her to consider space to be given, as the universal, unchangeable, and ever present environment of things, just as much as things and their qualities are recognised to be given as changeable and alternating.

In this sense I may couple Kant's assertion with another saying of his; 'space is imagined as an infinite given magnitude[1].' It has been objected against this too, that an infinite magnitude cannot be imagined as given; but no one knew this better than Kant. A reasonable exposition can only take his expression to mean, that space is above all things *given*, and is not like a universal of which there can be a doubt whether it applies to anything or not; and that further, in every actual limited perception space is given, as a magnitude whose nature demands and permits, that, as extending uniformly beyond every limit, it should be pursued to infinity. Hence, the infinity of space clearly is given; for there is no limit such that progress beyond it, although conceivable, yet would not be real in

[1] [Trans. Aesth. 2. (4).]

the same sense as the interval left behind ; every increment of extension, as it is progressively imagined, must be added to the former
quantity as equally a given magnitude.

Finally, all these observations strictly speaking do nothing but
repeat and depict the impression under which we all are in every-day
life. The moment we exert our senses, nothing seems surer to us
than that we are environed by Space, as a reality in whose depths the
actual world may lose itself to our sight, but from which it can never
escape ; therefore while every particular sense-perception readily falls
under suspicion of being a purely subjective excitement in us, to doubt
the objectivity of Space has always seemed to the common apprehension an unintelligible paradox of speculation.

105. The motives to such a startling transformation of the ordinary
view were found by Kant not in the nature of space itself, but in contradictions which seemed to result from its presupposed relation to the
real world. The attempt of the Transcendental Aesthetic, to demonstrate our mental picture of space to be an *a priori* possession of our
mind, does not in itself run counter to common opinion. For
suppose a single space to extend all round us and to contain within
it ourselves and all things ; precisely in that case it is of course impossible that the several visions of it, existing in several thinking
beings, could be the space itself ; they could not be more than subjective representations of it in those beings : so whether they belong
to us originally, or arise in us by action from without, there is no
prima facie hindrance to their being, *qua* images belonging to cognition, similar to a space which exists in fact.

Nothing short of the antinomies in which we become entangled, if
we attempt to unite our ideas of the entirety of the world or of its
ultimate constituent parts with this presupposition of an actual Space,
decided Kant for his assumption that the space-perception was
nothing but a subjective form of apprehension with which the nature
of the real world that had to be presupposed had nothing in common.
With this indirect establishment of his doctrine I cannot agree ;
because the purely phenomenal nature of space does not properly
speaking remove any of the difficulties on account of which Kant felt
compelled to assert it. It is quite inadmissible, after the fashion
especially of popular treatises of the Kantian school which exulted in
this notion, to treat Things in themselves as utterly foreign to the
forms under which they were nevertheless to appear to us ; there must
be determinations in the realm of things in themselves prescribing the
definite places, forms, or motions, which we observe the appearances

in space to occupy, sustain, or execute, without the power of changing them at our pleasure. If Things are not themselves of spatial form and do not stand in space-relations to one another, then they must be in some network of changeable intelligible relations with one another; to each of these, translated by us into the language of spatial images, there must correspond one definite space-relation to the exclusion of every other, How we are in a position to apply our innate and consequently uniform perception of space, which we are said to bring to our experiences ready made, so that particular apparent things find their definite places in it, is a question the whole of which Kant has left unanswered; the results of this omission, as I think it worth while to show briefly, encumber even his decision upon the antinomy of Space.

106. The real world, it is said, cannot be infinite in space, because infinity can only be conceived as unlimited succession, and not as simultaneous. Now how is our position bettered by denying all extension to the real world, while forced, with Kant, to admit that in all our experience space is the one persistently valid form under which that world appears? I cannot persuade myself that this so-called empirical reality of space is reconcilable with the grounds which cause the rejection of its transcendental validity for the world of Things in themselves.

In this world, the world of experience, if we proceed onwards in a straight line, we shall, admittedly, never come to the end of the line; but how do we suppose that our perceptions would behave during our infinite linear progress? Would there always be something to perceive, however far we advanced? And if there was, would there be some point after which it would be always the same or would it keep changing all through? In both of these cases there must be precisely as many distinguishable elements in the world of things in themselves as there are different points of space in this world of perception; for all the things that appear in different places, whether like or unlike, must be somehow different in order to have the power of so appearing, and so must at least consist in a number of similar elements, corresponding to the number of their distinguishable places. Consequently, on this assumption, space could only possess its empirical reality if there were conceded to the real world that very countlessness or infinity the impossibility of admitting which was the reason for restricting space to an empirical reality. I trust that it will not be attempted to object that in fact the infinite rectilinear progression can never be completed. Most certainly it cannot, and doubtless we are secure against advancing so far in space as to give practical urgency to the question how our perceptions will behave : but in treating of the formation of our idea of the world, we must consider the distances which we know we shall never reach as in their nature simultaneously existent, just as much as those which we have actually traversed are held simultaneously persistent ; it is impossible for us to assume that the former are not there till our perception arrives at them, and that the latter cease to be, when we no longer perceive them.

Now, one would think, the other assumption remains ; suppose at a definite point reached in our advance, the world of perception came to an end, and with it, all transmission of perceptions arising from the actually existing contents of the distances previously traversed. This would give the image of a finite actual world-volume floating in the infinite extension of empty space. Kant thinks it impossible ; his idea is that in such a case we should have not merely a relation of things *in* space, but also one of things *to* space ; but as the world is a whole, and outside it there is no object of perception with which it can stand in the alleged relation, the world's relation to empty space would be a relation of it to no object. The note [1] which Kant subjoins here, shows clearly what his only reason is for scrupling to admit this relation of a limitation of the real world by space : he starts with his own assumption that space is only a form to be attached to possible things, and not an object which can limit other objects. But the popular view, which he ought not to disregard as up to this point [2] he has not explicitly disputed it, apprehends space to be a self-existent form such as to include *possible* things, but clearly in treating it thus by no means takes it for a form which can only exist in attachment to things as one of their qualities, or for a simple non-entity. Rather it is held to be a something of its own enigmatic kind, not indeed an object like other objects, but with its peculiar sort of reality, and such therefore as could not be known without proof to be incapable of forming the boundary of the real world. But in any case we should have no occasion to expect of empty space a restricting energy, which should actively set limits to the world, as if it were obvious that in default of such resistance the world must extend into infinity. The fact is rather that the world must stop at its limit, because there is no more of it ; we may call this a relation of the world ' to no object,' but such a relation is at least nothing mysterious or suspicious; moreover, it would have to remain true even of our unspatial world of things in themselves; this also, the totality of existence, would be in the same way bounded by Nothing. So if in our progression through the world of experience, the coherent whole of our observations convinced us that at any point the real world came to an end, this fact alone would not cause us the difficulty by which Kant was impelled to overthrow the common idea; were it but clear what is meant by saying of things that they are in space, we should not be disturbed at their not being everywhere.

On the other hand it cannot be denied, that this boundedness of the world in space would also be reconcilable with Kant's doctrine, if this were once accepted, and supplemented in the way I suggest. If the world of things in themselves were a completed whole; if they all stood to each other in graduated intelligible relations, which our perception had to transform into spatial ones; then the pheno-menal image of such a world would be complete when all these actually existing relations of its elements had found their spatial expression in our apprehension. But beyond this bounded world-picture there would appear to extend an unbounded empty space; all conceivable but unrealised continuations or higher intensities of those intelligible conditions would like them enter into our percep-tion, but only as empty possibilities. To indicate it briefly; every pair of converging lines a b and c d whose extremities we found attached to impressions of real things, would require their point of intersection to be in the infinite void, supposing them not to find it within the picture of the real world. The boundedness of the real world is therefore admissible both on Kant's view of space and on the popular view, and so the choice between them is undetermined; it is equally undetermined if we assume the unboundedness of the world, as neither of the views in question by itself removes the difficulties which are found in the conception of the infinity of exist-ing things.

107. I intend merely to subjoin in a few words the corresponding observations on the infinite divisibility, or the indivisibleness, of the ultimate elements of real existence. If we abide strictly by the em-pirical reality of space, then in thinking of the subdivision of extended objects as continued beyond the limits attainable in practice, we must come to one of two conclusions about the result; either we must arrive at ultimate actual shapes, indivisible not only by our methods but in their nature; or else the divisibility really continues to infinity.

If real things were infinitely divisible the difficulty which we should see in the fact would be no more removed by assuming space to be purely phenomenal, than was the similar difficulty in the idea of infinite extension: every real Thing, which presented itself phenomenally to our perception as something single and finite occupying space, would have to be itself infinitely divisible into unspatial multiplicities; for every part of the divisible space-image, must, as it appears in a different point of space from every other part, be dependent on a real element which has an existence of its own and in its unspatial fashion is distinct, somehow, from all other points.

If on the contrary we arrived at the conviction, that definite minimum volumes of real things were indivisible, while the space they occupied of course retained its infinite geometrical divisibility, we might still think it obscure what could be meant at all by saying that real things occupy space: but if we assume this as intelligible, we should not be astonished that in virtue of its nature as a particular kind of unit, each real thing should occupy just this volume and no other, and allow no subdivision of it. Here once more the obscure point remarked upon is made no clearer by the assumption that space is merely phenomenal. We should have to represent to ourselves that every Thing in itself, though in itself unspatial, yet bore in its intelligible nature the reason why it is forced to present itself as a limited extension to any perception which translates it into spatial appearance. This idea involves another; that the real Thing, though indivisibly one, is yet equivalent to an indissolubly combined unity of moments, however to be conceived; every point of its small phenomenal volume, in order to distinguish itself from every other and form an extension with their help, presupposes a cause of its phenomenality in the Thing-in-itself, distinct from the corresponding cause of every other point, and yet indissolubly bound up with those causes.

How to satisfy these postulates we do not yet know; common opinion, which says that the Thing is actually extended in an actual space, probably thinks that it is no less wise, and much more clear, about the fact of the matter than the view of the unreality of space, which common opinion holds to be at all events not more successful in comprehending it.

Here, as in the last section, I dismiss the objection that there is a practical limit; that we can never get so far in the actual subdivision of what is extended, as to be enabled to assert either infinite divisibility or the existence of indivisible volumes. One of the two must necessarily be thought of as taking place as long as the empirical reality of space is allowed universal validity; that is as long as we assume that however far we go in dividing the objects of our direct experience, spatial ideas will find necessary application to all the products of this subdivision; that there would never be a moment when the disruption of what is in space would suddenly present us with non-spatial elements.

108. The foregoing discussions have brought me to the conviction that the difficulties which Kant discovers by his treatment of the Antinomies, neither suffice to refute the ordinary view of the objectivity of space, nor would be got rid of by its opposite; but that other motives are forthcoming, though less noticed by Kant, which nevertheless force us to agree with him.

The want of objective validity in the spatial perception is revealed before we come to apply it to the universe or to its ultimate elements. We have only to ask two other and more general questions; how can space, such as it is and must be conceived whether occupied or not, have ascribed to it a reality of its own, in virtue of which it exists before its possible content? And how can what we call the existence of things in space be conceived, whether such occupation by real things concerns its entire infinite extent, or only a finite part of it?

The first of our questions, more especially, but the second as well, require a further introductory remark. We must give up all attempt to pave the way for answering the two questions by assigning to space a different nature from that which we found for it in our former description. There is obvious temptation to do so in order to make the substantive existence of space, and its limiting action on real things, seem more intelligible. Thus we are inclined to supply to space, which at first we took for a mere tissue of relations, some substratum of properties, undefinable of course, but still such as to serve for a substantive support to these relations. We gain nothing by doing so; we do not so much corrupt the conception of space, as merely throw the difficulty back, and that quite uselessly. For the second of our questions was, how real things can at all stand in relation to space. Precisely the same question will be raised over again by the new substratum in which space is somehow to inhere. Therefore we must abide by this; there is simply nothing behind that tissue of relations which at starting we represented to ourselves as space; if we ask questions about its existence, all that we do or can want to know is, what kind of reality can belong to a thing so represented, to this empty and unsubstantial space.

109. No doubt, when so stated, the question is already decided in my own conviction by what I said above concerning the nature of all

' relations ¹'': that they only exist either as ideas in a consciousness
which imposes them, or as inner states, within the real elements
of existence, which according to our ordinary phrase stand in the
' relations.'

Still I do not wish to answer the present question merely by a
deduction from this previous assertion of mine; but should think it
more advantageous if I could succeed in arriving at the same result
by an independent treatment. But I do not hide from myself how liable
such an attempt is to fail; it is a hard achievement to expound by
discursive considerations the essential absurdity of an idea which
appears to be justly formed because it is every moment forming itself
anew under the overpowering impression of a direct perception; an
idea too, which never defines precisely what it means, and which
therefore escapes, impalpably, all attempts at refutation.

This is our present case. It is an impression which we all share
that space extends before our contemplating vision, not merely as an
example of external being independent of us, but as the one thing
necessary to making credible to us the possibility and import of
any such being. The idea that it would still remain there, even if
there were no vision for it to extend before, is an inference hard to
refute; for it does not explain in what the alleged being of that space
would any longer consist if it is to be neither the existence of a thing
which can act, nor the mere validity of a truth, nor a mental representation in us. It is vain to repeat, that space itself teaches us with
dazzling clearness that there are other and peculiar kinds of reality
besides these; this is only to repeat the confusion of the given perception with the inference drawn from it; the former does find space
appearing in its marvellous form of existence; but perception cannot
go outside itself and vouch that there corresponds to this reality which
is an object of perception a similar reality which is not; this notion
can only be subjoined by our thought, and is *prima facie* a questionable supposition.

I now wish to attempt to show how little this hypothesis does to
make those properties intelligible, which we can easily understand to
be true of space if we conceive it merely as an image created by our
perceptive power, and forthcoming for it only.

110. Every point p of empty space must be credited with the same
reality, whatever that may be, which belongs to space as a whole;
for whether we regard this latter as a sum of points, or as a product
of their continuous confluence with one another, in any case it could

¹ [§ 81, end.]

not exist, unless they existed. Again, we find every point p exactly like every other q or r, and no change would be made if we thought of p as replaced by q or by r. At the same time such an interchange is quite impossible, only real elements can change their relations (which we are not now discussing), to empty space-points; but these latter themselves stand immovable in fixed relations, which are different for any one pair and for any other.

Of course, no one even who holds space to be real, regards its empty points as things like other things, acting on each other by means of physical forces. Nevertheless, when we say 'Space exists,' it is only the shortness of the phrase that gives a semblance of settling the matter by help of a simple 'position [1]' or act of presenting itself, easily assigned or thought of as assigned to this totality, which we comprehend under the name of space. But, in fact, for space to *exist*, everything that we have alluded to must *occur ;* every point must exist, and the existence of each, though it is like every other, must consist in distinguishing itself from every other, and determining an unalterable position for itself compared with all, and for all compared with it. Hence the fabric of space, if it is to exist, will have to rest on an effectual reciprocal determination of its empty points ; this can in any case be brought under the idea of action and reaction, whatever distinction may be found between it and the operation of physical force, or between empty points and real atoms.

This requirement cannot be parried by the objection that as we have not to make space, but only to consider it as existing, we have no occasion to construct its fabric, but may accept it, and therefore the position of all its points, as given. True, we do not want to make space, as if it had not existed before, but this very act, the recognition of it as given, means presupposing that precise action and reaction of its points which I described. No points or elements, unless thought of as distributed in an already existing space, could conceivably be asserted simply *to be* in particular places, without being responsible for it themselves, and to share in the relation subsisting between these places ; but the points of empty space cannot be taken as localised in turn in a previous space, so as to have their reciprocal relations derived from their situation in it ; it must be in consequence of what they themselves are or do, that they have these relations, and by their means constitute space as a whole. Hence, if the two points p and q exist, their distance pq is something which would not be there without them, and which they must make for themselves.

---

[1] ['Position,' v. § 10.]

I can imagine the former objection being here repeated in another shape; that we did not conceive the spatial relations as prior, in order to place the points in them afterwards; and so now, we are not to assume the points first, so that they have to create the relations afterwards; the two together, thought in complete cohesion, the points *in* these relations, put before us, at once and complete, the datum which we call existing space. Granting then, that I could attach any meaning to points being in relations simply as a fact, without either creating or sustaining them by anything in themselves; still I should have to insist on the circumstance that every reality, which is merely given in fact, admits of being done away and its non-existence assumed at least in thought. Now not only does no one attempt to make an actual hole in actual empty space; but even in thought it is vain to try to displace one of the empty space-points out of that relation to others which we are told is a mere datum of fact; the lacuna which we try to create is at once filled up by space as good as that suppressed. Now of course I cannot suppose that anyone who affirms the reality of space will set down this invulnerability only to his subjective perception of it, and not to existing space itself; obviously this miraculous property would have to be ascribed to real extension as well.

This property is very easily intelligible on the view of the purely phenomenal nature of space. If a consciousness which recollects its own different acts or states, experiences a number n of impressions of any kind in a succession which it cannot alter at pleasure; if, in the transition from each impression to the next, it experiences alterations, sensibly homogeneous and equal, of its own feeling; if, again, it is compelled to contemplate these differences not merely as feelings, but owing to a reason in its own nature, as magnitudes of a space whose parts are beside each other; and if, finally, after frequently experiencing the same kind of progression, it abstracts from the various qualities of the impressions received and only calls to mind the form under which they cohered; then, for their consciousness, and this only, there will arise before the mind's eye the picture of an orderly series or system of series, in each of which between the terms $m-1$ and $m+1$ it is impossible for m to be missing. If there were no impression to occupy the place m, still the image of the empty place in the series would be at once supplied by help of the images of the two contiguous places and by means of the single self-identical activity of the representing consciousness.

All is different if we require an existing space, and conceive the absence of this consciousness, which combines its images, evokes some to join others, and never passes from one to the others without also representing the difference which divides them. Then, the empty points of space would have to take upon themselves what the active consciousness did; they would have to prescribe their places to each other by attraction and repulsion, and to exert of themselves the extraordinary reproductive power by which space healed its mutilations. And in spite of all we should at once get into fresh difficulties.

111. For, the relation or interval $p \, q$, which the two existing points p and q would be bound according to their nature to establish between them, ought at the same time to be different from every other similar relation which p and r or q and r for similar reasons would set up between *them*. But the complete similarity of all empty points involves, on the contrary, an impossibility of p and q determining any other relation between themselves, than any other pair of points could between themselves; even N, a number of connected points, conceived with determinate relations already existing between them, could assign no place in particular to another point s which we might suppose thrown in, because any other, t or u, would have as good a right to the same place.

It is easy to foresee the answer that will at once be made; that it is quite indifferent, whether the point is designated by s or t or u; it is in itself a yet undefined, and therefore, in strictness, a nameless point; it is only after N has assigned it a particular place that it becomes the point s, which is now distinct from the points t and u, which are differently localised by N. But this observation, though quite correct in itself, is out of place here. It would only apply if we were regarding s as the mere idea of an extreme term belonging to a series N begun in our consciousness; such an idea of s would be created by our consciousness, in the act of requiring it, in the particular relations to N which belonged to it; there would be no inducement to the production of any other image which had not these relations. Or again; our consciousness may not restrict itself to its immediate problem, but recalling previous experiences may first form the idea of an extreme term, *e.g.* for two series which converge, without being aware what place it will hold in a system of other independent terms which is to serve as the measure of its position; then we have a term x, which has as yet no name, and which is not particularised as s, t, or u, till we come accurately to consider the law according to which each series progresses, and so the simultaneous determining equations are both solved.

Such a productive process of determination, realising what it aims at, is explained in this case by the nature of our single consciousness, which connects with each other all the particular imagined points of its content; but if instead of mental images of empty points we are to speak of actual empty points, then we should really be compelled to assume, *either* that every existing number of points N is constantly creating new points, which by the act of their production enter into the relations appropriate to them; *or* that by exerting a determining activity N imposes these relations on points already existing whose own nature is indifferent to them. Obviously we should not conceive either of these constructions as a history of something that had once taken place, but only as a description of the continually present unmoving tension of activities which sustains in every moment the apparently inactive nature of space. Having once got so far into this region of interesting fancies I wish to pursue the former of these hypotheses one step further; the second, my readers will gladly excuse me from considering.

112. We cannot seriously mean to regard a particular ready-made volume N as the core round which the rest of space crystallises. Not merely any N whatever, but ultimately every individual empty point, would have the same right to possess this power of propagation, and we should arrive at the idea of a radiant point in space, fundamentally in the same sense in which it is known to geometry. Then, the radiant point p would produce all the points with which its nature makes a geometrical relation possible, and each of them in the precise relation which belongs to it in respect of p; among others the point q, which is determined by the distance and direction $p\,q$. All this is just as true of any other empty point; it would still hold good if among them was a q, and then among the innumerable points which q would create there would be one standing to q in the relation $q\,p$, the same which was above designated, in a different order, by $p\,q$. And now it might be supposed that we had done what we wanted, and obtained a construction of space corresponding to its actual nature; for it seems obvious that $p\,q$ and $q\,p$ indicate the same distance between the same points, and that thus the radiant activities of all points coincide in their results, so as to produce ordinary extension with its geometrical structure.

But this expectation is founded on a subreption. Before we completed our construction we knew nothing more of the empty points

from which it was to start, than that they are all similar to one
another, and that the same reality attaches to all of them; but beyond
this they had no community with each other. It is therefore by no
means self-evident, that the pencil of rays which starts from the
existing point p will ever meet the other, emitted by the independent
point q; both of them may, instead of meeting, extend as if into two
different worlds, and remain ever strange to each other, even more
naturally than two lines in space which not being in the same plane,
neither intersect nor are parallel. The point q, generated by the
radiant point p, is not obviously the same q, with that which, as given
independently, we expected to generate p; the second p generated
by the given q need not coincide with the first p, nor the line $q\,p$ with
the previous line $p\,q$; in a word, what is generated is not a single
space, in which all empty points would be arranged in a system, but
as many reciprocally independent spaces, as we assumed radiant
points; and from one of these spaces there would be absolutely no
transition into another. Our anticipation of finding that only a single
space is generated, started with the tacit assumption that space was
present as the common all-comprehending background, in which the
radiations from the points could not help meeting.

Still, if all the resources of a disputatious fancy are to be exerted
in defence of the attempted construction; there might be this escape.
Suppose there are countless different spaces, it might be said; still, just
because they do not concern each other, for that very reason they do
not concern us; excepting that particular one in which we and all our
experiences are comprehended, and with which alone, as the others
never come in contact with us at all, Metaphysic has to do. Then let
us confine ourselves to the space which is generated by the radiant
point p. The point q which it creates, has equal reality with p, and
so shares its radiant power; it must, in its turn, determine a point
towards which it imposes on itself the relation $q\,p$; and this point p
will certainly be no other than, but the same with, that which first
imposed on itself towards q the relation $p\,q$; therefore the lines $q\,p$
and $p\,q$ will certainly coincide.

But even this does not give us the result aimed at. As we cannot
regard a particular point p exclusively, but are able to regard any
whatever, as the starting-point of this genesis of space, the result of
our representation translated from the past tense of construction into
the present of definition, is simply this; that it is the fact that in
existing space every point has its particular place, and that a line $p\,q$
of determinate direction and magnitude, taken in the opposite direc-

tion $q\,p$, returns to its starting-point. No doubt this is correct; but no one will affirm that this last construction fulfils its purpose of explaining such a condition of things; there is something too extraordinary in the notion that an existing point generates out of itself an infinite number of points with equally real existence, and something too strange in the result that every existing empty point has as it were an infinite density, being created and put in its place by every other point, not merely by one; and finally, the whole idea is too empty a fiction, with its radiant power which if it is not to lead to a purely intensive multiplication of being into itself, but to an Extension, must in any case presuppose a space, in which its effect may assume this very character of radiation.

Nevertheless all these incredibilities appear to me to be unavoidable, as long as we persist in thinking of empty space with its geometrical structure as actually existing; but the doctrine of its purely phenomenal nature avoids them from the beginning; and it is hardly requisite to prove this by a protraction of this long exposition.

One can understand how, for a consciousness which remembers its previous progression through the terms $p\,q\,r$, there arises the expectation of a homogeneous continuance of this series in both directions, which implies an apparent power of radiation, as above, in those points; only what takes place here is not a self-multiplication of something existent, but a generation of ideas out of ideas, i.e. of fresh states of a single subject out of its former states, in accordance with the laws of its faculty of ideas and the movement of its activities which was in progress before. It is on this hypothesis equally easy to understand, that the converse march of the movement returns from q to the same p, i.e. reproduces the identical image p from which it started; for the image q has only such radiant power as it derives from representing to the mind the purport of the series; so that q by itself, as long as it is represented as a term in the series, can never induce a divergence from the direction of that series.

On the other hand, starting with a qualitatively determined impression π, which fills the geometrical place of the term p, there may be an advance to other impressions κ and ρ, such that the differences $\pi - \kappa$, $\kappa - \rho$, may be comparable with each other, though not comparable with the difference of the series p, q, r. Then we have the case which we mentioned above; π radiates too, but, so to speak, into another world, and the series π, κ, ρ, finds in fact no place in space-perception, and in respect to its relations within itself can only be metaphorically or symbolically represented by constructions in space, but cannot be shown to have a spatial situation.

113. I am sure that the whole of this account of the matter has only convinced those who were convinced before, and will not have done much to shake the preference for an existing space. Let us therefore ask once more where in strictness the difference of the two views lies; and what important advantage there is that can only be secured by the assumption of this enigmatic existence, so constantly reaffirmed, of an empty extension, and that must be lost by conceding that its import is purely phenomenal? The clearness and self-evidence, with which our perception sees space extended around us, is equally great for both views; we do not in the least traverse this perception, which is endowed with such self-evidence; but only the allegation of a being that underlies it, which must be inaccessible to perception and so cannot share its self-evidence. No doubt for common opinion every perception carries a revelation of the reality of what is perceived; but in the world of philosophy Idealism claims the first hearing, with its proof that what is perceived, in this case, space, is given to begin with merely as the subjective perception of our minds. Now of course in common life we do not need to go through the long toil of inference from perception before *attaining* the idea that what is perceived is real; but in the world of philosophy this investigation is essential, to decide whether we may *retain* this idea; for I repeat that in this region it is not the primary *datum*, but remains problematic till it is proved to be necessary.

Such a proof, in strictness, has never been attempted; the burden of disproof has been thrown on the opposite view, and its opponents have taken their stand on the probability of their own opinion as importing a valid presumption of its truth. The probability seems to rest on this; that a space, which exists by itself with all the properties ascribed to it by our perception, makes the origin of this perception seem much more natural than does our more artificial doctrine; according to which it arises from a combination of inner states of our consciousness wholly dissimilar to it. But the artificiality here objected to must be admitted, even if space were as real as could be wished. The pictures which are made of it in the countless minds which are all held to be within space, could not be more than pictures of it, they could not *be it*; and as pictures they could only have arisen by means of operations on the mind which could not be extensions, but could only be inner states corresponding to the nature of the subject operated upon. In every case our mental representation of space must arise in this way; we cannot get it more cheaply, whether we imagine beneath the picture presented to our mind an existence like it outside us, or one entirely disparate.

What can be gained then by maintaining the view which we oppose? Men will go on repeating the retort; that it is impossible to doubt the reality of space, which is so clearly brought home to us by immediate perception. But are we denying this reality? Ought not people at length to get tired of repeating this confusion of ideas, which sees reality in nothing but external existence, and yet is ready to ascribe it to absolute vacuity? Is pain merely a deceitful appearance, and unreal, because it subsists only for the moment in which it is felt? Are we to deny the reality of colours and tones because we admit that they only shine and sound while they are seen and heard? Or is their reality less loud and bright because it only consists in being felt and not in a self-sustained being independent of all consciousness? So then space would lose nothing of its convincing reality for our perception if we admitted that it possesses it only in our perception.

We long ago rejected the careless exaggeration which attaches to this idea; space is not a mere semblance in us, to which nothing in the real world corresponds; rather every particular feature of our spatial perceptions corresponds to a ground which there is for it in the world of things; only, space cannot retain the properties which it has in our consciousness, in a substantive existence apart from thought and perception. In fact, there is only one distinction forthcoming, and that of course remains as between the two views; for our view all spatial determinations are secondary qualities, which the real relations put on for our minds only; for the opposite view space as the existing background which comprehends things is not merely secondary but primary as a totality of determining laws and limits, which the Being and action of things has to obey, so that the things and ourselves are in space; while our view maintains that space is in us. This brings us naturally to the second of the questions, which were proposed[1] above.

114. When I want to know what precisely we mean by saying that things are in space, I can only expect to meet with astonishment, and wonder what there is in the matter that is open to question; nothing, it will be said, is plainer. And in fact this spatial relation is given so clearly to our perception, that we find all other relations, in themselves not of the spatial kind, expressed in language by designations borrowed from space. We even meet with philosophical views which

[1] [Sect. 108.]

not only demand constructions in space by way of sensuous elucidation of abstract thought, but prefer to regard the problem of cognition as unsolved till such constructions are found. I have no hope of making clear the import of my question to such a ' scientific mind.' But the assumption of a purely phenomenal space has little difficulty in answering it.

Only I feel compelled to repeat the warning, that this assumption does not any more than the other aim at denying or modifying the directness of the overwhelming impression which makes space appear to us to include things in it ; it only propounds reflections on the true state of the facts, which makes this impression possible ; and we expressly admit of our reflections that they are utterly foreign to the common consciousness. The power of our senses to see colours and forms or to hear sounds, seems to us quite as simple ; we need, we think, only to be present, and it is a matter of course that sensations are formed in us, which apprehend and repeat the external world as it really is ; the natural consciousness never has an inkling of the manifold intermediate processes required to produce these feelings ; and one who has gained scientific insight into their necessity does not feel them a whit more noticeable in the moment of actual sensation.

It is the task of psychology to ascertain these intermediate processes for the case in hand ; its solution will not point to an image of empty space, formed prior to all perceptions, into which the mind had subsequently to transplant its impressions ; it is rather the series of peculiar concomitant feelings of homogeneous change of its condition, experienced in the transition from the impression p to the other impression q, that is felt by it as the distance $p\,q$; and from the comparison of many such experiences there arises, as I indicated just now, by help of abstraction from the content of the various impressions, the picture of empty extension. After it has arisen, to localise an impression q in a particular point of this space simply means : taking an impression p as the initial state from which the movement of consciousness starts, to contemplate the magnitude of the change which consciousness felt or must feel in order to reach q, under the form of a distance $p\,q$.

These different concomitant feelings, which distinguish the impressions p and q, are independent of the qualitative difference of their content, and may attach to like as well as to unlike impressions. Therefore metaphysic can only derive the feelings from a difference in the effects produced on the soul by the real elements which correspond to them, in conformity with a difference of actual relations in which the realities stand to the soul, and consequently, with a determinate actual relation in which they stand to each other. I reserve for a moment my further explanations concerning these intelligible relations, as we may call them, of the realities, which we regard as causes of our perceived relations of space; I only emphasise here the fact that they consist in actual relations of thing and thing, not of things and space; and that it is not they, as merely subsisting between the things, but the concentration in the unity of our consciousness of effects of the things varying in conformity with them, that is the proximate active cause of our spatial idea in which we picture their locality, and their distance from each other.

115. From this point we may obtain a conspectus of the difficulties which spring from the opposite view, that space has an existence of its own, and that things are in it. If space exists, and consequently the point p exists, what is meant by saying that a real element π is in the point p? Even if p itself is not to be taken to be a real thing, still, between it as something existent, and the reality π, some reciprocal operation must be conceivable by the subsistence of which the presence of π in p is distinguished from its not being present in p. But as regards π we do not believe that its place does anything to it; on the contrary, it remains the same in whatever place it may be; therefore there is nothing which takes place in it by which its being in p is distinguishable from its being in q; the two cases would only be distinguishable to an observer, who had reason on the one hand to distinguish p from q, and on the other to associate the image of π in the moment of perception only with p and not with q.

If we go on to ask what happens to the point p when π is in it, we should suppose that the nature of p would be just as little changed as that of π; but no doubt the answer will be: the very fact that p is occupied by π distinguishes it from q, which is now not the place occupied by π. Against this answer I am defenceless. It is indeed unassailable if we can once conceive, and accept as a satisfactory solution, that between two realities, the point p and the actual element π, there should be a relation as to which neither of the related points takes note of anything except that it, the relation, subsists, while in every other respect the two things are exactly as they would be if it did not subsist. I might add, that p would not be permanently filled by π, but, in turn, by other real elements κ or ρ; surely the one case ought somehow to distinguish itself from the other, and the point p to be different when occupied by π from what it is when occupied by κ.

But this would be unavailing; I should be answered with the same acuteness : that in all these cases p remains just the same in every other respect, and the distinction between them is constituted by the simple fact, that the occupation of p, which does not affect it in itself, is carried out by π in one case and by κ in another. As all this more-over is as true of q as of p, I can only meet this reassertion by reas-serting the opposite notion ; that the whole state of things alleged is inconceivable to me as in real existence, and only conceivable as in the thought of an observer, who, as I indicated, has reason to dis-tinguish p from q and, at the moment, to combine either π or κ with p or q—to make one combination and not another.

Finally, taking $p\,q$ as the distance between the real elements π and κ which occupy the points p and q, we do not in fact treat this localisation as unimportant in our further investigation of things; for we believe the intensity of reciprocal action between π and κ to be conditioned according to the magnitude of the distance. But their action cannot be guided by this changeable distance unless it is somehow brought home to them; how are we to suppose this to be done? The distance $p\,q$ is not in the points p and q but between them ; if we suppose the empty point q represented at p by some effect produced by q on p, which makes the distance $p\,q$ always present to p, and consequently, though I can see no reason for the inference, present also to the element π in p and determining its behaviour, still this would hold equally good of any other empty point r or s. All of them would be represented at p, consequently they would all have an equal right to determine the behaviour of the element π at p; the pre-eminence of q which is at the moment occupied by the real element κ, could only depend on the latter, and would have to be accounted for thus : the empty point q must undergo a change of state by becoming filled, must transmit the change to p through $q\,p$ and there transfer it to the element π; a reaction between real existence and the void, which would be as inevitable as it is inexplicable. The argument might be pursued farther, but I conclude here, hoping that the mass of ex-travagances in which we should be involved has persuaded us of the inconceivability of the apparently simple assumption that space has independent existence and that things have their being in space.

116. The opposite view which I am now maintaining leads to a series of problems which I will not undertake to treat at present ; it is enough to characterise their import as far as is requisite to establish the general admissibility of the doctrine. We may begin by ex-pressing ourselves thus ; that we regard a system of relations between

the realities, unspatial, inaccessible to perception, and purely intelligible, as the fact which lies at the root of our spatial perceptions. When these objective relations are translated into the subjective language of our consciousness, each of them finds its counterpart in one definite spatial image to the exclusion of all others. I should avoid calling this system of relations an 'intelligible space' and discussing whether it is like or unlike the space which we represent to ourselves by help of our senses. I start from the opposite conviction, that there exists no resemblance between the two; for it would transfer to the reality of the new condition of things all the difficulties which we found in the reality of empty space.

However, it is not worth while to keep up the idea of such a system of relations, which was only of use as a brief preliminary expression of the fact ; we now return to the conviction expressed above ; it is not relations, whether spatial or intelligible, *between* the things, but only direct reactions which the things are subject to from each other, and experience as inner states of themselves, which constitute the real fact whose perception we spin out into a semblance of extension. Let P and Q be two real elements thought of as unrelated ; let $P\kappa$ and $Q\pi$ indicate them when in the states of themselves which are set up by a momentary mutual reaction ; these states of theirs contain the reason why P and Q, or at the moment $P\kappa$ and $Q\pi$, appear in our perception in the places p and q, separated by the interval pq. It need hardly be observed that the mere fact of the reaction subsisting between P and Q cannot by itself set up our perception ; but can only do so by means of an action of P and Q upon us, conformable to their momentary states κ and π ; and therefore other than it would have been in the moment of a different mutual reaction. The meeting of these two actions in our consciousness causes, first, in virtue of its unity, the possibility of a comparison and reciprocal reference of the two ; secondly, in virtue of its peculiar nature the necessity that the result of this comparison should assume the form of distance in space to our perception ; and finally, the magnitude of the difference which is felt between the two actions on us, determines, to put it shortly, the visual angle by which we separate the impressions of the two elements.

Thus the theory attaches itself to a more general point of view, which I adopt in opposition to a predominant tendency of the philosophic spirit of the age ; holding that thought should always go back to the living activities of things, which activities are to be considered as the efficient cause of all that we regard as external

relation between things.   For in calling these latter ' relations ' we are
in fact using a mere name ; we cannot seriously conceive them to be
real and to subsist apart from thought.   I regret that there is an increasingly widespread inclination in the opposite direction, namely, to
apprehend everything that takes place as the product of pre-existing
and varying relations ; overlooking the circumstance that ultimately,
even supposing that such relations could exist by themselves, nothing
but the vital susceptibility and energy which is in Things could
utilise them, or attach to any one of them a result different from that
attaching to the others.

117. As an elucidation, and more or less as a caution, I add what
follows. If the arrangement of perceivable objects in space were
always the same, we might think of them as the image of a systematic order in which every element had a right to its particular
place, in virtue of the essential idea of its nature. It would not be
necessary that the elements which presented a greater resemblance of
nature should occur in closer contiguity in space, or that dissimilar
things should be more widely separated ; the entire scheme of *M*,
which realises itself in the simultaneously combined manifold of
things, might easily necessitate a multitude of crossing relations or
reactions between them, of such a kind that similar elements should
repeatedly occur as necessary centres of relation at very different
parts of the whole system, while very dissimilar ones would have to
stand side by side, as immediately conditioning each other.

The movability of things makes it superfluous to go deeper into
this notion ; the ground of localisation is clearly not in the nature of
the things alone, but in some variable incident which occurs to them,
compatible with their nature, but not determined by it alone. This
might lead to the idea, that it was simply the intensity of the subsisting reaction between them which dictated the apparent situation of
things in space ; whether we presume that in all things what takes
place is the same in kind and varies only in degree ; or, that the
inner states produced in things by their reactions are different in
kind, but so far comparable that their external effects are calculable as
degrees of one and the same activity.

It would be no objection to this that it is observed that there often
are elements contiguous in space which seem quite indifferent to each
other, while distant ones betray a lively reciprocal action. No
element must be torn from its connexion with all others, and none of
its states from their cohesion with previous ones ; contiguous elements
which are indifferent are together not because they demand one

another, but because their relations to all others deny them every other place, and only leave them this one undisputed; the remote elements in question act powerfully on one another, because the ceaseless stream of occurrence has produced counteractions, which hinder the two elements from attaining the state towards which they are now striving.

However, it is not my intention to continue the subject now, or to show by what general line of thought my view of space might be reconciled with the particular facts of Nature. The following sections will compel us to make this attempt, but they would entirely disappoint many expectations unless I began by confessing that the theory of a phenomenal space when applied to the explanation of the most general relations of nature will by no means distinguish itself for facility and simplicity in comparison with the common view. On the contrary; the latter is a gift which our mental nature gives us as a means to clearness and vivid realisation. But I insist upon it that my view is not propounded for its practical utility, but simply because it is necessary in itself, however much it might ultimately embarrass a detailed enquiry were we bound to keep it explicitly before us at every step. We shall see that we are not obliged to do so; but at present I maintain with a philosopher's obstinacy, that above all things that must hold good which we find to be in its nature a necessary result of thought, though all else bend or break. In no case may we regard other hypotheses as definitive truth (convenient as they may be for use and therefore to be admitted in use), if they are in themselves as unthinkable as the indefinite species of reality, which the ordinary view attributes to empty space.

CHAPTER II.

Deductions of Space.

118. AMONG the commonest undertakings of modern philosophy are to be found attempted deductions of Space; and they have been essayed with different purposes. Adherents of idealistic views, convinced that nothing could be or happen without being required by the highest thought which governs reality, had a natural interest in showing that Space was constrained to be what it is, or to be represented as it is represented to us, because it could not otherwise fulfil its assigned purpose. Self-evident as the belief fundamentally is, that everything in the world belongs to a rational whole, there are obvious reasons why it should be equally unfruitful in the actual demonstration of this connexion in a whole; and even the deduction of Space has hardly given results which it is necessary to dwell on.

The solidarity of the whole content of the universe was maintained, in the dawn of modern philosophy, by Spinoza; but in a way which rather excluded than favoured the deduction of Space. The reason lay in an enthusiasm, somewhat deficient in clearness, for the idea of Infinity, and for everything great and unutterable that formal logical acumen combined with an imagination bent on things of price could concentrate in that expression. Hence he spoke of infinitely numerous attributes of his one infinite substance, and represented it as manifesting its eternal nature by means of modifications of each of them. Our human experience, indeed, was restricted to two only of them, consciousness and extension, the two clear fundamental notions under which Descartes had distributed the total content of the universe; and the further progress of the Spinozistic philosophy takes account of these two only. But it adheres to the principle laid down at its starting about all attributes; each of them rests wholly on itself, and can be understood by us only by means of itself; we find it expressly subjoined, that though it is one and the same substance which expresses its essence as well in forms of extension as in forms of thought; yet the shape which it assumes in one of these attributes can never be derived from that which it has assumed in the other. This prohibits any attempt to deduce the attributes of Space from what is not Space; but at the same time Consciousness and Extension are considered to be as manifestations of the absolute quite on the same level; in assuming the shape of extension, it does a positive act as much as in giving existence to forms of consciousness; neither of these is the mere result or semblance of the other.

119. These notions influenced Schelling. After Kant had destroyed all rational cohesion between things-in-themselves and spatial phenomena, it was natural to make the attempt to restore Space to some kind of objective validity. If we may here eliminate the many slight alterations which Schelling's views underwent, the following will be found a pretty constant series of thoughts in him. *Empty* Space is for him too only the subjectively represented image, which remains to our pictorial imagination when it disregards the definite forms of real existence in Space, that is, of matter ; it is not a prior creation of the absolute which goes before the production of the things to be realised in it, but matter itself is this first production, and spatial extension is only real in matter, but in it is actually real and not a mere subjective mode of the spectator's apprehension. How he represents the creation of matter as coming to pass, we need not describe here ; but in general it is easy to see how the desire to explain by one and the same root the distinction which experience presents between the material and spiritual world might lead to denying the primary presence of the characteristic predicates of these two worlds in the Absolute, the root required ; while conceiving, in the complete indefiniteness thus obtained of this absolute Identity, two eternally co-existent impulses, tendencies, or factors, out of which the distinction that had been cancelled might again arise. Some interest attaches to the different expressions which Schelling employs to designate them ; he opposes to the real objective producing factor, which embodies the infinite in forms of the finite, the ideal subjective defining factor which re-moulds the finite into the infinite ; it is the former whose predominance creates Nature, the latter that creates the world of Mind ; though the two are so inseparably united that neither can produce its result without the co-operation, and participation as a determining factor, of the other.

This account admits of no idea of a deduction proper of Space ; still I think that the equal rank assigned to the above designations

contains an indication of the reason which made the space-generating
activity of the absolute appear indispensable to the idea of it. It became
obvious not only that nothing could be generated out of the void of
absolute Identity, but it was also impossible for the determinations
which might have been held to be included in it as merely ideal, to
be more than unrealisable problems failing one condition ; that
something should be forthcoming, given, with content, and for perception ; such as the ideal forms could never create, and as applied
to which, *qua* forms of its relations, and so only, they would possess
reality.　Thus, not without a reminiscence of Kant's construction of
matter out of expanding and contracting forces, Schelling makes the
one, that is the productive factor, provide above all things for the
creation of that which the ideal factor has only to form and to determine ; it is only by the activity of the first that results are made *real*,
which for all the second could do, would never be more than a
postulate, that is, an idea.　Even the actual form which the creation
assumes is determined by the character of the productive factor ; for
it is only this character that can, though under the control and
guidance of the other factor, create such shapes of reality as are
within its range.

120. The indefiniteness of the absolute Identity has disappeared
in Hegel, and the position of the two factors has altered ; the comprehensive system of notions which forms his Logic may be regarded
as the interpretation of what the ideal factor, now the proximate and
primary expression of the Absolute, demands ; the consciousness,
how strongly all these determinations involve and postulate that as
determinations of which they must be presented in order to be real,
appears as the urgency of the ideal factor or hitherto purely logical
idea, to pass over into its form of otherness ; that is, into a shape
capable of direct or pictorial presentation, such as can only exist in the
forms by which a multiplicity whose parts are outside one another is
connected into a whole.　Therefore the logical idea, doing away its
own character as logical, produces Space as 'the abstract universality
of its being outside itself[1].'　Hegel says on this point[2], 'As our procedure is, after establishing the thought which is necessitated by the
notion[3], to ask, what it looks like in our sensuous idea[4] of it ; we go
on to assert, that what corresponds in direct presentation to the
thought of pure externality is Space.　Even if we are wrong in this,

---

[1] ['Die abstracte Allgemeinheit ihres Aussersichseins.']
[2] Naturphilosophie.　Sämmtliche Werke, Bd. VII. § 47.
[3] ['Begriff.']　　　　　　　　　　　　[4] ['Vorstellung.']

that will not interfere with the truth of our thought.' I refer to this remarkable passage in order to indicate the limits which such speculative constructions of Space as this is can never overstep. They may of course derive in a general way, from the thought in which they conceive themselves to express the supreme purpose of the world, a certain postulate which must be fulfilled if the end is to be fulfilled; but they are not in a position to infer along with the postulate what appearance would be presented by that which should satisfy it. In the passage quoted Hegel admits this; in pronouncing Space to be the desired principle of externality he professes to have answered a riddle by free conjecture; the solution might be wrong, but the problem, he asserts, would still be there.

Just in the same way Weisse says [1] 'That primary quality of what exists, the idea of which arises from quantitative infinity being specified and made qualitative by the specific character of triplicity— is Space;' only that he, although in this sentence expressly separating enigma and answer by a mark of interruption, yet regards the latter as a continuous deduction of the space which is present to perception from his abstract and obscure postulate. It can never be otherwise; after, on the one hand, we feel justified in making certain abstract demands which reality is to satisfy, and after, on the other hand, we have become acquainted with Space, then it is possible to put the two together and to show that Space, being such as it is, satisfies these demands. But it is impossible to demonstrate that only it, and no other form, can satisfy them; we are confined to a speculative interpretation of space, and any deduction of it is an impossibility on this track. One would think that the opinion Hegel expresses could not but incline him *prima facie* to the view of the mere phenomenality of the sensuous idea of space; but what he adds on the subject can make no one any wiser as to his true meaning; as a rule the views of his school have adhered to extension as a real activity of the Absolute.

121. Philosophical constructions, it was held, were under the further obligation, to demonstrate not merely of Space as a whole, but further of each and every property by which geometry characterises it, that it is a necessary consequence of ideal requirements. Attempts have been made on obvious and natural grounds to conceive the infinite divisibility and the homogeneousness of an infinite extension, as antecedent conditions of that which the idea sets itself to realise within space; but the most numerous and least fortunate endeavours

[1] Metaphysik, p. 317.

have been devoted to the three dimensions. There are two points in these innumerable attempts that have always been incomprehensible to me.

The first is, the entire neglect of the circumstance that space contains innumerable directions starting from every one of its points, and that the limitation of their number to three is only admissible under the further condition that each must be perpendicular to the two others. Accessory reasons, which are self-evident in the case of geometry and mechanics, have no doubt led to the habit of tacitly understanding, by dimensions of space, such *par excellence* as fulfil this condition; but the philosophical deductions proceed as if the only point was to secure a triplicity, and as if it was unnecessary to find among the abstract presuppositions from which space is to be deduced, a special reason why the dimensions which are to correspond to three distinct ideal moments (however these may be distinguished), should be at right angles to one another.

The second point which I cannot understand is the fastidiousness with which every demonstration partaking of mathematical form, that a fourth perpendicular dimension must necessarily coincide with one of the other three, is always rejected as an external and unphilosophical process of proof. I think, on the contrary, that if we once supposed ourselves to have deduced that certain relations which we postulated in an abstract form must take the shape of lines and angles between them, then the correct philosophical progress would consist in the demonstration that these elementary forms of space being once obtained were completely decisive of its whole possible structure. As a whole subject to law it can have no properties but those constituted in it by the relations of its parts; if its properties are to correspond besides to certain ideal relations then it ought to have been shown that this correspondence demanded just those primary spatial relations from which the properties must proceed as inevitable result. However, it is not worth while to go at greater length into these unsuccessful undertakings, which are not to the taste of the present time, and, we may hope, will not be renewed.

122. Our attention will be much longer detained by other investigations which are sometimes wrongly comprehended under the name of Psychological Deductions of Space. In virtue of the title ' Psychological' they would not claim mention till later; but they treat in detail or touch in passing three distinct questions, the complete separation of which seems to me indispensable.

1. The first, were it capable of being solved, would really belong

to Psychology: it is this: what is the reason that the soul, receiving from things manifold impressions which can only be to begin with unextended states of its own receptive nature, is obliged to envisage them at all under the form of a space with parts outside each other? The cause of this marvellous transfiguration could only be found in the peculiar nature of the soul, but it never will be found; the question is just as unanswerable as how it comes to pass that the soul brings before consciousness in the form of brightness and sound the effects which it can only experience by means of light and sound vibrations transmitted through the senses. It is important to make clear to ourselves that these two questions are precisely alike in nature; and that to answer the first is neither more essential nor more possible than to answer the second, which every one has long desisted from attempting. All endeavours to derive this elementary and universal character of ideas of space, this externality, which appears to us in the shape of an extended line, from any possible abstract relations, which are still unspatial, between psychical affections, have invariably led to nothing but fallacies of subreption; by which space, as it could not be made in this way, was brought in at some step of the deduction as an unjustified addition.

2. On the other hand, if we postulate as given the capacity and obligation of the soul to apprehend an unspatial multiplicity as in space, then there arises the second problem, which I hold to be capable of being solved though a long way from being so; What sort of multiplicity does the soul present in this peculiar form of its apprehension?—for there are some which it does not treat thus. And under what conditions, by what means, and following what clue, does it combine its occasional particular impressions in the definite situation in space in which they are to us the express image of external objects? As no perception of this variable manifold can take place but by the instrumentality of the senses, the solution of this question concerning the localisation of sensations belongs wholly to that part of psychology which investigates the connexion of sensations, and the associations of these remembered images; which latter are partly caused by the conjoint action of nervous stimuli, partly by the activity of consciousness in creating relations.

3. There remains a third question, that of the geometrical structure of extension which arises if we develope all the consequences that the given character of the original externality necessitates or admits; and which is wanted to complete the totality of the Space-image in whose uniformly present environment we are obliged to set in array the

various impressions of sensation. This investigation, which has fallen
to the share of Mathematics, has hitherto been conducted by that
science in a purely logical spirit; it took no account of the play of
psychical activities, which bring about in the individual apprehending
subject a perception of the truth of its successive propositions, a play
of which in these days we think we know a great deal, and really know
nothing; it attached the convincingness of their truth purely to the
objective [1] necessity of thought with which given premisses demand
their conclusions. But the premisses themselves, as well as that
combination of them on which the conclusion has to rest, were
simply accepted by Mathematics from what it called Direct or Intuitional Perception [2]. Nor could the word perception [2] be held to
designate any psychical activity, which could be shown to possess a
peculiar and definite mode of procedure; every impartial attempt to
say what perception [2] does, must end with the admission that it really
does nothing, that there is no visible working or process at all as a
means to the production of its content; but that on the contrary it is
nothing but a direct receptivity, with an entirely unknown psychical
basis, which merely becomes aware of its object and the peculiar
nature of that object. Obviously, an investigation cannot begin before
the matter is given to which it is to refer; but again, it will only
consist, even when the matter is forthcoming, in presenting one by
one to this receptivity all the details which do not fall at once in the
line of our mental vision; and defining their differences or similarities
by help of marks which make it possible to transfer from one to the
other of these features the judgments about them made by direct perception, and to connect all such features systematically together.

I shall return later on to what it is indispensable to say on this
head; I will only add now that it was possible for the Euclidean
geometry, which arose in the above way, to remain unassailed as long
as no doubt was raised of the objective validity of space; while it was
believed, that is, that we had in it if not a real thing, at least the
actual and peculiar form attaching to real things. It was not indeed
solely, as we shall see, but chiefly, the modern notion which sees in it
only a subjective mode of perception, that disturbed this unsuspicious
security and raised such questions as these; of how much that is
true about the world can we properly be said to get experience by
help of this form of apprehension; could there not be other species
of perception that might teach us the same truth about Things better,
or other truths quite unknown; and finally, may not the whole fabric

---

[1] ['Sachliche.']   [2] ['Anschauung.']

of our spatial perceptions be incomplete, perhaps charged with inner contradictions which escape our notice for want of the empirical stimuli which would bring them to light ?   The diversity of opinions propounded in relation to the above matters compels me in my metaphysic to enter upon the essential nature of space in its geometrical aspect ; and I begin my task by a very frank confession.   I am quite unable to persuade myself that all those among my fellow-students of philosophy, who accept the new theories with applause, can really understand with such ease what is quite incomprehensible to me ; I fear, that from over-modesty they do not discharge their office, and fail, on this borderland between mathematics and philosophy, to vindicate their full weight for the grave doubts which they should have raised in the name of the latter against many mathematical speculations of the present day.   I shall not imitate this procedure ; but while on the contrary I plainly say that the whole of this speculation seems to me one huge coherent error, I am quite happy to risk being censured for a complete misapprehension, in case my remarks should have the good fortune to provoke a thorough and decisive refutation.

123. I begin with the first inference suggested by the doctrine that space is only the subjective form of apprehension which is evolved from the nature of our souls, though not deducible by us. Then, there is nothing to interfere with our thinking of beings endowed with mental images as differing in nature within very wide limits ; or with our assigning to each of these kinds a mode of apprehension of its own, which, as is commonly said, it holds in readiness to apply to its future perceptions. Meantime we have convinced ourselves how little use such forms could be to these minds, if they were only a subjective manner of behaviour and destitute of all comparability with the things. In short, things would not be caught in nets whose meshes did not fit them ; far less could there be in purely subjective forms any ground of distinction which could compel things to prefer one place to appear in rather than another. We must therefore necessarily give a share in our consideration to the connexion in which the forms of apprehension are bound to stand with the objects which they are to grasp. The following cases will have to be distinguished.

Let X and Z be two of those modes of perception, different from our space S, which we arbitrarily assign to two kinds of beings endowed with mental images, and organized differently from us. This assumption would cause us no difficulty as long as, (i.) we supposed the worlds which are to be perceived by their means, to differ from the world M accessible to our experience, but to be such as to admit of apprehension in the forms X and Z as easily as the world M lends itself to our apprehension in the form of our space S. Only, this assumption would not interest us much; though free from internal contradiction, in fact, strictly, a mere tautology, it has no connexion whatever with the object of our doubt; the interest of our question depends entirely on a different presupposition; (ii.) that this same world M, which we represent to ourselves as enclosed in the frame of Euclidean space S, appears to other intellectual beings in the utterly heterogeneous systematic forms X or Z. On this supposition also there are two cases to be kept separate. The actions and reactions which the things of this world M reciprocate with each other may be extremely various; it is neither necessary nor credible that they only consist in such activities as cause us to localise the things in spatial relations in accordance with them; on the contrary, much may go on within the things that is not able to find expression in their appearance in space, even with the help of motion. Therefore there is still this alternative; either, (a.) the forms of perception X and Z reproduce relations of things which cannot be represented in our space S and do not occur in it; about this assumption we can have no decisive judgment, but only a conjecture, which I will state presently; or (β.) we assert that the same relations of things which appear to us as relations in space S are accessible to other beings under the deviating modes of perception X or Z; and on this point we shall have something more definite to say.

124. Let us begin with the former alternative (ii. a). We are justified in subordinating the idea of space S to the more universal conception of a system of arrangement of empty places, within which the reciprocal position of any two terms is fully determined by a number n of relations of the two to others. And there is nothing to prevent us, as long as no other requirements are annexed, from conceiving many other species of this genus, in which the reciprocal definition of the terms might be effected by other rules than those valid for the space S, or might require a greater or smaller number of conditions than are required in it. Still, it seems to me unfruitful to refer for further illustration of such ideas to the well-known attempts to arrange in a spatial conspectus either the whole multiplicity of sensations of musical sound, with reference to strength, pitch, quality, and harmonic affinity; or the colours in all their variety on similar grounds. Nothing indeed is more certain than that (1) we here have

before us relations of the terms to be arranged for the adequate
representation of which our space S is unfitted ; but at the same time
I think nothing can be more doubtful than the implied idea by which,
whether furtively or explicitly, we console ourselves, that (2) there
may be other modes of perception X or Z which permit to beings of
different organization the feat which we cannot perform. I must
speak more fully of both parts of my assertion.

125. (1) We may arrange musical notes in a straight line according
to their rise of pitch ; but as there appears to be an increasing divergence from the character of the keynote up to the middle of the
octave, and from that point again an increasing approximation to it,
having regard to this we may represent the notes still more clearly, by
arranging them as Drobisch does in a spiral, which after every circuit
corresponding to an octave returns to a point vertically above the
starting-point. But in doing so we should bear in mind that all this,
like any other appropriate device which might be added to the scheme,
is still a *symbolical* construction ; the notes *are* not in the space in
which we localise them for the convenience of our perception, nor is
the increment-element Δp of the pitch p really the element Δs of a
line in space s, to which, for the purpose of our perception, we treat
it as equivalent. No one refuses this concession; but it is not precisely in this that the ground of my difficulty lies. Seeing that I have
asserted the phenomenal nature of space there is no longer any meaning for me in distinguishing Things as in space, from sounds as only
to be projected into it by way of symbolism. When Things appear
to us in space, what we do to them is just the same as the treatment
to which we submit the ideas of notes in the above constructions ;
like them, things have neither place nor figure in space, nor spatial
relations ; it is only within our combining consciousness and only to
its vision that the living reactions which Things interchange with each
other and with us expand into the system of extension, in which
every phenomenal element finds its completely definite place. So if
the innumerable mental representations of sounds compelled us as
unambiguously to place each of them in definite spatial relations to
others, I should not be able to see how such an arrangement must be
less legitimate for them than for things, for which also it remains a
subjective apprehension in our minds.

It will further be observed, and quite correctly, that Things are
movable in space, and their place at any time only expresses the sum
of relations in which they stand to other things, which subsists at the
moment but is essentially variable ; it tells nothing of the Thing's own

P 2

nature; whereas such constructions of the realms of colour or sound aim at a completely different result; they attempt to assign to each one of these sensations conformably with the peculiar combination in which each unites definite values of the universal predicates of colour and sound, a systematic position between all others which it can never exchange for another place. No doubt this difference is important as regards the nature of the elements which it is proposed to systematise in the two cases; still there is no essential obstacle to copying the eternal and permanent articulation of a system of contents[1] fixed in the shape of ideas by means of the same mode of perception which is used to represent the variable arrangement of real Things. In fact, for every single indivisible moment the existing arrangement of real things in space would be precisely the total expression of the complete systematic localisation appropriate to the individual things in virtue of the actions which intersected each other in them at that moment. The circumstance that within things there is motion, which will not admit of being represented for ever by the same fixed system, is a fact with its own importance, but not a proof that the space form is inadequate to express systematic relations. Therefore the felt inadequacy of the space-form S can only rest on the fact that its articulation, though fitted for what we perceive in it, is not fitted for such matter as these sensations which we project into it.

126. Things then obviously do not arrange themselves in space according to a constant affinity of their natures, but according to some variable occurrence within them, consisting of the reactions which they interchange. We are not justified in assuming an entirely homogeneous form of event as produced in all of them by these actions; but we cannot help regarding as homogeneous all that part of such events which has its effect in fixing their place in space; in designating it by the name of 'mechanical relations' of things we approach the common view of physical science, which considers that in every moment the place which a body occupies abandons or tends to, is determined by the joint action of entirely comparable forces and impulses.

Now it is just this comparability which is wanting to the musical properties of sounds; that is, the felt properties, for we are only speaking of them, not of the comparable physical conditions of their production. The graduated series of loudness[2] i and of pitch p may

---

[1] ['Inhaltsystem.']

[2] ['Tonstärken.' I have retained the i because it probably stands for 'Intensität' (intensity).]

no doubt be formed, each separately, by addition of homogeneous increments; but when we come to the series of qualities q we find it cannot be exhibited in this way; and in any case Δi, Δp, and Δq would remain quite incomparable with each other. The lines i, p, and q, though we might suppose that each could be constructed by itself, yet would diverge from any point in which they were united, as it were into different worlds; and if one of them were arbitrarily fixed in space still there would be nothing to determine the angles at which the others would cross it or part from it.

It will of course be said that this as well as the difficulties raised in the last section, was known long ago; but that no one can be sure that (2) beings different from us have not at command forms of apprehension X or Z, which attach themselves to the content to be arranged just as unambiguously and perfectly, as our space S does to its matter, the mechanical relations of things. Yet I cannot see how this should be supposed possible as long as we ascribe to those beings the same achievement as that in which we fail. If instead of the qualitatively different colours and tones which we see and hear, they perceived only uniform physical or psychical actions, from a mixture of which those sensations arose in us, I do not dispute that in that case they might have for such actions an adequate perceptive form X or Z; but the relations which they would have to arrange would again be purely mechanical, only mechanical in a different way from those which we reproduce in our space S.

On the other hand, if those beings are supposed to feel the same difference between red and blue as we do, or to feel the pitch of a note as independently of loudness and quality as we feel it, then the different progressions i, p, q, would be as incomparable for them as for us; though they might arbitrarily reduce the relations of tones and colours to the forms X and Z by way of symbolism, with the same sort of approximation as we obtain in our space S. But I hold that a special colour-space X or tone-space Z is an impossibility; an impossibility that is, as an endowment of the supposed beings with two faculties of the nature of empty forms of apprehension, prior to all content and so having none of their own, but able to dictate particular situations to disparate elements subsequently received into them, solely in virtue of the rules of connexion between individual places which they contain. No form of perception X, be it what it may, can enable elements which remain disparate even for it to prescribe their places in it definitely and unambiguously to each other. And conversely; there may no doubt be rules of criticism for variously combined values of disparate predicates, which, being based on an estimate of the efficient causes which produce such combinations, show how to exclude impossible terms and to arrange possible ones in series according to their various aspects; but a form of perception X such as to unite all these different series of ideas *about* the material into a single image *of* the material seems to me impossible.

I cannot see how we lose much if we admit this; the many-sided affinities, resemblances, and contrasts of colours and tones are not lost to us because we cannot satisfactorily symbolise them in space; we have the enjoyment of all of them when we compare the impressions with each other. Now it seems to me that no being can get beyond this discursive knowledge in respect of elements which in their sum of predicates combine different properties that remain disparate even for that being; a form of perception, in the sense of an ordered system of empty places, can only exist for such relations of elements as are completely comparable, and each of which is separated from a second by a difference of the same kind as separates this second from any third or fourth. It is possible that things contain some system of uniform occurrences which escape us, but form the object of perception for other beings, and are in fact apprehended by them in forms of perception which differ from our space-form S and adapt themselves to the peculiar articulation of the occurrences; but this idea being motived by no definite suggestion need not be pursued further, at least for the moment.

127. We are much more interested in the other of the cases distinguished above (ii. β) [1]. If the same relations of things which are imaged by us as in space were supposed to meet with forms of a different kind in other beings; at least we know that there is nothing in the nature of these relations to make them intractable to combination before the mind's eye into one entire image; such an X or Z undoubtedly might bear the character of perceptive forms. They would not need to be in the least like our space S; the difference between two places of the system which appears to us in our space as the line s, would represent itself in them in the form x or z; both of which would be as disparate from s as the interval between two notes from the distance between two points. As long as we maintain these postulates, we have no reason to deny the possibility of these perceptions X and Z; but as we do not possess them their assumption remains an empty idea, and we know absolutely nothing further

[1] [§ 123, end.]

of how things present themselves and what they look like under those forms. Only we must not require more of them than our own space-apprehension can achieve; not, therefore, that the beings which enjoy them shall be enabled by them in each individual perception to apprehend the true relations of what is perceived. This is more than even our space S does for us; for instance we have to assign ourselves a place in it, with the change of which the whole constellation of our impressions is displaced; even to us, owing to the laws of the optical impressions made on us, parallel lines inevitably appear to converge at a distance, magnitudes to diminish, and the horizon of the sea to rise above the level of the shore. As we require the comparison of many experiences to enable us to apprehend the true relations in despite of the persistent semblance of the false, no more than this ought to be demanded of the nature of X and Z; that is, that combined experiences should give criteria for the elimination of the contradictions and mistakes of isolated ones. We may say then, subject to such conditions, that the same relations of things as appear to us in space admit of other kinds of perception completely unknown to us but leading to equally true cognition. Still even this is by no means what is as a rule in people's minds; it is expressly other *space*-perceptions than ours that it is hoped to make conceivable in this way. It is to be taken as settled that the relation of two elements presented to perception is given by perception the shape of the extended line s, and the relation of two such relations that of the angle a; and still even so there is to be a possibility that by help of other combinations this s and a may form not our space S but a different one S^1 or S^2, like ours in respect of the character of its elements s and a as pictured to the mind, but unlike in the fabric of the whole which they generate. Perhaps it will not be too painful to the feelings of philologists if I propose for these forms S^1 or S^2 the name of Raumoids [1] ['quasi-spaces']. I know no shorter way of expressing the difference between these forms and our previous forms X and Z; and as I mean to maintain that there cannot be Raumoids, their name will soon disappear again supposing I am right; if I am wrong, I make a present of it to my antagonists as the only thing I can do for their cause. For I shall hardly myself be brought to surrender my conviction that to accept s and a as elements of space is to decide its total form and inner structure, fully, unambiguously, and quite in the sense of the geometry which has hitherto prevailed.

126. I hold it, strictly speaking, unreasonable to require any other

[1] [From 'Raum,' 'Space.']

proof of this than that which lies in the development of the science down to the present time. That assuming the elements *s* and *a* they admit of other modes of combination than can be presented in our space *S*; and that these other combinations do not remain mere abstract names, but lead to kinds of perception S^1 and S^2; all this could only be proved by the actual discovery of the perceptions in question. But it is admitted that our human mode of representation cannot discover S^1 and S^2; nothing but *S* can be evolved out of it; therefore if the logical sequence of this evolution were established, and we still believed in other beings who could form divergent perceptions out of the same elements *s* and *a*, we should have to credit them with other laws of thought than those on which the truth of knowledge rests for us. Such an assumption would destroy our interest in the question; though no doubt it would not in the least run counter to the taste of an age whose tendency is so indulgent as to take anything for possible, which cannot be at a moment's notice demonstrated impossible.

But there is a point at which our geometry has long been thought deficient in consecutiveness of deduction; that is in the doctrine of parallel lines and of the sum of the angles of a triangle. Still it appears to me as if philosophical logic could neither advance nor properly speaking admit the peculiar claims to strictness of procedure made at this point by the logic of mathematics. After all, discursive proof cannot make truth, but only finds it; the perception of space with the variety of its inner relations faces us as the given object of inner experience; one which, if not so given, we should never be able to construct by a logical combination of unspatial elements, or even of those elements of space which we assumed; all demonstrations can but serve to discover certain definite relations between a number of arbitrarily chosen points to be implied in the nature of the whole. For such discovery perfect strictness of reasoning is indispensable; and elegance of representation may also require that the multiplicity of relations shall be reduced to the minimum number of directly evident and fundamental ones; but it will always be fruitless to assume fewer independent principles than the nature of the facts requires, and always erroneous to presuppose that it does not require a considerable number. We convinced ourselves in the Logic that all our cognition of facts rests on our application of synthetic judgments; the law of Identity will never tell us more than that every *A* is the same as itself; there is no formal maxim which gives us any help about the relation of *A* to *B*, except the one law which simply

disjoins them because they are not the same ; every positive relation
which we assert between *A* and *B* can only express a content which
is given us, a synthesis ; such as could be derived neither from *A* nor
from *B*, nor from any other relation between them which was not
itself in turn given to us in the same way. It is impossible to pursue
this here in its general sense, but it will be useful to elucidate it in
relation to space in particular.

129. The first consequence of what has been referred to is that a
case is possible in which we are unable to give adequate definitions
either of *A* or of *B* without involving the relation *C* in which they are
given to us, and equally so to define this relation apart from *A* and *B*.
It would be impossible to say what a point of space is and how
distinguished from a point of time, unless we include in our thought
the extension in which it is, and treat it, for instance, as Euclid does,
as the extremity of a line ; no more could we construct this line out
of points without a like presupposition. Two precisely similar and
co-existent points may have innumerable different relations of the kind
which we know as their greater or less distances from one another ;
but how could we guess or understand this unless the space in which
they are distributed, being present to the mind's eye, taught us at
once that the problem is soluble and what the solution looks like ?
Just as little can a line be generated by motion ; it can only be
followed ; for we could not set about to describe the track left behind
us without the idea of a space in general which furnishes the place
for it ; again any definite line could only be generated in space if in
every point which we pass through the further direction which we
mean to take were already present to our imagination.

Again, in any line when we compare it with others we shall be able
to distinguish its length from its direction ; but we cannot make the
simplest assertions about either property without learning them from
perception. That the addition of two lines of the length *a* gives a
line of the length 2 *a* seems a simple application of an arithmetical
principle ; but strictly arithmetic teaches only that such an addition
results in the sum of two lines of the length *a*, just as putting together
two apples weighing half an ounce each gives only the sum of these
two, not one apple twice the weight. The possibility of uniting the
one line with the extremity of the other so that it becomes its unbroken continuation and the two lengths add up into one only follows
from the mental portrayal of a space within which the junction can
be effected. I say expressly, ' of a *space* '; for not even the consideration that the things to be united are two lines is sufficient ; on the

contrary, we know that a thousand lines [1], if thought of as between the same extremities, will form no more than one and the same line; they must be put together lengthways, and to do this the image of the surrounding space which gives the necessary room is indispensable. Geometry only expresses the same thing in another form, when it says that every line is capable of being produced to infinity.

As regards direction, it is easily seen that it is a delusion to suppose that we have a conception of it to which straightness and curvedness can be subordinated as co-ordinate species; its conception is only intelligible as completely coinciding with that of the straight line which is called from another point of view, in relation to its extremities, the distance between them; every idea of a curve includes that of a deviation from the straight direction of the tangents and can only be fixed in the particular case by the measurement of this deviation. Thus we can it is true assign a criterion for any extended line which is security for its straightness; the distance between its extremities must be equal to the sum of the distances between all pairs of points by which we may choose to divide the line; but of course we do not by this get rid of the conception of straightness in principle; the distance between the extremities and each of these intermediate distances can only be conceived under that conception. So in fact it is not proper to say that the straight line is the shortest distance between two points; it is rather *the distance* itself; the different circuits that may be made in going from *a* to *b* have nothing to do with this distance which is always one and the same; but their possibility calls our attention to the circumstance that perception is in that fact telling us something more than would follow from its teaching up to that point taken alone.

130. If a straight line can be drawn between *a* and *b* and another between *a* and *c*, it does not in the least follow from these isolated premisses that the same thing can or must take place between *b* and *c*; the two lines might diverge from *a* as if into different worlds, and their extensions have no relation to each other. But they have one; our spatial perception and nothing else reveals to us the angle *a*, and shows us that space extends between the two lines and allows a connexion between the points *b* and *c* by means of a straight line *bc* of the same kind as *ab* and *ac*; it teaches us at the same time that there is this possibility for all points of *ab* and *ac*, and so creates the third element of our idea of space, the plane *p*. This, after having so discovered it, we are able to define as the figure in space any point of

[1] ['*Straight* lines' of course.]

which may be connected with any other point by a straight line lying wholly in that figure. This definition however, though I should think it a sufficient one, contains no rule for construction according to which we could produce for ourselves the plane *p* without having had it before ; for what is really meant by requiring all connecting lines to be contained in the spatial figure which is to be drawn is only made clear by the spatial perception of the plane. Now I will not deny that it may be of use in the course of scientific investigations to demonstrate even simple conceptions as the result of complicated constructions; in cases, that is to say, in which it is our object to show that the complicated conditions present in a problem must have precisely this simple consequence; but I cannot comprehend the acumen which seeks as the basis of geometry to obtain the most elementary perceptions by help of presuppositions, which not only contain of necessity the actual elements in question but also more besides them.

It is possible to regard the straight line as a limiting case in a series of curves ; but it would not be possible to form the series of these curves without in some way employing for their determination and measurement the mental presentation of the straight line from which they show a measurable deviation. Whoever should give it as a complete designation of a straight line, that it was the line which being rotated between its extremities did not change its place, would plunge us into silent reflexion as to how he conceived the axis of that rotation ; and by what, without supposing a straight line somewhere, he would measure the change of place which the curve experienced in such a rotation.

I hold it quite as useless to construct the plane *p* over again, after it has once been given by perceptive cognition ; no doubt it is also the surface in which two spheres intersect, and reappears as the result of countless constructions of the kind ; but every fair judge will think that it is the perception of the plane which elucidates the idea of the intersection and not *vice versa*.

And now, if we may let alone these attempts to clear up what is clear already, we are invited to a more serious defence of the rights of universal Logic by the dazzling play of ambiguities which endeavours to controvert and threatens to falsify the perception itself. A finite arc of a circle of course becomes perpetually more like a straight line as the radius of the circle to which it belongs is increased ; but the whole circle never comes to be like one. However infinitely great we may conceive the radius as being, nothing can prevent us from conceiving it to complete its rotation round the centre; and till such rotation is completed we have no right to apply the conception of a circle to the figure which is generated; discourse about a straight line which, being in secret a circle of infinite diameter, returns into itself, is not a portion of an esoteric science but a proof of logical barbarism. Just the same is shown by phrases about parallel lines which are supposed to cut each other at an infinite distance; they do not cut each other at any finite distance, and as every distance when conceived as attained, would become finite again, there simply is no distance at which they do so; it is utterly inadmissible to pervert this negation into the positive assertion, that in infinite distance there is a point at which intersection occurs. Here again, however, I am not denying that in the context of a calculation good service may be rendered within certain limits by modes of designation which rest on assumptions like these; so much the more useful would be a precise investigation within what limits they may be employed in every case, without commending to notice absolute nonsense by help of pretentious calculation.

131. It is obvious that according to the above general discussion, I cannot propose to solve the dispute about parallels by the demonstrative method commonly desiderated; I am content with expressing my conviction by saying that in presence of direct perception I can see no reason whatever for raising the dispute. We call parallel the two straight lines a and b which have the same direction in space, and we test the identity of their direction by the criterion that with a third straight line c in the same plane p, the straight lines a and b form on the same side of them s, the same angle a. In saying this I do not hesitate to presuppose the plane p and side s as perfectly clear data of perception; still they might both be eliminated by the following expression; a and b are parallel if the extremities a and β of any equal lengths $a\,a$ and $b\,\beta$ taken on the two straight lines from their starting points a and b, are always at the same distance from one another. It follows from this as a mere verbal definition, that $a\,b$ will also be parallel to $a\,\beta$; and at the same time from the matter of the definition, that a and b, as long as they are straight lines, must remain at the same distance from each other, measured as above; every question whether to produce them to infinity would make any change in this is otiose, and contradicts the presupposition which conceives identity of direction to infinity as involved in the direction of a finite portion of a straight line. That the sum of the interior angles which $a\,a$ and $b\,\beta$ make with $a\,b$ or with $a\,\beta$ is equal to two right angles, only requires the familiar elucidation.

Now if a triangle is to be made between aa and $b\beta$ both the lines must change their position, or one of them *its* position relatively to the other. If we suppose aa to turn about the point a so that the angle which it forms with ab is diminished, our spatial perception shows us that the interval between its intersection with $a\beta$ and the extremity β of that line must also diminish; if the turning is continued this interval is necessarily reduced to zero, and then ab, $a\beta$, and $b\beta$, enclose the required triangle. When this has been done the line $a\beta$ and the line of its former position aa make an angle, which is now excluded from the sum of the angles which were before the interior angles between the parallels aa and $b\beta$; but the vertical angle opposite to this angle, and therefore the angle itself, is equal to the new angle which $a\beta$ produces by its convergence with $b\beta$; the latter forms a part of the sum of the angles of the triangle which is being made, which sum as it loses and gains equally, remains the same as it was in the open space between the parallels; that is, in every triangle, whatever its shape may be, it is equal to two right angles. If this simple connexion between the two cases will not serve, still we could attach no importance to any attempt to postulate a different sum for the angles of a triangle, except on one condition; that it should not only proceed by strictly coherent calculations but should also be able to present the purely mathematical perception of the cases which corresponded to its assumption with equal obviousness and lucidity. For in fact it is not obvious, why, if the sum of the angles of a triangle were generally or in particular cases different from what we made it, this state of things should never be discovered to exist or be demonstrated to be necessary. But here we plainly have misunderstandings between philosophy and mathematics which go much deeper. Philosophy can never come to an understanding with the attempt which it must always find utterly incomprehensible, to decide upon the validity of one or the other assumption by external observations of nature. So far these observations have agreed with the Euclidean geometry; but if it should happen that astronomical measurements of great distances, after exclusion of all errors of observation, revealed a less sum for the angle of a triangle, what then? Then we should only suppose that we had discovered a new and very strange kind of refraction, which had diverted the rays of light which served to determine the direction; that is, we should infer a peculiar condition of physical realities in space, but certainly not a real condition of space itself which would contradict all our perceptive presentations and be vouched for by no exceptional presentation of its own.

132. However all this is the special concern of geometry, without essential importance for metaphysic. There is another set of ideas in which the latter has a greater interest. I admitted above that a being endowed with ideas would not evolve forms of space-perception which no occasion was given him to produce. Others have connected with such an idea the conjecture of a possibility that even our geometry may admit of extensions the stimulus to which in human experience is either absent or as yet unnoticed.

Helmholtz (Popular Scientific Lectures, III) in his first example supposes the case of intelligent beings living in an infinite plane, and incapable of perceiving anything outside the plane, but capable of having perceptions like ours within the extension of the plane, in which they can move freely. It will be admitted that these beings would establish precisely the same geometry which is contained in our Planimetry; but their ideas would not include the third dimension of space.

Not quite so obvious, I think, are the inferences drawn from a second case, in which intelligent beings with the same free power of movement and the same incapacity of receiving impressions from without their dwelling-space, are supposed to live on the surface of a sphere. At least, I suppose I ought to interpret as I did in the last sentence the expression that they [1] 'have not the power of perceiving anything outside this surface'; the other interpretation that even if impressions came to them from without the surface, they nevertheless are unable to project them outside it, would give the appearance of an innate defect in the intelligence of these beings to what according to the import of such descriptions ought only to result from the lack of appropriate stimuli. Under such conditions the direct perceptions of these beings would certainly lead in the first place to the ideas which Helmholtz ascribes to them; but I cannot persuade myself that the matter would end there, supposing we assume that the mental nature of such beings has the tendency with which our own is inspired, to combine single perceptions into a whole as a self-consistent and complete image of all that we perceive.

For shortness' sake I take two points N and S as the North and South poles of the surface of the sphere, and suppose the whole net of geographical circles to be drawn upon it. Suppose first that a being B moves from a point a along the meridian of this point. We must assume then that B is not only capable of receiving quali-

---

[1] ['Popular Lectures on Scientific Subjects;' Atkinson's translation, 2nd series, p. 34.]

tatively different or similar impressions from East and West; it must be informed by some feeling, by whatever means produced, of the fact of its own motion, and at the same time have capacity to interpret this feeling into the fact of its motion, that is, into the change of its relation to objects which for the time at least are fixed; it must finally have equally direct feelings which enable it to distinguish the persistent and similar continuance of this motion or change from a change of direction or a return in the same direction. However these postulates may be satisfied in the being B, it is certain that if we are to count upon any definite combination of the impressions it receives, it can experience no change of its feeling of direction in its continuous journey along the meridian; for by the hypothesis it is insensible to the concavity of its path towards the centre of the sphere. So if having started from a it passes through N and S and returns to a, keeping to this path, such a fact admits of the following interpretations for its intelligence.

As long as a only distinguishes itself from b or c by the quality of the impression it makes on B it will remain unestablished that the a which has recurred is that from which its movement started; it may be a second, like the first but not identical with it. On the other hand, the feelings which arise in B from its actual movement may prove to it a change in its own relation to objects, but as long as this is all it is not self-evident that the feelings can only indicate a change of *spatial* relation to them; the feelings are simply a regular series of states, the repeated passage through which is always combined with the recurrence of one and the same sensation a; very much, though not exactly, like running up the musical scale, when we feel a continuous increase in the same direction of our exertion of the vocal organs, which brings us back in certain periods not indeed to the same note, but to its octave which resembles it.

If B can feel no more than this, no space-perception can be generated; in order that it should be, a further separate postulate is required; B must be forced by the peculiar nature of its intelligence to represent to itself every difference between two of its felt states as a distance in space between two places or points. Under this new condition the interpretation of the experience gained is still doubtful, until the identity of the two a's is determined; as B does not experience a deviation to East or West, and by the hypothesis does not feel the curvature of its path inwards, it might suppose itself to have moved along an infinitely extended straight line, furnished at definite equal intervals with similar objects a.

But it is not worth while to spend time on this hypothesis; let us suppose at once that *B* moves freely on the surface and is able to compare in its consciousness innumerable experiences acquired in succession; then it will find means to establish not only the exact resemblance [1] but the identity of the two *a*'s. If this has taken place its journey along the meridian from *a* by *N* and *S* back to *a* will appear to it to establish the fact that by following a rectilinear movement in space, without change of direction or turning back, it has returned to its starting-point. At least I do not know how its path could appear to it other than rectilinear; as it can measure the whole distance from *a* to *a* by nothing but the length of the journey accomplished, it is of course equal to the sum of all the intermediate distances from point to point of this journey and so falls under the conception of straightness which was determined above; and on the other hand we cannot assume that *B* would detect in every element that made part of his journey, therefore in each of the minimum distances from point to point, the character of the arc of a circle; it would then possess the power denied to it of perceiving convexity in terms of the third dimension; and therein it would at once have a basis for the complete development of the idea of that dimension, its possession of which is disputed.

But such an idea must undoubtedly arise in its mind, not on grounds of direct perception, but by reason of the intolerable contradiction which would be involved in this straight line returning into itself, if this apparent result of experience were allowed to pass as an actual fact. For a power of mental portrayal which has got so far as to imagine manifold points ranged beside each other in a spatial order the content of the experience which has been acquired is nothing but the definition of a curve, and indeed, all things considered, of the uniform curve of the circle; but as it cannot turn either East or West, there must necessarily be a third dimension, out of which immediate impressions never come, and which cannot therefore be the object of a sense-perception for the being *B* in the same way as the two other dimensions; but which nevertheless would be mentally represented by *B* with the same certainty with which we can imagine the interior of a physical body although hidden by its surface. As soon as this conception of the third dimension is established the being *B* would evolve from the comparison of all its experiences according to the most universal laws of logic and mathematics precisely the same geometry that we acquire

[1] ['Gleichheit.']

more easily, not having to call to our aid a dimension which for our sensuous perception is imaginary, to reduce things to order; the being B would by this time understand its dwelling-space to be what it is, a figure in space which is extended in three dimensions; and would be in a position to explain the extraordinary phenomena which its experience of motion had presented to it by help of this form of idea.

133. Parallel lines, Helmholtz continues, would be quite unknown to the inhabitants of the sphere; they would assert that any two lines, the straightest possible, would if sufficiently produced, cut one another not merely in one point but in two. It depends somewhat on the definition of parallelism and on the interpretation of the assumptions which are made whether we are forced to agree to the former assertion. Movements along the meridians could of course not lead to the idea of parallel lines; but still, in case of free power to move, B might traverse successively two circles of the same north and south latitude; it would find that these circles have equal lengths to their return to the starting-point, that they never either cut or touch each other; but that counting from the same meridian the extremities of equal segments of the two have always the same distance from each other. This seems to me sufficient ground for calling them parallel, and in fact we use the term parallel of the circumferences of similarly-directed sections of a cylinder, which in this case the two circles would really be.

But that would be, as I said, merely a question of names; I mention these movements here for a different reason. The tangential planes of the successive points of the southern circle cut each other in straight lines which converge to the south; the corresponding sections for the northern circle do the same to the north; the question is whether the being B would be aware of this difference or not. If it were not, then B would really suppose itself to traverse two paths of precisely the same direction, which would in fact be parallel in the same sense as the above cylinder-sections; and then it might, as long as no other experiences contradicted the idea, conceive both paths to be in one plane as circles, the centres of which are joined by a straight line greater than the sum of their radii.

This would not be so in the other case, which we must anyhow regard as the more probable hypothesis. Of course it is hard to obtain a perfectly clear idea of what we mean by calling B sensitive only to impressions in the surface of the sphere; but we may assume that it would become aware of the slope of the tangential planes to

North and South from the fact that the meridians, known to it from other experiences, make smaller angles with its path on the side on which the plane inclines to the pole, and greater on the opposite side. However this might produce its further effect on *B*'s feelings of motion, the only credible result would be that it would think its path along the southern parallel concave to the south; and that along the northern parallel concave to the north; in other respects it would take them for circles, returning into themselves. These two impressions given by this second case would not be capable of being reconciled with the experience above mentioned of the constant distance maintained between equal segments of the two paths, taking these latter as transferred into a plane; and this case also would necessitate, in order to reconcile the contradiction it involves, the invention of the third dimension though not directly perceptible.

134. This result must guide us in forming our opinion on the vexed question of the fourth dimension of space. I omit all reference to fancies which choose to recommend to notice either time, or the density of real things in space, or anything else as being this fourth dimension; if we do not intend an unmeaning play upon words we must take it for granted at least that any new dimension is fully homogeneous and interchangeable with those to the number of which it is added; moreover if it is to be a dimension of *space*, it must as the fourth be perpendicular to the three others, just as each of them is to the remaining two.

It is conceded that for our perception this condition cannot be fulfilled; but the attempt is made to invalidate this objection by referring to the beings which have been depicted, whose knowledge stops short even of the third dimension of space because perception affords them no stimulus to represent it to their minds. Therefore, it is argued, a further development of our receptivity might perhaps permit to us an insight into a fourth dimension, now unknown to us from lack of incitement to construct it. The possibility that some beings content themselves with a part of the space-perception attainable can of course be no proof by itself that this form of perception is not in itself a whole with certain limits; or that it admits of perpetual additions even beyond the boundary we have reached; but we must admit that for the moment the appeal to these imaginary cases at least obscures the limit at which we may suppose the mental image to have reached such a degree of completeness as forbids any further additions. This makes it all the more necessary to see what that appeal can really claim. The imaginary beings which could only receive perceptions from a single plane, would have been in the most favourable situation, supposing changed life-conditions to bring them impressions from outside it, for the utilisation of such new perceptions; they would have been able to add the geometry of the newly discovered direction to the Planimetry which they possessed without having to change anything in their previous perceptions.

When we came to the beings on the sphere-surface, we at once found a different situation; they were forced to devise the third dimension by the contradictions in which the combination of their immediate perceptions entangled them; but yet they never found a direct presentation of it given, and could not do so without remodelling all their initial ideas of space.

If we mean to use this analogy to support the possibility in our own case of a similar extension of our perceptive capacity, I hope that attention will be given to the differences which exist between our position and that of those imaginary beings. In particular; they were compelled precisely by the contradictions in their observations to postulate the new dimension; we have no contradiction present to us, of a kind to force us as in their case to regard our space-image as incomplete, and to add a fourth to its three dimensions. At the same time we are not, at all events just now, in the position of the beings in the plane, who were unsuspectingly content with their Planimetry and never even conjectured the third dimension, which we know; for the idea of a fourth dimension which is now mooted on all sides is so far a substitute for the absent incitements of experience that it does not leave us quite unsuspicious of the enlargement of our space-perception which may be possible, but draws our attention to it, more seriously than in fact is worth while. If such an enlargement were possible, things would have to go on very strangely for the examination of space as we picture it to ourselves not to reveal it to us even without suggestions on the part of observation; on the other hand if the required observations came to us, without the possibility of remoulding our space-image so as to reconcile their contradictions, we should simply have to acquiesce in the contradictions. Now the following difference subsists; the beings on the sphere-surface were no doubt compelled by observations to alter their initial geometrical images, but then they found the alteration practicable; we are not in any way compelled to make the attempt, and besides, we find it utterly impracticable; in our space S it is admittedly impossible to construct a fourth dimension perpendicular to the other three and coincident with none of them. This seems to me to settle

the matter; for no one should appeal to the possibility that the space
S, without itself becoming different, may still admit of a different
apprehension, exhibiting a fourth dimension in it. As long as the
condition is maintained that the dimensions must be at right angles
to each other, such an apprehension is impossible; if it is dropped,
what we obtain is no novelty; for in order to adapt our formulæ to
peculiar relations of what exists or can be constructed in space it has
long been the practice to select a peculiar and appropriate system of
axes. Nothing would prevent us from assigning to the plane alone
three dimensions cutting each other at angles of 60°; which would
give a more convenient conspectus of many relations of points distributed in space than two dimensions at right angles.

Therefore only the other question remains provisionally admissible; whether there can be another form of apprehension X or
Z, unlike the space S, which presents four or more dimensions,
perfectly homogeneous, interchangeable, and having that impartial
relation to each other which appears in the property of being
at right angles as known in the space S. I shall return to it
directly; meantime I must insist upon the logical objection for which
I have been censured; it is absolutely unallowable to transfer the
name and conception of a space S to formations which would only
be co-ordinate with it under the common title of a system of arrangement capable of direct presentation to the mind; but whose special
properties are entirely incompatible with the characteristic differentia
of the space S, that is with the line s, the plane p, the angle a, and the
relations which subsist between these elements. It is this dangerous
use of language that produces the consequences which we have before
us; such as the supposition that the space S in which we live really
has a fourth dimension over and above its three, only is malicious
enough not to let us find it out; but that perhaps in the future we
may succeed in getting a glimpse of it; then by its help we should
be able to make equal and similar bodies coincide, as we now can
equal and similar plane figures. This last reason for the probability
of the fourth dimension is moreover one which I fail to understand;
what good would it do us to be occupied with folding over each other
bodies of the same size and shape, and what do we lose now by
being unable to do it? and further; must everything be true which
would be a fine thing if it were? No doubt it would be convenient if
the circumference of the circle or any root with index raised to any
power in the case of any number could be expressed rationally; but
no one hopes for an extension of arithmetic which would make this

possible. What have we come to? Has the exercise of ingenuity killed all our sense of probability? The anticipation of such trans-figurations of our most fundamental kinds of perception can only remind us of the dreams of the Fourierists, who expected from the social advance of man a corresponding regeneration of nature, ex-tending to the taming of all savageness and ferocity in its creatures. But perhaps the two processes may help each other; it will be a fine thing when we can ride on tame whales through the fourth dimension of the *eau sucré* sea.

135. To return to the above question; I am convinced, certainly, that the triplicity of perpendicular dimensions is no special property of our space S; but the necessary property of every perception R which presents, however differently from our space, a background or comprehending form for all the systematic relations of a co-existent multiplicity. Still I could wish that I had a stronger argument to sustain my conviction than what I am now going to add. To avoid all confusion with ideas taken from existing space which of course press upon us as the most obvious symbols to adopt, let us con-ceive a series of terms X, between which, putting out of sight their qualitative character which we treat therefore as wholly uniform, there are such relations, homogeneous in nature but now not otherwise known, that every term is separated from its two next neighbours by a difference x. How in such a system of arrangement R this differ-ence x would be imagined, or pictured to the mind, we leave quite out of the question; it is merely a form or value of an unknown r, and corresponds to what appears in our space-perception as the straight line s or as the distance in space between two points. Now let O be the term of the series X from which we start; then the differences between its place in the series and that of any other term, that is the differences between the particular elements of the re-quired perception R itself measured in the unknown form r, will be of the form $\pm mx$, where m is to be replaced by the numbers of the natural series. Now O may be at the same time a term of another series Y of precisely similar formation, whose terms we will designate by $\pm my$ so that each my is not merely like in kind but also equal to mx.

There are two conditions which these two series X and Y would have to satisfy in order to stand in a relation corresponding to that of two lines in space at right angles to each other. First, progression in the series Y, however far continued, should bring no increment of one-sided resemblance in the terms my so arising to $+mx$ or $-mx$, but every

my should have its difference from $+mx$ equally great with that from
$-mx$ in whatever such difference consists. Secondly, this difference
should not consist in any chance quality, but should be comparable
both in kind and in magnitude both with x and with y. This second
condition must be remarked; obviously countless series like Y can be
conceived, starting from a term O common to it and X and extending, so to speak, into different worlds, whose terms would approach
neither $+x$ nor $-x$ because quite incomparable with either; but
such suppositions would have nothing to do with our subject. In our
space S the difference between *my* and *mx* is a line s, just as *mx* and
my themselves are lines of the kind s; in the other system of places
R which we are here supposing this difference is of the otherwise
unknown kind r, just as *mx* and *my* are comparable forms or values
of r.

From this point we might proceed in different ways. We might
attempt to form the idea, still problematic, of several series Y, all of
which satisfy these conditions; but against this suggestion it is rightly
urged, that as long as we are without the conception of a space whose
plainly presented differences of direction would show us how to keep
asunder these several Y's, so long they are all in their relation to
X, (and so far they are defined by nothing else), to be considered as
one single series; they would not be many, till the same difference
should subsist between them, as between them and X, and that
without interfering with their common difference from X. Now let us
consider one of these Y's as given; the others, which, in the abstract
sense which we explained, are as well as the given Y perpendicular
to the series X, may have the most diverse relations to the former;
their progressive terms may approximate more or less to the $+my$
or $-my$ of the first given series; but among all these series there can
conceivably be only one which we will call Z, whose successive terms
mz though commensurable with $\pm my$ still have equally great differences from the positive and from the negative branch of Y. It is true
too of this third series Z as long as it is defined by nothing but its
relation to Y, that it is only to be regarded as one; but of it too we
may form the problematic idea that it is forthcoming in a number of
instances, all of which stand in the same relation of being perpendicular to Y. If we now choose one of these many Z's, then the rest
may stand to it again in the most diverse relations; but again only
one, which we will call V, could be such that its progressive terms
mv would have always equal differences from the $+mz$ and the
$-mz$ of that one determinate Z. Observations of this kind might be

continued for ever; but there is an absolutely essential and decisive point which as they stand, they just omit.

We have so far only supposed the Y's perpendicular to X, the Z's to Y, and the V's to Z, but have not decided the question, how far the relation of Z as at right angles to Y brings this Z into a necessarily deducible relation with X, or that of V to Z has a similar effect upon V as regards Y or X. If we really added nothing further this would be a case of what I have more than once expressed in metaphor; the Z's would no doubt have the same relation to the Y's that the Y's have to the X's; only the relation of the Z's as perpendicular to Y would as it were point into another world from that of the Y's as perpendicular to X; and though we should be able to have a perception of each particular one of these relations, that of the Y's to the X's and that of the Z's to the Y's, yet we should not bring together these two instances of one and the same relation into any definite mental picture at all, in spite of the common starting-point O.

Therefore in this way we shall never obtain the collective perception R, which we were looking for and within which we hoped to distribute in determinate places all the points we met with in its alleged n dimensions; only the accustomed perception of space S, which we introduce unawares, misleads us into the subreption that it is self-evident that these successive perpendicular branchings of the X's from Y, of the Y's from Z, and of the Z's from V take place in a common intuitional form R. But in fact, to secure this, the particular condition must be added to which I drew attention above. A Z which is perpendicular to a Y, or deviates in a measurable degree from the perpendicular to it, must by this circumstance enter also into a perfectly definite relation with X, to which that Y is perpendicular. At present we have only to do with one of these various relations; which is this; among the Z's perpendicular to Y, that one which is also to be perpendicular to X must necessarily be one among the many Y's, as they included all the series that had this relation to X; therefore even this third dimension cannot exist in R without its coinciding with one, and taking X as given, with a particular one of the many instances of the second dimension all perpendicular to X; still less can there be a fourth dimension V, at once perpendicular to X, Y, and Z, and yet distinct from the one particular Z which stands alone in answering to the two conditions of being perpendicular to X and at the same time to Y. I maintain therefore that in no intuitional form R, however unlike our space

S, provided that it really is to have the character of a comprehensive
intuitional form for all co-existing relations of the content arranged
in it, can there be more than three dimensions perpendicular to each
other; taking the designation 'perpendicular' in the abstract meaning
which I assigned it, and which refers not only to lines *s* and angles *a*
but to every element *r*, however constituted, in such a form of perception *R*. Of course this whole account of the matter is, and
in view of the facts can be, nothing but a sort of retranslation
from the concrete of geometry into the abstract of logic; perhaps
others may succeed better in what I have attempted. I believe that
I am in agreement with Schmitz-Dumont on this question as well as
on some of the points already discussed, but I find it hard to
adopt the point of view required by the whole context of his exposition.

136. Among the properties which our common apprehension
believes most indispensable to Space is the absolute homogeneousness of its infinite extension. The real elements which occupy it or
move in it may, we think, have different densities of their aggregation and different rules for their relative positions at different points;
space itself, on the other hand, as the impartial theatre of all these
events, cannot possess local differences of its own nature which might
interfere with the liberty of everything that is or happens at one of its
points to repeat itself without alteration at any other. Now if we
conceive a number of real elements either united in a system at rest,
or set in motion, by the reactions which their nature makes them
exert on one another, then there arise surfaces and lines, which can
be drawn in space, but are not a part of its own structure; they
unite points in a selection which is solely dependent on the laws of
the forces which act between the real things. Mathematics can
abstract from the recollection of these causes of special figures in
space and need not retain more than the supposition of a law,
(disregarding its origin,) according to which definite connected series
of points present themselves to our perception out of the infinite
uniformity of extension as figures, lines, or surfaces.

So far ordinary ideas have no difficulty in following the endeavours
of geometry when in obedience to the law of combination of a multiplicity given in an equation it searches for the spatial outlines which
unite in themselves the particular set of spatial points that correspond
to this law. But in the most recent speculations we meet with a
notion, or at least imagine we meet with it, which we cannot understand and do not know how to justify. It is possible that the diffi-

culties which I am going to state are based on a misconception of the
purposes aimed at by the analytically conducted investigations of this
subject; but then it is at least necessary to point out plainly where
the need exists for intelligibility and explanation which has not been
in the least met by the expositions hitherto given.

To put it shortly, I am alluding to the notion that not only may
there be in infinite uniform extension innumerable surfaces and lines
whose structure within the particular extent of each is very far from
uniform, that is, *variously formed figures in space;* but that also there
may be *spaces* of a peculiar structure, such that uniformity of their
entire extension is excluded. It is clear to us what we are to think
of as a spherical or pseudo-spherical surface, but not clear what can
be meant by a spherical or pseudo-spherical space; designations
which we meet with in the discussion of these subjects without any
help being given to us in comprehending their meaning. In the
following remarks I shall only employ the former of these designations; the mention of 'pseudo-spherical space,' which is harder to
present definitely to the mind, could only reinforce our impression of
mysteriousness, without contributing to the explanation of the matter
any more than the allusion to the familiar spherical figure. The idea
of a spherical surface, being that of a figure in space, presupposes the
common perception of space; the situation of its points is determined,
at least has been hitherto, by some system of co-ordinates which
measures their distance and the direction of that distance from an
assumed point of origin according to the rules which hold for a
uniform space. To pass from the spherical surface to a spherical
space, one of two assertions seems to me to be needed; either
this surface *is* the whole space which exists, really or to the mind's
eye; or this totality of space *arises* out of the spherical surface by
making the co-ordinates pass continuously through the whole series of
values compatible with the law of their combination. If we do the
latter there arises by the unbroken attachment of each spherical surface to the previous one, the familiar image of a spherical Volume,
which we may either limit arbitrarily at a particular point or conceive
as growing to infinity, as the equation of the surface remains capable
of construction for all values of the radius; in this way we attain to
nothing more than the admissible but purely incidental aspect, that
.the infinite uniform extension of space is capable of a complete
secondary construction, if from any given point of origin we supposed a minimum spherical surface to expand in all directions conformably to its equation. But in the interior of this spherical volume

there is no further structure revealed than that of uniform space, on the basis of which the co-ordinates of the boundary-surface at each particular moment had been determined : the interior does not consist permanently and exclusively of the separate spherical shells out of which in this case our representing faculty created its representation ; the passage from point to point is not in any way bound to respect this mode of creation of the whole, as though such a passage could take place better or more easily in one of the spherical surfaces than in the direction of a ·chord which should unite any places in the interior. The conception of a measure of curvature has its proper and familiar import for each of the surfaces, distinguishable in this space by thought, but wholly obliterated in the space itself ; but it is impossible to conceive a property of space itself to which it could apply.

In the case of the sphere its law of formation permitted the continuous attachment of surface to surface ; but equations are conceivable which if constructed as a system of positions in space would produce either a series of discrete points or one of discrete surfaces, perhaps partially connected or perhaps not at all. We know such constructions primarily as figures in space and nothing else, and conceive their production as conditioned by equations between co-ordinates whose power of being reciprocally defined by each other corresponds to the nature of uniform space, now known as Euclidean space ; but let us assume that we had escaped from that postulate and had employed co-ordinates which themselves partook of the special nature of the variously formed space which is to be obtained. It may then be difficult to project an image of these strange figures within our accustomed modes of space-perception ; I attach more weight to another difficulty, that of determining what we properly mean when we speak of them as spaces. Let us assume that the fundamental law, being capable of algebraical expression, which prevails in a system of related points not yet explicitly apprehended as spatial, conditions a systematic order of them which could only be represented in our space S by a number of curved sheets not wholly attached to one another ; then the fact, form, and degree of their divergence could only be observed by us through the medium of distance measured according to the nature of the space S, as existing between particular points in the different sheets.

However, let us even put out of the question all idea of a space S as the neutral background on which the figure X was constructed, and attempt to regard this X as the sole represented space ; still the different sheets of it could not possibly extend as if into different worlds, so as to prevent there being any measurable transition from one to another ; just as little could that which separates them and makes them diverge be a mere nothingness when compared to the space X itself, and capable of no measurable degrees whatever ; even in this case that which gave the reason for their being separate could not but be a spatial magnitude or distance, uniform and commensurable with the magnitudes which formed the actual space X. Thus our attempt would be a failure ; we should not be able to regard that X as space, but only as a structure in a space ; we might no doubt assume, for the moment, of this space that in each of its minutest parts it had a structure other than that of our space S, but we should have to admit at once that it formed a continuous whole with the same inner structure in every one of its parts. For, provided that this tentatively assumed space X is not to be regarded as something real, but as the empty form of a system for the reception of possible realities, there can be no difference of reality or value between the points contained in those sheets and the other points by the interposition of which their divergence arises ; they would all accordingly have equal claims to be starting-points of the construction in question, and from the intersection of all these constructions there would once more be formed the idea of a space uniform through an infinite extension, and indifferent to the structure of the fabrics designed in it. Not even a break in the otherwise uniform extension is possible ; such a break is only conceivable if in the first place there is a something between the terms which keeps them asunder, and if moreover that something is comparable in kind and magnitude with what it bounds on both sides of itself ; hence space cannot consist of an infinite number of intersecting lines which leave meshes of what is not space between them ; it uncontrollably becomes again the continuous and uniform extension which we supposed it to be at first ; and the manifold configurations of the kind X are conceivable in it only as bounded structures, not as themselves forms of space.

137. I feel myself obliged to maintain the convictions which I have expressed even against Riemann's investigations into a multiplicity extended in n directions. My objections are on the whole directed to the point, that here again the confusion which seems to me to darken the whole question has not been avoided ; the confusion of the universal localisation-system of empty places presented to the mind, a system in which structures of any shape or any extent can be arranged, with the structure and articulation belonging to that which has to be arranged

in this system; or to repeat the expression employed above, the confusion of space with structures in space. In II. § 4 of his treatise on the hypotheses which lie at the foundation of Geometry, Riemann expresses himself as follows: 'Multiplicities whose measure of curvature is everywhere zero, may be treated as a particular case of multiplicities whose measure of curvature is everywhere constant. The common character of multiplicities whose degree of curvature is constant may be expressed by saying that all figures can be moved in them without stretching. For, obviously, figures could not be made to slide or rotate in them at pleasure, unless the degree of curvature were constant. On the other hand, by means of the constant degree of curvature the relations of measurement of the multiplicity in question are completely determined; accordingly in all directions about one point the relations of measurement are exactly the same as about another, and therefore the same constructions are practicable starting from the one as from the other and consequently in multiplicities with a constant measure of curvature figures can be given any position.'

Now I have no doubt at all that by analytical treatment of more universal formulæ the properties of space indicated may be deduced as a special case; but I must adhere to my assertion that it is only with these special properties that such an 'extended multiplicity' is a space, or corresponds to the idea of a system of arrangement for perception; all formulæ which do not contain so much as these determinations, or which contain others opposed to them, mean either nothing, or only something which as a special or peculiar formation may be fittingly or unfittingly reduced to order in that universal frame. A system of places which was otherwise formed in any one of its parts than in another, would contradict its own conception, and would not be what it ought to be, the neutral background for the manifold relations of what was to be arranged in it; it would be itself a special formation, 'a multiplicity extended in *n* directions' instead of being the *n*-dimensional multiplicity *of extension*, about which the question really was.

I cannot believe that any skill in analysis can compensate for this misconception in the ideas; alleged spaces of such structure that in one part of them they would not be able to receive, without stretching or change of size, a figure which they could so receive in another, can only be conceived as real shells or walls, endowed with such forces of resistance as to hinder the entrance of an approaching real figure, but inevitably doomed to be shattered by its more violent impact. I trust that on this point philosophy will not allow itself to be imposed upon by mathematics; space of absolutely uniform fabric will always seem to philosophy the one standard by the assumption of which all these other figures become intelligible to it. This may be illustrated by the analogy of arithmetic. The natural series of numbers with its constant difference 1, and its direct progression, according to which the difference of any two terms is the sum of the differences of all intermediate terms, may be treated as a special case of a more general form of series just as much as can uniform space. But, by whatever universal term it might be attempted to express the law of formation of this series, it could have no possible meaning without presupposing the series of numbers. Every exponent or every co-efficient which this universal formula contained, would be of unassign-able import unless it had either a constant value in the natural series of numbers, or else a variable one, depending in particular cases on the value, measurable only in this series of numbers, of the magnitudes whose function it might be. Every other arithmetical series only states in its law of formation how it *deviates from* the progression of terms of equal rank which forms in the series of numbers; no other standard can be substituted for this, without standing in need in its turn of the simple series of numbers to make it intelligible. Precisely the same seems to me to be the case in the matter of space; and I cannot persuade myself that so much as the idea of multiform space or of a variable measure of curvature in space could be formed and defined, without presupposing the elements of uniform space, recti-linear tangents and tangential planes, in fact uniform space in its entirety, as the one intelligible and indispensable standard, from which the formation of the other, if it could be pictured to the mind at all, would present definite deviations.

CHAPTER III.

Of Time.

THE Psychologist may if he pleases make the gradual development
of our ideas of Time the object of his enquiry, though, beyond some
obvious considerations which lead to nothing, there is no hope of his
arriving at any important result. The Metaphysician has to assume
that this development has been so far completed that the Time in
which, as a matter of fact, we all live is conceived as one comprehensive form in which all that takes place between things as well as our
own actions are comprehended. The only question which he has to
ask is how far Time, thus conceived, has any application to the Real
or admits of being predicated of it with any significance.

188. In regard to the conception I must in the first place protest
against the habit, which since the time of Kant has been prevalent
with us, of speaking of a direct perception of Time, co-ordinate with
that of space and with it forming a connected pair of primary forms
of our presentative faculty. On the contrary we have no primary and
proper perception of it at all. The character of direct perception
attaching to our idea of Time is only obtained by images which are
borrowed from Space and which, as soon as we follow them out,
prove incapable of exhibiting the characteristics necessary to the
thought of Time. We speak of Time as a line, but however large
the abstraction which we believe ourselves able to make from the
properties of a line in space in order to the subsumption of Time
under the more general conception of the line, it must certainly be
admitted that the conception of a line involves that of a reality belonging equally to all its elements. Time however does not correspond to this requirement. Thought of as a line, it would only
possess one real point, namely, the present. From it would issue two
endless but imaginary arms, each having a peculiar distinction from
each other and from simple nullity, viz. Past and Future. The distinction between these would not be adequately expressed by the

opposition of directions in space. Nor can we stop here. Even though we leave out of sight the relation in which empty Time stands to the occurrences which fall within it, still even in itself it cannot be thought of as at rest. The single real point which the Present constitutes is in a state of change and is ceaselessly passing over to the imaginary points of the Past while its place is taken by the realisation of the next point in the Future.

Hence arises the familiar representation of Time as a stream. All however that in this representation can be mentally pictured originates in recollections of space and leads only to contradictions. We cannot speak of a stream without thinking of a bed of the stream : and in fact, whenever we speak of the stream of Time, there always hovers before us the image of a plain which the stream traverses, but which admits of no further definition. In one point of it we plant ourselves and call it the Present. On one side we represent to ourselves the Future as emerging out of the distance and flowing away into the Past, or conversely — to make the ambiguity of this imagery more manifest—we think of the stream as issuing from the Past and running on into an endless Future. In neither case does the image correspond to the thought. For this never-ending stream is and remains of equal reality throughout, whether as it already flows on the side where we place the future or as it is still flowing on that which stands for the past ; and the same reality belongs to it at the moment of its crossing the Present. Nor is it this alone that disturbs us in the use of the image. Even the movement of the stream cannot be presented to the mind's eye except as having a definite celerity, which would compel us to suppose a second Time, in which the former (imaged as a stream) might traverse longer or shorter distances of that unintelligible background.

139. Suppose then that we try to dispense with this inappropriate imagery, and consider what empty time must be supposed to be, when it is merely thought of, without the help of images presented to the mind's eye. Nothing is gained by substituting the more abstract conception of a series for the unavailable image of a line. It would only be the order of the single moments of Time in relation to each other that this conception would determine. It is, no doubt, involved in the conception of Time that there is a fixed order of its constituents and that the moment m has its place between $m+1$ and $m-1$: also that its advance is uniform and that the interval between two of its members is the sum of the intervals between all the intervening members. Thus we might say that if Time is to be compared with a line

at all, it could only be with a straight line. Time itself could not be spoken of as running a circular course. There may be a recurrence of events in it, but this would not be a recurrence if the points of Time, at which what is intrinsically the same event occurs, were not themselves different. So far the conception of a series serves to explain what Time is, but it does so no further. Time does not consist merely in such an order as has been described. That is an order in virtue of which the moment m would have its place eternally between $m + 1$ and $m - 1$. The characteristic of Time is that this order is traversed and that the vanishing m is constantly replaced by $m + 1$, never by $m - 1$. Our thoughts thus turn to that motion of our consciousness in which it ranges backwards and forwards at pleasure over a series which is in itself at rest. If Time were itself a real existence, it would correspond to this motion, with the qualification of being a process directed only one way, in which the reality of every stage would be the offspring of the vanished or vanishing reality of the preceding one and itself in turn the cause of its own cessation and of the commencing reality of the next stage. We might fairly acquiesce in an impossibility of learning what the moments properly are at which these occurrences take place and what are the means by which existence is transferred from one to the other. In the first place it would be maintained that Time is something *sui generis*, not to be defined by conceptions proper to other realities : and secondly we know that the demand for explanation must have its limit and may not insist on making a simplest possible occurrence intelligible by constructions which would presuppose one more complex. But without wanting to know how Time is made, it would still be the fact that we were bringing it under the conception of a process and we should have to ask whether to such a conception of it any complete and consistent sense could be given.

We cannot think of a process as occurring in which nothing proceeds, in which the continuation would be indistinguishable from the beginning, the result produced from the condition producing it. This however would be the case with empty Time. Every moment in it would be exactly like every other. While one passed away, another would take its place, without differing from it in anything but its position in the series. This position however it would not itself indicate by a special nature, incompatible with its occupying another. It would only be the consciousness of an observer, who counted the whole series, that would have occasion to distinguish it by the number of places counted before it was reached from other moments with

which it might be compared. But if so, there would not in Time itself
be any stream, bringing the new into the place of the old. Nor can
appeal be made to the view previously stated, according to which
even the unchanged duration of a certain state is to be regarded as
the product of a process of self-maintenance in constant exercise and
thus as a permanent event, though there would be no outward change
to make this visible. If this view were applied to Time, it would only
help us to the idea of a Time for ever stationary, not flowing at all.
A distinction of earlier and later moments in it would only be possible
on the basis of the presentation to thought of a second Time, in which
we should be compelled to measure the extent in a definite direction
of the first Time, the Time supposed to be at rest.

140. Such is the obscurity which attaches to the notion of a stream
of empty Time, when taken by itself. The same obscurity meets us
when we enquire into the relation of Time to the things and events
which are said to exist and take place *in* it. Here too the convenient
preposition only disguises the unintelligibleness of the relation which
it has the appearance of enabling us to picture to the mind. There
would be no meaning in the statement that things exist in Time if
they did not incur some modification by so existing which they would
not incur if they were not in Time. What is this? To say that the
stream of Time carries them along with it would be a faulty image.
Not only would it be impossible to understand how empty Time could
exercise such a force as to compel what is not empty but real to a
motion not its own. The result too would be something impossible
to state. For even supposing the real to be thus carried along by the
stream of Time, it would be in just the same condition as before, and
thus our expression would contradict what we meant it to convey.
For it is not a mere change in the place of something which throughout retains its reality, but an annihilation of one reality and an
origination of another, that we mean to signify by the power at once
destructive and creative of the stream of Time. But, so understood,
this power would involve a greater riddle still. Its work of destruc-
tion would be unintelligible in itself, nor would it be possible to conceive the relation between it and that vital power of things to which
must be ascribed the greater or less resistance which they offer to
their annihilation. Empty Time would be the last thing that could
afford an explanation of the selection which we should have to suppose it to exercise in calling events, with all their variety, into existence
in a definite order of succession.

But if, aware of this impossibility, we transfer the motive causes of this variety of events to that to which they really belong, viz. to the nature and inner connexion of things, what are we then to make of the independent efflux of empty Time, with which the development of things would have to coincide without any internal necessity of doing so? There would be nothing on this supposition to exclude the adventurous thought that the course of events runs counter to time and brings the cause into reality after the effect. In short, whichever way we look at the matter, we see the impossibility of this first familiar view, according to which an empty Time has an existence of its own, either as something permanent or in the way of continual flux, including the sum of events within its bounds, as a power prior to all reality and governed by laws of its own. But the certainty with which we reject this view does not help us to the affirmation of any other.

141. Doubts have indeed been constantly entertained in regard to the reality which is commonly ascribed to Time and many attempts have been made, in the interests of a philosophy of religion, to establish the real existence of a Timeless Being as against changeable phenomena. A more metaphysical basis was first given to this exceptional view by the labours of Kant. He was led by the contradictions, which the supposition of the reality of Time seemed to introduce even into a purely speculative theory of the world, to regard it equally with space as a merely subjective form of our apprehension. This is not the line which I have myself taken. It seemed to me a safer course to show that Time in itself, as we understand it and as we cannot cease to understand it without a complete transformation of the common view, excludes every attribute which would have to be supposed to belong to it if it had an independent existence prior to other existence. On the other hand I cannot find in the assumption of its merely phenomenal reality a summary solution of difficulties, which only *seem* to arise out of the application of Time to the Real but in truth are inseparable from the intrinsic nature of the Real. On this subject I may be allowed to interpose some remarks.

142. Were it intrinsically conceivable that an independent existence of any kind should belong to Time, and were it further possible to conceive any way in which the course of the world could enter into relation to it, then the difficulties which Kant found in the endlessness of time would cause me no special disturbance. That the world has of necessity a beginning in Time, is the Thesis of his antinomy, and this according to the method of ἀπαγωγή he seeks to prove by disproving the antithesis. It may be noticed in passing that for those who do not, to begin with, find something unthinkable in empty Time as having an existence of its own, the reference to the world which fills Time is even here really superfluous. The Thesis might just as well assert of Time itself that it must have a beginning, and then proceed as it does. 'For[1] on supposition that *Time* has no beginning, before any given moment of Time there must have elapsed an eternity, an endless series of successive *moments*. Now the endlessness of a series consists in this, that it can never be completed by successive synthesis. An endless past lapse *of Time* is therefore impossible and a beginning *of it* necessary.'

I confess to having always found something questionable in the relative position which Kant here assigns to the thought of the endlessness of Time on the one hand, and that of the impossibility of completing the endless series by synthesis on the other. He thinks it obvious that the latter constitutes a reason against the former, whereas one might be tempted on the contrary to consider it merely an obvious but unimportant consequence of this thesis. For undoubtedly, in contemplating an endless lapse of time, we suppose that a regress from the present into the past would never come to an end, and that accordingly we could not exhaust the elapsed time by a successive synthesis of the steps taken in this regress. The two thoughts are thus perfectly consistent, and the endlessness of the past would not be found to involve any contradiction until we could succeed in discovering a last stage in the regress. Presumably indeed Kant merely meant by the second thought to exhibit more clearly an absurdity already implicit in the first. But it is just on this point that I cannot accord him an unqualified assent.

143. To begin with, I propose to put my objection in the following general form: the right and duty to admit that something is or happens does not depend on our ability by combining acts of thought to *make* it in that fashion in which we should have to present it to ourselves as being or happening, *if* it were to be or to happen. It is enough that the admission is not rendered impossible by any inner contradiction, and is rendered necessary by the bidding of experience. By no effort of thought can we learn how the world of Being is made ; but there was no contradiction in the conception of it, and experience compelled us to adopt the conception. We have had no experience how the world of Becoming is made, on the contrary, the attempt to construct it in thought constantly brings us to the edge of inner con-

---

[1] [Altered from Kant's Kritik d. r. Vernunft, p. 304 (Hartenstein's ed.). The words in italics are Lotze's alterations.]

tradictions, and it is only experience that has shown us that there may
happen in reality what we cannot recreate in thought. We cannot
make out how the operation of one thing on another is brought about,
and in this case we found it impossible to overcome the inner contradiction implied in the supposition that independent elements, in no
way concerned with each other, should yet concern themselves with
each other so far that the movement of one should be regulated by
that of the other. This conception of operation, accordingly, we
could not admit without discarding the supposition of the obstructive
independence of things, and so rendering possible that mutual regulation of their motions, which experience shows to be a fact. Could
the ascription to empty space of an existence of its own, independent
of our consciousness, be carried out without contradiction, the infinite
extension, inseparable from its nature, would not have withheld us
from recognising its reality, although we were aware that we could
never exhaust this infinity by a successive addition of its points or of
the steps taken by us in traversing it. It was no business of ours to
make Space. It is the concern of Space itself how it brings that to
pass which the activity of our imagination cannot compass. Certainly,
if a self-sustained existence, it was not bound to be small enough for
us to be able to find its limits. In its infinity no contradiction was
involved. From every limit, at which we might halt for the moment,
progress to another limit was possible, which means that such progress
was always possible. A contradiction would only have arisen upon a
point being found beyond which a further progress would not have
been allowed, without any reason for the stoppage being afforded by
the law which has governed the process through the stages previously
traversed, and against the requirement of that law. From this infinity
of Space the impossibility of exhausting it by successive synthesis
would have followed as a necessary, but at the same time, unimportant
consequence: unimportant, because the essence of Space, as a complex of simultaneous not successive elements, would have been quite
unaffected by the question whether a mode of origination, which is
certainly not that of Space, is possible.

In this respect the case is undoubtedly quite different in regard to
Time. It is by the succession of moments that every section of Time
comes into being. Therefore no wrong is done it by the question
whether its infinity is attainable by the method of successive synthesis,
which ceases in this case to be merely the subjective method followed
by our thought. But even here the impossibility of coming to an end
cannot be regarded as disproving the endlessness of Time. Kant

speaks expressly of a *successive* synthesis, and of the certainty that the infinite series can *never* be exhausted by it. .If we insist on these expressions, it is clear that the course of Time, the infinity of which is alone ostensibly impugned, is itself already regarded as a real condition antecedent of that activity of imagination, which attempts the synthesis said to be fruitless. The several steps of this activity follow each other. Now whatever the celerity with which this task of adding moment to moment may be supposed to be carried on, no one will maintain that it is achieved more quickly than the lapse of the moments which it counts. The mental reconstruction of Time in time by means of the successive synthesis of its moments will take as much time as Time itself takes for its own construction ; therefore an endless Time, if Time is endless. And this is in fact, as it seems to me, the real meaning of the word *never* in the above connexion. It cannot have the mere force of negation, *not.* It only asserts what is in itself intelligible, that no succession in Time, neither that of our mental representation of Time nor that of Time itself, can measure an infinite Time in a finite Time. But no inner contradiction lies in this progress from point to point. This is the more apparent from the consideration that the progress must be supposed really to take place if we are to conceive the possibility of the successive synthesis, by which we are said to learn that it continues so endlessly as never to be completed. It is not with itself therefore that the endlessness of Time is in contradiction, but only with our effort to include its infinite progress in a finite one of the same kind.

144. In writing thus, I am not unaware of the possible objection that this view admits of unforced application only to the Future, which no one would seriously doubt to be without limits. It may be said that the Future, as we conceive it, contains that which is coming to be but has not yet taken shape, and the endlessness of its progression agrees with this conception : whereas the Past (if Infinity is to be ascribed to it) would compel us to assume a finished and ready-made Infinity. I cannot help thinking, however, that we have here a confusion of ideas.

In the first place, I would dispose of the difficulty which may be suggested by Kant's expression, that '*up to* any moment of the present an infinite series of Time must have elapsed.' It seems to me improper to represent the Present as the end of this series. It is not the stream of Time of which the direction can be described by saying that it flows out of the Past, through the Present, into the future. It is only that which fills Time—the concrete course of the world—that conditions what is contained in the later by what is contained in the earlier. Empty Time itself, if there were such a thing, would take the opposite direction. The Future would pass unceasingly into the Present and this into the Past. In presenting it to ourselves we should have no occasion to seek the source of this stream in the past.

This correction, however, only alters the form of the above objection, which might be repeated thus :—If the Past is held to be infinite, then there must be considered to have elapsed an infinite repetition of that mysterious process, by which every moment of the empty Future becomes the Present, and again pushes the Present before it as a Past. The true ground, however, of the misunderstanding is as follows. Future and Past alike are *not;* but the manner of their not-being is not the same. It is true that in regard to empty Time, though we would fain make this distinction, we cannot show that it obtains, for one point of the elapsed void is exactly like every point of the void that has still to come. But if we think of that course of the world which fills Time, then the Future presents itself to us as that which, for us at any rate, is shapeless, dubious, still to be made, while the Past alone is definitely formed and ready-made. Only the Past—which indeed *is* not, but still has known what Being is—we take as given, and as in a certain way belonging to reality. For every moment of what has been the series of conditions is finished—the conditions which must have been thought or must have been active in order to make it the definite object which it is. This character of what has been, since it belongs to every moment of the past, is shared by the whole past of the world's history, and is transferred by us to empty Time. Thus, as a matter of course, when we speak of an endless Past, we take it to be the same thing as saying that this endless Past '*has been*'[1]. But it is quite a different notion that Kant conveys by his expression 'gone by'[2]. This is the term used of a stream, of which it is already known or assumed that it has an end and exhausts itself in its lapse. But there is nothing in the essential character of the Past to justify this assumption. Nothing is finished but the sum of conditions which made each single moment what it has been. To say, however, that this determination is in each case finished is quite a different thing from saying that the series of repetitions of this process is itself closed, and must be held to be given as a closed series or to have gone by, if it is to be equivalent to the series of what has been. The latter is indeed the assertion of Kant, but the

---

[1] ['Sei gewesen.']   [2] ['Verflossen.']

thought so expressed is not one necessarily involved in the conception of that which has been, so as to be alleged as a disproof of the assumption of an infinite past. All that can be said is that whoever thinks of an infinite past, thinks of an infinite that has been. Why he should not think this does not appear. He will simply deny that the conception of what has been contains a presumption of its being finite. But that, on supposition of an infinite past, we should never come to an end in an attempt to reconstruct the past by the successive synthesis of a process of imagination, is not anything to surprise us. It is the natural result of our assumption. A contradiction would only arise if the infinity asserted broke off anywhere.

145. The doctrine that our imagination can only approach the infinitely great by a progress which can be continued beyond every limit that may be fixed for the moment may be met with elsewhere than in Kant. I do not dispute the correctness of this doctrine. But if it is meant to convey a definition of the infinite I must object, that it would be a definition of the object only by one of its consequences which may serve as a mark of it, not by the proper nature from which these consequences flow. For that the progress in question admits of being continued beyond every limit is something that cannot have been learned by any actual experiment. Any such experiment must necessarily have stopped at some finite limit without any certainty that the next step in advance, which had unfortunately not been taken, might not have exhausted the infinite. Rather we derive this certainty, that the imagination with its posterior constructions will not exhaust it, from a prior conception which does exhaust it, were it only the simple recognition that the infinite has not an end, and that therefore, as a matter of course, such an end cannot be found.

The above definition by consequences may, notwithstanding, have its use. What must, on the contrary, be disputed is the conclusion connected with it, that in the range of our thoughts about the real a case can never occur in which we might recognise the infinite as actually present and given ; or, to put it otherwise, that an infinite can never possess the same reality which we ascribe to finite magnitudes of the same kind. If we continue the series of numbers by the addition of units, the infinite cannot, it is true, be found as a number. To require that it should be so found would be to contradict our definition of it. But to every further number admitted beyond the last which we presented to ourselves, we have to ascribe the same validity as to this last. The series does not so break off where our synthesis comes to an end as that the further continuation should be in any way

distinguishable from the piece already counted, as the merely possible
or imaginary from something real or given. On the contrary, to our
conception the series has undiminished validity as an infinite one,
although on the method of addition of units it could never be begotten
for our imagination. The Tangent of an angle increases with the
increase of the angle. Not only, however, do we continuously approximate to the case in which its value becomes infinite; we actually
arrive at it if the angle is a right angle and the Tangent parallel to the
Secant. This infinite length remains throughout unmeasurable by
successive synthesis of finite lengths: but we are at the same time
forced to admit that as the concluding member of a series of finite
Tangent-values, which admit of being stated, this infinite inexhaustible
Tangent presents itself with just the same validity as those that are
exhaustible. We say with equal *validity*, and that is all that we can
say, for none of these lines are realities, but only images which we
present to the mind's eye. But I find nothing to prove that in the
conception of reality, as such, there is anything to hinder us from
recognising, beside finite values which we are forced to admit, the
reality of the infinite, as soon as the necessary connexion of our
thoughts compels us to do so.

Now for those who consider a stream of empty Time, as such,
possible, such a necessity lies not merely in the fact that no moment
of this time has any better title than another to form the beginning.
On the contrary, try as we may, an independent stream of Time
cannot be regarded as anything but a process, in which every smallest
part has the condition of its reality in a previous one. There thus
arises the necessity of an infinite progression—a necessity equally unavoidable if, on the other hand, we look merely to the real process of
events and regard this as producing in some way the illusion of there
being an empty Time. It is impossible to think of any first state of
the world, which contains the first germ of all the motion that takes
place in the world in the form of a still motionless existence, and yet
more impossible to suppose a transition out of nothing, by means of
which all reality, together with the motive impulses contained in it,
first came into being.

146. All these remarks, however, have only been made on supposition that a stream of empty Time is in itself possible. Since we
found it impossible, we will try how far we are helped by the opposite
view, that Time is merely a subjective way of apprehending what is
nQt in Time. A difficulty is here obvious, which had not to be encountered by the analogous view of Space. Ideas, *ex parte nostra*, do

not generally admit of that which forms their content being predicated of them. The idea of Red is not itself red, nor that of choler choleric, nor that of a curve curved. These instances make that clear and credible to us which in itself, notwithstanding, is most strange; the nature, namely, of every intellectual presentation, not itself to be that which is presented in it. It may indeed be difficult for the imagination, when the expanse of Space spreading before our perception announces itself so convincingly as present outside us, to regard it as a product, only present for us, of an activity working in us which is itself to no conditions of Space. Still, in the conception of an activity there is nothing to make us look for extension in Space on the part of the activity itself as a condition of its activity. On the contrary, had we believed that the impressions of Space in our inner man could themselves have position in Space, we should have been obliged to seek out a new activity of observation which had converted this inner condition into a knowledge of it, and to look to this activity for that strange apprehension of what is in Space which must do its work without being in Space itself.

If, on the other hand, we try to speak in a similar way of a timeless presentation of what is in time, the attempt seems to break down. The thought that Time is only a form or product of our presentative susceptibility, cannot take away from the presentation itself the character of an activity or at least of an event, and an event seems inconceivable without presupposition of a lapse of time, of which the end is distinguishable from the beginning. Thus Time, unlike Space, is not merely a product of the soul's activity, but at the same time the condition of the exercise of the activity by which Time itself as a product is said to have been obtained, and the presentation to consciousness of any change seems impossible without the corresponding real change on the part of the presenting mind. Now it must be borne in mind that in no case could Time be a subjective form of apprehension in such a sense as that the process of events, which we present to ourselves in it, should be itself opposed to the form of apprehension as being of a completely alien nature. Whatever basis in the way of timeless reality we may be disposed to supply to phenomena in Time, it must at any rate be such that its own nature and constitution remain translateable into forms of Time. To this hidden timeless reality, it may be suggested, that activity of thought would itself belong, of which the product in our consciousness would be that course of occurrences and of our ideas which is seemingly in Time. Of it, and by consequence of every activity as such, it must be sought to show,

according to the view which takes Time to be merely our form of
apprehension, that while not itself running a course in a time already
present, it may yet present itself to sense in its products as running
such a course. Let us pursue the consideration by which it may be
attempted to vindicate this paradoxical notion.

147. No one will maintain that the stream of empty Time brings
forth events in the sense of being that which determines their character and the succession of the various series of them. It would be
admitted that all this is decided by the actual inner connexion of
·things. But although that which happens at one moment contains
the ground G of that which at the next is to appear as consequence
F, it may be fancied that the lapse of Time is a *conditio sine qua non*
which must be fulfilled if the grounded consequence is really to follow
from its ground. A reference to the general remarks previously
made, upon the several kinds of cause distinguished in common
parlance, may meanwhile suffice to convince us that what we call a
conditio sine qua non can stand in no other relation to the effect resulting than does every other co-operative cause. The mere presence of
that which in each case is so called is never sufficient to draw a
distinct event in the way of consequence after it. The case rather is
that the presence of such a complementary condition must always
manifest itself by an effect exercised on the other real elements
which without it would not have sufficed for the production of the
consequence F.

Now if upon such a supposition we assume first that at a certain
moment a state of things, G, is really given which forms the complete
ground of a necessary consequence, F, there is no conceivable respect
in which the lapse of an empty Time, T, should be necessary, or
could contribute, to bring about the production of F by G. Granted
that, during the time T, G has continued without change, neither
producing F nor a more immediate consequence, f, preliminary to
the other, then at the end of the interval T everything will be just as
at the beginning, and the lapse of time T will have been perfectly
barren. If, on the other hand, during the same interval G has passed
into the series of consequences $f_1, f_2, f_3 \ldots$, each related to the next
following as ground to consequence, the same remark is applicable to
any two proximately related members of this series. If f_2 is the sole
ground of f_3, then the lapse of the smaller interval of empty Time $t_3 - t_2$
can be neither contributory nor essential to the production by f_2 of its
effect f_3. It will no doubt be objected that the flaw of our argument consists in this, that we fix a certain momentary state of things,

G, and consider this fixed state of things, in complete identity with itself, to act as the operative cause of an effect ; whereas in fact G only becomes such a cause through a lapse of Time during which it is itself in continuous process of becoming. For this reason, it will be said, the series of determinate causes and effects unfolds itself as a process of events, while on our supposition it remains out of Time and just for that reason cannot form more than a system of members which stand to each other eternally in graduated relations of dependence without ever moving in these relations. It must be admitted that whoever puts this objection strikes a most essential point. He is perfectly right in insisting upon ceaseless motion or uninterrupted becoming as constituents of the real. For undoubtedly, if once the perfectly unchanging fact G were recognised as given, then the consequent F, of which it contains the sufficient reason, would as speculatively valid truth, subsist permanently along with G, while considered as reality it would either always exist along with it or never come into being out of it. For then the addition of the lapse of an empty Time t would not produce the motion absent from G at all, at any rate not produce it more or less than would the lapse of $0 . t$ or $\infty . t$.

This shall be more fully considered below. For the present my concern is to show that for the very process of Becoming in question the mere lapse of Time can afford no means, any possible application of which could be necessary to bringing it about. The proof of this, however, I hold to be involved in what has been already said. For here it comes to the same thing in effect whether we only speak of a series of distinct causes which produce their several effects, so to speak, by jumps, or whether taking the case of continuity we understand by f_1, f_2, f_3 constituents of a continuous stream of causation—constituents which are only arbitrarily fixed in thought but of which really each in turn moves. On the latter supposition it would be just as impossible that the internal motion, which results in the emission of f_3 from f_2, should be dependent on the lapse of the empty time t_3-t_2 in such a way as that it could not take place unless this lapse of time preceded. Such an influence is unintelligible unless we suppose that the lapse of empty time can announce itself to f_2—nay that the completion of the period t_3-t_2 makes itself felt as different from that of the longer period t_4-t_2, in order that in the former there may be occasion for the advance of the process of becoming from f_2 only to f_3, in the latter to f_4. But the ends of the two periods are completely like each other and like every other moment of empty time. The entry of the one has no such distinction from that of the other as can give to f_2 the signal for this or that amount of advance. For that reason the sum of the continuously flowing moments, which forms the duration of each period, cannot make itself felt by the operative power f_2 as a measure of the work which it has to do in the way of production of Becoming. On the contrary, it will only be in the same way in which we measure a period of Time for purposes of our knowledge that the length of this period can announce itself to f_2 so as to determine the magnitude of the change which f_2 has to undergo. This is by the enumeration of the repetitions of a similar process, which at the end of some period of Time exhibits a different state of reality from what it did at the beginning. So far as our knowledge is concerned, the perception of the different positions which a pendulum, for instance, occupies at the beginning and at the end of its vibration, would suffice for the purpose. For a reality, which was to take account of the lapse of Time in order to direct its becoming accordingly, there would be needed the constant summing of the impressions received by it from another real process, by means of which it itself or its own condition had been so changed as to be able to serve as indicator of the length of Time elapsed. The conclusion plainly is that a process of becoming, B, which required a lapse of time in order to come about, must have already traversed in itself a succession of different stages, in order to feel in that succession the lengths of the periods according to which it is supposed to direct itself, and which it is supposed to employ for the purpose of effecting the transition from one stage to another.

148. These considerations do not lead us at once to the end of our task. For the present I may put their result, which I shall not again discuss, as follows. It is quite unallowable to put the system of definite causes and effects, which gives its character to any occurrence, on one side and on the other side to suppose a stream of empty Time, and then to throw the definitely characterised event into the stream in expectation that its fabric of simultaneous conditions will in the fluidity of this stream melt into a succession, in which each of the graduated relations of dependence will find its appropriate point of time and the period of its manifestation. It is only in the actual content of what happens, not in a form present outside it into which it may fall, that the reason can be found for its elements being related to each other in an order of succession, and at the same time for the times at which they succeed each other.

The other view therefore begins to press itself upon us—the view that it is not Time that precedes the process of Becoming and

Activity, but this that precedes Time and brings forth from itself either the real course of Time or the appearance in us of there being such a thing. The constant contradiction to this reversal of the habitual way of looking at the matter which our imagination would present, we could no more get rid of than we could of the habit of saying that the sun rises and sets. What we might hope to do would be to understand one illusion as well as the other. It is also our habit to speak of general laws, standing outside things and occurrences and regulating their course; yet we have been forced to the conviction that these have no reality except in the various particular cases of their application. Only that which happens and acts in determinate forms is the real. The general law is the product of our comparison of the various cases. After we have discovered it, it appears to us as the first, and the realities, out of the consideration of which it arose, as dependent on its antecedence. In just the same way, after the manifold web of occurrence has in countless instances assumed for us forms of succession in Time, we misunderstand the general character of these forms, which results from our comparison of them—the empty flowing Time—and take it for a condition antecedent, to which the occurrence of events must adjust itself in order to be possible. That we are mistaken in so doing and that the operation of such a condition is unthinkable— this 'reductio ad impossibile,' which I have sought to make out, is, it must be admitted, the only thing which can be opposed to this unavoidable habit of our mental vision.

149. The positive view, which we found emerging in place of the illusion rejected, is still ambiguous. Is it a real Time that the process of events, in its process, produces or only the appearance of Time in us? In answering this question we cannot simply affirm either of the alternatives. One thing is certainly clear, that the production of Time must be a production *sui generis*. Time does not remain as a realised product behind the process that produces it. As little does it lie before that process as a material out of which the process can constantly complete itself. Past and Future are *not*, and the representation of them both as dimensions of Time is in fact but an artificial projection, which takes place only for our mind's eye, of the unreal upon the plane which we think of as containing the world's real state of existence.

Undoubtedly therefore Time, conceived as an infinite whole with its two opposite extensions, is but a subjective presentation to our mind's eye; or rather it is an attempt, by means of images borrowed from space, to render so presentable a thought which we entertain as to the inner dependence of the individual constituents of that which happens. What we call Past, we regard primarily as the condition ' *sine qua non* ' of the Present, and in the Present we see the necessary condition of the Future. This one-sided relation of dependence, abstracted from the content so related and extended over all cases which it in its nature admits of, leads to the idea of an infinite Time, in which every point of the Past forms the point of transition to Present and Future, but no point of Present or Future forms a point of transition to the Past. That this process must appear infinite scarcely needs to be pointed out. The condition of that which has a definite character can never lie in a complete absence of such character. Every state of facts, accordingly, of which we might think for a moment as the beginning of reality, would immediately appear to us either as a continuation of a previous like state of facts, or as a product of one unlike; and in like manner every state of facts momentarily assumed to be an end would appear as the condition of the continuance of the same state of facts, or in turn as the beginning of a new one. If finally the course of the world were thought of as a history, which really had a beginning and end, still beyond both alike we should present to ourselves the infinite void of a Past and Future, just as two straight lines in space which cut each other at the limit of the real, still demand an empty extension beyond in which they may again diverge.

150. It will be felt, however, that we have not yet reached the end of our doubts. It will be maintained that though the process of Becoming does indeed make no abiding Time, it yet does really bring into being or include the course of Time, by means of which the various parts of the content of what happens, standing to each other in the relation of dependence described above, having been at first only something future, acquire *seriatim* the character of the Present and the Past. If we chose to confine ourselves simply to highly developed thought, and to regard the dimensions of Time merely as expressions for conditioned-ness or the power of conditioning, then the whole content of the world would again change into a motionless systematic whole, and everything would depend on the position which a consciousness capable of viewing the whole might please to take up facing, so to speak, some one part of it, m. From this point of departure, m, the contemplator would reckon everything as belonging to the Past, $m-1$, in which he had recognised the conditions that make the content of m what it is, while he would assign to the Future, $m+1$, all the

consequences which the necessities of thought compelled him to draw from it : and this assignment of names would change according as *m* or *n* might be made the point of departure for this judgment. This however does not represent the real state of the case. This capacity of tracing out the connexion of occurrences in both directions—forwards and backwards—would only be possible to a consciousness standing outside the completed course of the world. It belongs to us only in relation to the past, so far as the past has become known to us through tradition. Immediate experience is confined to a definite range, and neither does the recollection of the past reproduce for experience its actual duration, nor does the sure foresight of the future, in the few cases where it is possible, take the place for experience of the real occurrence of the foreseen event.

What then is the proper meaning of the Reality, which in this connexion of thought we ascribe only to the Present? Or conversely, what constitutes this character of the present, which we suppose to belong successively in unalterable series to the events of which each has its cause in the other, and to be equivalent to reality? I will not attempt to prepare the way for an answer to this question, or to lead up to it as a discovery. I will merely state what seems to me the only possible answer to it. It is not the mere fact that they happen which attaches *this* character to the content of events. On the contrary the import of the statement that they happen is only explained by the expression ' the Present,' in which Language aptly makes us aware of the necessity of a subject, in relation to which alone the thinkable content of the world's course can be distinguished either as merely thinkable and absent on the one hand, or on the other as real and present. To explain this, however, I am obliged to go into detail to an extent for which I must ask indulgence and patience.

151. Let us consider one of the finite spiritual beings like ourselves, which shall be called *S*. In the collective content of the world, *M*, which to begin with we will think of as we did before, merely as a regularly arranged whole of causes and effects, *S* has its proper place in the system at *m* between a past $m-1$, which contains its conditions, and a future $m+1$, of which it is itself a joint condition. We will first assume that the place *m*, which *S* holds in *M*, is without extension. By this I mean that it is only in this single plane of a section *m* through the manifold interlacing series of causes and effects which forms the content of *M*—not in any other $m-1$ or $m+1$—that there lie the conditions of *S* : while at the same time

every element of $M-S$ among others—may be supposed to have
knowledge, immediately and not by gradual acquisition, as to the whole
structure and content of M. All that would be implied in this supposition would be that S would no longer be able ·at its pleasure to
seek out positions indifferent as concerned itself for its survey of the
whole of M. Being only able to plant itself in the position m, everything in which it recognises a joint condition of its own being will
appear to belong to a different branch, $m-1$, of the world's content,
from that in which it finds reactions from its own existence—that
existence which is confined to m. At the same time this knowledge
on the part of S, that it is merely co-ordinated in this entire system
of conditions with the other parts of the world's content that are included in m, would remain a mere speculative insight, which would
excite in S no stronger interest in this m, and one of no other nature,
than the interest in the fact of the dependence of m upon $m-1$ and
of $m+1$ upon m. Thus, although S would distinguish according to
their import the two branches of the system of conditions that have
their point of departure in m, it would yet have no occasion to
oppose them both to m as what is unreal and absent to what is real
and present. And this would still be the case, though we so far
altered our assumption as to suppose S to be not only contained in
the one section-plane m of M, but also to be co-ordinated with the
contents of other planes $m-a$ and $m+a$, without undergoing any
change in itself. To us indeed, who are accustomed to the idea of
Time, this position of S in a system would present itself as a duration,
as the filling by S of the period of time, $2a$: but to S itself, if S
continued to possess the immediate knowledge supposed, it could
only convey the speculative impression that S is interwoven in an
extended section of M, while S would still have no occasion to
oppose this section as present to others as absent.

All this would be changed on one supposition only, which indeed
for other reasons must be made ; the supposition, namely, that the
place of S in the system contains not only the conditions of its existence but those of its knowledge. In this is implied that only those
elements of $m-1$ can be an object of its knowledge which not only
systematically precede it as conditions but of which the consequences
are contained in m, and only as far as their consequences are so contained. Of $m+1$ on the contrary all that will be knowable will be the
impulse, already present in m, which is the condition of $m+1$. Even
the entire content of m will not, merely as such, form an object of
knowledge to S. Even the fact of belonging to m is for each element

of it only the condition of a more special relation to S, which we may call its effect on S in the way of producing knowledge. If now we return to our supposition that m is a place without extension, then the knowledge possessed by S will be an unchangeable presentation to consciousness, without there being any occasion for the distinction of Present from Future in it. If on the contrary S found itself contained in the whole extended section $2\,a$ of M, then it would follow—since we are now supposing its knowledge to rest upon the effect produced in it by the content of this section—that S is no longer identical with itself in all points of $2\,a$, but has to be defined by s_1, s_2, s_3, corresponding to the various conditions to which it is subject in the various points of $2\,a$. But thus S would fall asunder into a multiplicity of finite beings, unless something supervened to justify us in adhering to the unity asserted of it, and this justification, if it is not merely to establish an accidental view about s in us but to constitute an essential unity on the part of s, can only consist in an action of its own on the part of s by which it unites the several s's.

This requirement however is not satisfied by the assumption of an S having unity, which distinguishes the several s's in itself as its states. S as thus constituted would still never live through any experience. The whole content of its being would be presented to it just in the same way as on our previous supposition. There would indeed be a clear insight into the plan upon which the elements are formed into a connected whole, but the whole would be presented simultaneously, just as is the frame-work of theoretic propositions which appear to us not as arising out of each other in a course of time but as always holding good at the same time, although we understand their dependence on each other. Only one of the s's can in any case be the knowing subject, but in it—in s_3, let us say—the content of s_2 must not only be contained by its consequences, through which it helps to constitute the nature of s_3, but this content as presented to consciousness must be distinguishable in the form of a recollection from that which belongs to s_3 as its own feeling or perception. On this condition only is it possible for s_3 to distinguish this latter experience as present from that represented content as absent, and on the same condition, since the same reproduction of s_1 in s_2 has already taken place, the whole series of these mutually dependent contents, as represented in consciousness, while preserving its inner order, will be pushed back to various distances of absence. The question indeed as to the foundation of this

faculty of distinguishing a represented absent object from one experienced as present is a question upon which any psychological
or physiological explanation may be thankfully accepted in its place.
Here however it would be useless. What we are now concerned
with is merely the fact itself, that we are able to make this distinction
and to represent to ourselves what we *have* experienced without
experiencing it again. This alone renders it possible for ideas of
a proper succession to be developed in us, in which the member n
has a different kind of reality from $n + 1$. It would have been more
convenient to arrive at this result otherwise than by this tedious process of development. I thought the process indispensable, however,
because it leads to some peculiar deductions, which require further
patient consideration.

152. For instance; what has been said will be found very intelligible—not to say, obvious—if only we allow ourselves to inter-
polate the thought that s_2 ceases to exist when it has produced s_1;
that thus there is a time in which those section-planes of M or of $2a$
succeed each other. But it will be thought to be as impossible after
our discussion as it was before it, to look upon the content of the
world as out of time, a whole of which the members are related
systematically but not successively, while yet there arises in parts of
it the appearance of there being a lapse of time on the part of the
periods which those parts observe. For if there is no successive
alternation of Being and not-Being, then, it will be said, every stage
of development, s_2, which a subject, s_3, believes itself to have experienced in the past, will possess, as a ground of s_3, the same
reality as the consequence s_3 itself. Accordingly we should be compelled, it would seem, to think of all that is past—all histories, actions,
and states of an earlier time—as still existing and happening; and
every individual being s_n, would have alongside of itself as many
doubles, s_1, s_2, s_3, completing themselves one after another, as it counts
various moments in the existence which it seems to have lived through.

Against this objection, however, we must maintain that such peculiar views would not be the logical consequence of our denial of the
lapse of Time, but on the contrary of the inconsistency of allowing
the succession that has been denied again to mix itself with our
thoughts. For only this habituation of our imagination to the idea of
Time could mislead us into treating the elements of the world,
which are of equal value—all, that is to say, equally indispensable to
the whole—as if they must be contemporaneous unless they are to
be successive, when all the while our purpose was to show that every

determination in the way of time is inapplicable to them, as such. We shall never succeed in ridding ourselves of this habit of fantasy. Only in thinking shall we be able to convince ourselves, in standing conflict with our demand for images presentable to the mind's eye, that adherence to the assumption of timelessness does not lead to the consequences in which we have just found a stumbling-block. There would not indeed on our view be that kind of past into which the conditioning stage of development would be supposed to vanish instead of illegitimately continuing in the present alongside of the consequence conditioned by it—that consequence to which it ought to have transferred the exclusive possession of the quality of being present. The histories of the past would not continue to live in this present, petrified in each of their phases, alongside of that which further proceeded to happen in the course of things. It would not be the case that s_1 really existed earlier than s_2 and strangely continued along with it, but rather that it had reality only so far as it was contained in s_2 and was presented by the latter to itself as earlier. It will be with Time as with Space. As we saw, there is no such thing as a Space in which things are supposed to take their places. The case rather is that in spiritual beings there is formed the idea of an extension, in which they themselves seem to have their lot and in which they spatially present to themselves their non-spatial relations to each other. In like manner there is no real Time in which occurrences run their course, but in the single elements of the Universe which are capable of a limited knowledge there developes itself the idea of a Time in which they assign themselves their position in relation to their more remote or nearer conditions as to what is more or less long past, and in relation to their more remote or nearer consequences as to a future that is to be looked for more or less late.

It is not out of wantonness that I have gone so far in delineating this paradoxical way of looking at things. It is what we must come to if we wish to put clearly before us the view of the merely subjective validity of Time in relation to a timeless reality. It is vexatious to listen to the mere asseveration of this antithesis without the question being asked whether, when adopted, it intrinsically admits of being in any way carried out, and whether it would be a sufficient guide to the understanding of that experience from which we all start. The description which has been given will be enough to raise a doubt whether the latter is the case. The reasons for this doubt however are not all of equal value. In regard to them again, while passing to the consideration of this contradiction, I must ask to be allowed some detail.

153. In order to find a point of departure in what is familiar, I will first repeat the objection which will always recur. Pointing to the external world the objector will enquire—' Is it not then the case that something is for ever happening? Do not things change? Do they not operate on each other? And is all this imaginable without a lapse of time?' Imaginable it certainly is not, and we have never maintained that it is so. But in what relation do the lapse of Time and this happening stand to each other, which might enable us to maintain the correctness of this imagination of ours? That it is only in what is contained in a sufficient cause, G, that there lies a necessity for the consequence, F—that the necessity, if otherwise lacking, could not supervene through lapse of a time, T— this we found obviously true. It was admitted also that, G being given, it would neither be intelligible where the hindrance should come from which should retard its transition into F, nor how the lapse of empty Time could overcome that hindrance. Thus constrained to confess that our habit of thinking the effect as *after* the cause does not point to anything which in the things themselves contributes to the production of the effect, what other conclusion can we draw than this, that succession in Time is something which our mode of apprehension alone introduces into things—introduces in a way absolutely inevitable for us, so that our thought about things remains constantly in contradiction with our habit of presenting them to the mind's eye?

One may attempt to make this thought clear to oneself by gradual approximation. To a definite period of Time it is our habit in common apprehension to ascribe a certain absolute quantity. If we ask ourselves, however, how long a century or an hour properly lasts, we at once recollect that the time filled by one series of events we always measure simply according to its relation to another series, with the ends of which those of the first series do or do not coincide. Our ordinary impression of the duration of periods of time is itself the uncertain result of such a comparison, in which we are not clearly conscious of the standard of our measurement. Hence the same period may appear long or short in memory. The multiplicity of the events contained in it gives it greater extent for the imagination. Poverty of events makes it shrink into nothing. It has itself no extensive quantity which is properly its own. Therefore no hindrance meets us in the attempt to suppose as short a time as we will for the collective course of events. However small we think it, still it is not in it but in the dependence of events on each other that the reason lies of the order in which events occur ; and the entire history which fills centuries admits of being presented in a similar image, as condensed into an infinitely small space of Time through proportional diminution of all dimensions.

With this admission however it will be thought necessary to come to a stop. However small, it will be said, still this differential of Time must contain a distinction of before and after, and thus a lapse, though one infinitely small. But we want to know exactly why. Undoubtedly the transition to a moment completely without extension would deprive History of the character of succession in Time ; but then our question is just this, whether the real needed this succession on its own part in order to its appearance as successive to us. And in regard to this we must constantly repeat what has been already said ; that neither could the order of events be constituted by Time, if it were not determined by the inner connexion of things, nor is it intelligible how Time should begin to bring that which already has a sufficient cause to reality, if that reality is still lacking to it. On the other hand, we believe that we do understand how a presentative faculty such as to derive from its own nature the habit of viewing the world as in time, should find occasion in the inner connexion between the constituents of that world, as conditioning and conditioned by each other, to treat its parts as following each other in a definite order and as assuming lengths—definite in relation to each other but, apart from such relation, quite arbitrary — of this imagined Time. Thus even upon this method, by help of the idea of an infinitely small moment, we should have mastered the thought of a complete timelessness on the part of what fills the world. For in that case we should certainly not go out of our way to think of that extension of time, within which this moment would seem of a vanishing smallness, and so bring on the world the reproach of a short and fleeting existence, as compared with the duration which expansion into infinite Time would have promised it.

154. After all, it will be objected, we have not yet touched the proper difficulty. If all that we had to take account of were an external course of the world, then it would indeed cost us little effort to regard all that it contains as timeless, and to hold that it is only in relation to our way of looking at it that it unfolds itself into a succession. But the motion, which we should thus have excluded from the outer world, would so much the more surely have been transferred into our Thought, which, on the given supposition, must itself pass from one of the elements which constitute the world to another, in

order to make them successive for its contemplation. For the unfolding, by which what is in itself timeless comes to be in time, cannot
take place in us without a real lapse of Time ; the appearance of succession cannot take place without a succession of images in consciousness, nor an apparent transition of *a* into *b* without the real transition
which we should in such a case effect from the image of *a* to that of *b*.

But convincing as these assertions are, they are as far from containing the whole truth. On the contrary, without the addition of
something further, the doctrine which they allege would be fatal to
the possibility of that which it is sought to establish. If the idea of
the later *b* in fact merely followed on that of the earlier *a*, then a
change of ideas would indeed take place, but there would still be no
idea of this change. There would be a lapse of time, but not an
appearance of such change to any one. In order to a comparison in
which *b* shall be known as the later it is necessary in turn that the
two presentations of *a* and *b* should be objects, throughout simultaneous, of a relating knowledge, which, itself completely indivisible,
holds them together in a single indivisible act. If there is a belief
on the part of this knowledge that it passes from one of its related
points to another, it will not itself form this idea of its transition
through the mere fact of the transition taking place. In order that
the idea may be possible, the points with which its course severally
begins and ends, being separate in time, must again be apprehended
in a single picture by the mind as the limits between which that
course lies. All ideas of a course, a distance, a transition—all, in
short, which contain a comparison of several elements and the relation between them—can as such only be thought of as products of
a timelessly comprehending knowledge. They would all be impossible, if the presentative act itself were wholly reducible to that
succession in Time which it regards as the peculiarity of the objects
presented by it. Nay if we go further and make the provisional
admission that we really had the idea of *a* before we had that of *b*,
still *a* can only be known as the earlier on being held together with *b* in
an indivisible act of comparison. It is at this moment, at which *a* is
no longer the earlier nor *b* the later, that for knowledge *a* appears as
the earlier and *b* as the later. In assigning these determinate places,
however, to the two, the soul can only be guided by some sort of
qualitative differences in their content—by temporal signs, if we like
to say so, corresponding to the local signs in accordance with which
the *non*-spatial consciousness expands its impressions into a system of
spatial juxtaposition.

Such could not but be the state of the case even *if* there were a lapse of Time in which our ideas successively formed themselves. The real lapse of Time would not, immediately as such, be a sufficient cause to that which combines and knows of the succession in Time which it presents to itself. It would be so only mediately through signs derived by each constituent element of the world from that place in the order of Time into which it had fallen. But such various signs could not be stamped on the various elements by empty time, even though it elapsed, since one of its elements is exactly like every other. They could only be derived from the peculiar manner in which each element is inwoven into the texture of conditions which determine the content of the world. But just for that reason there was no need of a real sequence in Time to annex them to our ideas as characteristic incidental distinctions. Thus it would certainly be possible for a presentative consciousness, without any need of Time, to be led by means of temporal signs, which in their turn need not have their origin in Time, to arrange its several objects in an apparent succession in the way of Time.

155. I am painfully aware that my reader's patience must be nearly exhausted. Granted, he will say, that in every single case in which a relation or comparison is instituted this timeless faculty of knowing is active: it remains none the less true that numberless repetitions of such action really succeed each other. Yesterday our timeless faculty of knowledge was employed in presenting the succession of *a* and *b*, to-day it presents that from *c* to *d*. There are thus, it would seem, many instances of Timeless occurrence which really succeed each other in Time. I venture, however, once again to ask, Whence are we to know that this is so? And if it were so, in what way could we know of it? That consciousness, to which the comparison made yesterday appears as earlier than that made to-day, must yet be the consciousness which we have to-day, not that which may have been yesterday and have vanished in the course of Time. That which appears to us as of yesterday cannot so appear to us because it is *not* in our consciousness, but because it is in it; while at the same time it is somehow so qualitatively determined, that our mental vision can assign it its place only in the past branch of apparent Time.

I will allow, however, that this last reply yields no result. The Past indeed, of which we believe ourselves already to have had living experience, one may try to exhibit as a system of things which has never run a course in Time, and which only consciousness, for its own benefit, expands into a preceding history in Time. But how then would the case stand with the Future, which we suppose ourselves still on the way to meet? Let s_3, according to the symbols previously used, stand for this Ego, which s_2 and s_1 never really preceded but always seem to have preceded, what then is s_4 which s_3 in turn will thus seem to have preceded? What could prevent s_3 from being conscious also of s_4, its own future, if the temporal signs which teach us to assign to single impressions their position in Time, depended only on the systematic position which belongs to their causes in the complex of conditions of a timeless universe? It may be that the content of s_4, which follows systematically upon s_3, is not determined merely by the conditions, which are contained in s_3 and previously in s_2 and s_1, but jointly by others, resting on the states of other beings which do not cross those of S till a later stage of the system. For that reason s_4 might be obscure to s_3 and this might constitute the temporal character which gives it in the consciousness of s_3 the stamp of something future. But if this were the case, the process would have to stop at this point. It would only be for another being s_4 that what was Future to s_3 could, owing to its later place in the system, be present. On the other hand in a timeless system there would be no possibility of the change by means of which s_4 would be moved out of its place into that of s_4: yet this would be necessary if to one and the same consciousness that is to become Present which was previously Future to it. If one and the same timeless being by its time-less activity of intellectual presentation gives to one constituent of its existence the Past character of a recollection, to another the significance of the Present, to a third unknown element that of the Future, it could never, if it is to be really timeless, change this distribution of characters. The recollection could never have been Present, the Present could never become Past and the Future would have to remain without change the same unknown obscurity. But if there is a change in this distribution of light; if it is the case that the indefinite burden of the Future gradually enters the presence of living experience and passes through it into the other absence of the Past; and finally if it is impossible for the activity of intellectual presentation to alter this order of sequence; then it follows necessarily that not merely this activity, but the content of the reality which it presents to itself, is involved in a succession of determinate direction.

This being so, we must finally decide as follows: Time, as a whole, is without doubt merely a creation of our presentative intellect. It neither is permanent nor does it elapse. It is but the fantastic image which we seek, rather than are able, to project before the mind's eye, when we think of the lapse of time as extended to all the points of relation which it admits of *ad infinitum*, and at the same time make abstraction of the content of these points of relation. But the lapse of events in time we do not eliminate from reality, and we reckon it a perfectly hopeless undertaking to regard even the idea of this lapse as an *a priori* merely subjective form of apprehension, which developes itself within a timeless reality, in the consciousness of spiritual beings.

153. Thus, at the end of a long and troublesome journey, we come back, as it will certainly appear, to complete agreement with the ordinary view. I fear however that remnants of an error still survive which call for a special attack—remnants of an error with which we are already familiar and which have here needed to be dealt with only in a new form, viz. the disintegration of the real into its content and its reality. We are unavoidably led by our comparison of the manifold facts given to us to the separation of that on the one hand which distinguishes one real object from another—its peculiar content which our thought can fix in abstraction from its existence—and on the other hand of that in which every thing real resembles every other— the reality itself which, as we fancy, has been imparted to it. For this is just what we go on to imagine—that this separation, achieved in our thoughts, represents a metaphysical history; I do not mean a history which has been completed once for all, but one which perpetually completes itself; a real relation, that is to say, of such a kind that that content, apart from its reality, is something to which this reality comes to belong. The prevalence of this error is evidenced by the abundant use which philosophy, not least since the time of Kant, has made of the conception of a 'Position,' which meeting with the thinkable content establishes its reality. In an earlier part of this work we declared ourselves against this mistake. We were convinced that it was simply unmeaning to speak of being as a kind of placing which may simply supervene upon that intelligible content of a thing, without changing anything in that content or essence or entering as a condition into its completeness. As separate from the energy of action and passion, in which we found the real being of the thing to consist, it was impossible even to think of that essence, impossible to think of it as that to which this reality of action and passion comes from without, as if it had been already, in complete rest, the same essence which it is under this motion.

It is the same impossible separation that we have here once again,

Of Time.

in consideration of the prevalence of the misunderstanding, carefully pursued to its consequences in the form of the severance of the thing which happens from its happening. It was thus that we were led to the experiment of seeking the essence[1] of what happens—that by which the actual history of the world is distinguished from another which might happen but does not—in a complex system of relations of dependence on the part of a timeless content of thought; while the motion in this system, which alone constitutes the process of becoming and happening, was regarded as a mode of setting it forth which might simply be imposed on this essential matter, or on the other hand, might be wanting to it without changing the distinctive character of the essence. We could not help noticing, indeed, the great difference between reality and that system of intelligible contents. In the latter the reason includes its consequence as eternally coexisting with it. In the former the earlier state of things ceases to be in causing the later. Then began the attempts to understand this succession, which imposes itself like an alien fate on the system in its articulation. They were all in vain. When once the lapse of empty time and the timeless content had been detached from each other, nothing could enable the set nature of the latter to resolve itself into a constant flux in the former. It was clear that in this separation we had forgotten something which forced that content—involving as it did, *if* it moved, the basis of an order of time—to pass in fact into such a state of motion. I will not suppose that crudest attempt to be made at supplying the necessary complement—the reference to a power standing outside the world which laid hold on the eternal content of things, as on a store of material, in order to dispose its elements in Time in such a way as their inner order, to which it looked as a pattern, directed it to do. Let us rather adopt the view that in the content itself lies the impulse after realisation which makes its manifold members issue from each other. Still, even on that view it would be a mistake, as I hold, to think of the measure and kind of that timeless conditionedness, which might obtain between two elements of the world's content, as the antecedent cause which commanded or forbade that operative impulse to elicit the one element from the other. What I am here advancing is only a further appli-

---

[1] [This is still 'the content'—'that which distinguishes one real object from another.' A verbal difficulty is caused by the distinction being here, *per accidens*, between the actual world and an imaginary world, so that but for the context we might take 'essence' to be used in just the opposite sense to that of p. 265, and to refer to that which distinguishes what is real from what is unreal.]

cation of a thought which I have previously expressed. Every relation, I have said, exists only in the spirit of the person instituting the
relation and for him. When we believe that we find it in things themselves, it is in every case more than a mere relation: it is itself already
an efficient process instead of being merely preliminary to effects.

On the same principle we say—It is not the case that there is *first*
a relation of unchanging conditionedness between the elements of
the world, and that afterwards in accordance with this relation the
productive operation, even though it may not come from without but
may lie in the things themselves, has to direct itself in order to give
reality to legitimate consequences and avoid those that are illegitimate. On the contrary first and alone is there this full living operation itself. Then, when we compare its acts, we are able in thought
and abstraction to present to ourselves the constant *modus agendi*,
self-determined, which in all its manifestations has remained the same.
This abstraction made, we can subordinate each single product of the
operation, as we look backward, to this mode of procedure as to an
ordaining *prius* and regard it as determined by conditions which are
in truth only the ordinary habit of this operation itself. This process
of comparison and abstraction leads us in one direction to the idea
of general laws of nature, which are first valid and to which there
then comes a world, which submits itself to them. In another direction it leads to the supposition of an empty Time, in which the
series of occurrences succeed each other and which, in the character
of an antecedent *conditio sine qua non*, makes all operation possible. But
this last way of looking at the matter we have found as untenable
as would be the attempt to represent velocities as prior to motions
(somewhat as if each motion had to choose an existing velocity), and
to interpret the common expression, according to which the motion
of a body *assumes* this or that velocity, as signifying an actual fact;
whereas in truth the motion is nothing but the velocity as following a
definite direction.

In this sense we may find more correctness in the expressions that
may be often heard, according to which it is not Time that is the
condition of the operation of things, but this operation that produces
Time. Only what it brings forth, while it takes its course, is not an
actually existing Time as an abiding product, somehow existing or
flowing or influencing things, but only the so-called ' vision ' of this
Time in the comparing consciousness. Of this—the empty total
image of that order in which we place events as a series—it is thus
true that it is only a subjective form of apprehension ; while of the

succession belonging to that operation itself, which makes this arrangement of events possible, the reverse is true, namely that it is the most proper nature of the real.

157. I should not be surprised if the view which I thus put forward met with an invincible resistance from the imagination. The unconquerable habit, which will see nothing wonderful in the primary grounds of things but insists on explaining them after the pattern of the latest effects which they alone render possible, must here at last confess to being confronted by a riddle which cannot be thought out. What exactly happens—such is the question which this habit will prompt—when the operation is at work or when the succession takes place, which is said to be characteristic of the operative process? How does it come to pass—what makes it come to pass—that the reality of one state of things ceases, and that of another begins? What process is it that constitutes what we call perishing, or transition into not-being, and in what other different process consists origin or becoming?

That these questions are unanswerable—that they arise out of the wish to supply a *prius* to what is first in the world—this I need not now repeat : but in this connexion they have a much more serious background than elsewhere, for here they are ever anew excited by the obscure pressure of an unintelligibility, which in ordinary thinking we are apt somewhat carelessly to overlook. We lightly repeat the words 'bygones are bygones'; are we quite conscious of their gravity? The teeming Past, has it really ceased to be at all? Is it quite broken off from connexion with the world and in no way preserved for it? The history of the world, is it reduced to the infinitely thin, for ever changing, strip of light which forms the Present, wavering between a darkness of the Past, which is done with and no longer anything at all, and a darkness of the Future, which is also nothing? Even in thus expressing these questions, I am ever again yielding to that imaginative tendency, which seeks to soften the 'monstrum infandum' which they contain. For these two abysses of obscurity, however formless and empty, would still be there. They would always form an environment which in its unknown within would still afford a kind of local habitation for the not-being, into which it might have disappeared or from which it might come forth. But let any one try to dispense with these images and to banish from thought even the two voids, which limit being : he will then feel how impossible it is to get along with the naked antithesis of being and not-being, and how unconquerable is the demand to be able to think even of that which is not as some unaccountable constituent of the real.

Therefore it is that we speak of distances of the Past and of the Future, covering under this spatial image the need of letting nothing slip completely from the larger whole of reality, though it belong not to the more limited reality of the Present. For the same reason even those unanswerable questions as to the origin of Becoming had their meaning. So long as the abyss from which reality draws its continuation, and that other abyss into which it lets the precedent pass away, shut in that which is on each side, so long there may still be a certain law, valid for the whole realm of this heterogeneous system, according to the determinations of which that change takes place, which on the other hand becomes unthinkable to us, if it is a change from nothing to being and from being to nothing. Therefore, though we were obliged to give up the hopeless attempt to regard the course of events in Time merely as an appearance, which forms itself within a system of timeless reality, we yet understand the motives of the efforts which are ever being renewed to include the real process of becoming within the compass of an abiding reality. They will not, however, attain their object, unless the reality, which is greater than our thought, vouchsafes us a Perception, which, by showing us the mode of solution, at the same time persuades us of the solubility of this riddle. I abstain at present from saying more on the subject. The ground afforded by the philosophy of religion, on which efforts of this kind have commonly begun, is also that on which alone it is possible for them to be continued.

CHAPTER IV.

Of Motion.

THE perceived facts of motion are a particularly favourable subject matter for numerical calculation; but our present interest is not in the manifold results obtained by the mathematical treatment of accepted relations of proportion between intervals of space and of time; but solely in the question which phoronomic and mechanical investigations are able to disregard for their immediate purpose; the question what motion implies as taking place in the things that move.

158. Common apprehension takes motion, while it lasts, to be the traversing of an interval of space; and its result at every moment in which we conceive it as arrested to be a change of place on the part of the thing moved. We shall be obliged for the moment to invert this order of our ideas, in order to remain in agreement with our view of the merely phenomenal validity of space. Things cannot actually traverse a space which does not actually extend around them, and whose only extension is in our consciousness and for its perception; what happens is rather that just as the sum S of all the intelligible relations in which an element e at a given moment stands to all others assigns it a place p in our spatial image; exactly in the same way any change of that sum of relations S into Σ will demand the new place π for the impression which is to us the expression, image, or indication of e. Therefore change of place is the first conception to which we are led in this connexion; and from that point we do not arrive quite directly at the notion that a journey through space is essential to the change; even an apparent journey, that is, for we no longer think a real one possible.

It only follows from what was said just now that in every moment the thing's situation p or π in apparent space is determined by the then forthcoming sum S or Σ of its intelligible relations; it is still undecided in what way the transition takes place from one situation to another. However, it only happens in fairy-tales that a thing disappears in one place and suddenly reappears in another, without having traversed a path leading in space from the one place to the other; all observation of nature assumes as self-evident that the moving object remains in all successive moments an object of possible perception in some point of a straight or curved path, which unites its former and subsequent position without breach of continuity. We have no intention of doubting the validity of this assumption; it involves for us the further one, that in like manner the sum S of intelligible relations does not pass into another Σ without traversing all intermediate values that can be intercalated, without break though not necessarily with uniform speed. And this is what we really think of all variable states which are in things, as far as our modern habit of referring every event to an alteration of external relations will allow us to speak of such states at all. We do not believe that a sensation comes suddenly into being with its full intensity; nor that a body at a temperature t_1 passes to another t_2, without successively assuming all intermediate temperatures; nor that from a position of rest it acquires the velocity v, without acquiring in unbroken series all degrees of it between o and v. Thus we speak of a Law of Continuity to which we believe that all natural processes are subject; yet however familiar the idea may be to us, and however irresistible in most cases to which it is applied, still its necessity is not so self-evident to thought that all consideration of the ground and limits of its validity is wasted.

159. Of course the application of the law of Continuity is not attempted where disparateness between two extremes excludes all possibility of a path leading from one to the other in the same medium. No one conceives a musical note as changing continuously into colour; a transition between the two could only be effected by annihilation of the one and creation of the other anew; but that negation of the note would not have the import of a definite zero in a series such as could not but expand into colours on the other side of it; it would be a pure nothing, of which taken by itself nothing can come, but after which anything may follow, that we choose to say is to follow. On the other hand, in what relation to each other are Being and not-Being, the actual transition between which is put before us in every instance of change? Are we to assume that because this transition takes place it too must come to pass by continuous traversing of intermediate values between Being and not-Being? We unhesitatingly negative this suggestion, if it is to require for one and the same content a a gradation of existence such as without changing a itself to remove it by degrees from reality to unreality or *vice versa*; we could

attach no meaning to the assertion of a varying intensity of being
which should make a permanent unvarying[1] *a* partake of reality in
a greater or less degree. We should on the other hand assent to
this ; that the content of *a* itself could not disappear and could not
come into being without traversing all the values intermediate between o and *a*, which its nature made possible; the not-being of *a*
is always in the first place the being of an *a*, which is continuous
with *a* as the value immediately above or below it. Therefore the
transition from being to not-being of the same content is no continuous one, but instantaneous; still, no value *a* of a natural process
or state arises thus instantaneously out of absolute nothingness, but
always out of a reality of its own kind, whose value *a* is the proximate
increase or diminution of its own.

The case is different with the increase or decrease which property,
for instance, is exposed to in games of chance or in commerce. A
sum of money which we have staked on a cast of the dice becomes
ours or not ours in its whole amount at once, and is whichever it is
immediately in the fullest sense. It was no one's property as long as
the game was undecided; our hopes of calling it our own are a
matter of degree, and no doubt might rise *per saltus*, though not
continuously, as one die after another came to rest; but neither this
nor any other intermediate process, even if some of them were
continuous, can alter the essential state of the facts ; on the one hand
our complete right of ownership begins instantaneously on the aggregate result of the throw becoming quite certain, and so far from
existing to a less degree the moment before, had then no existence at
all. On the other hand, this suddenly created right applies at once
to the whole sum in question, without extending by degrees over more
and more of it. In this instance and in innumerable similar ones
presented by human intercourse based on contract, a perfectly arbitrary ordinance has attached to an absolutely peculiar case S a consequence F of which S is not the obvious producing cause; therefore
by an equally arbitrary ordinance all the cases $s_1 s_2 s_3$ which naturally
belong to the same series as S may be made completely ineffectual ;
and all equally so, irrespective of their greater or less approximation
to the favourable condition S. Such relations can only occur in
artificial institutions, in which a covenant, quite foreign to the nature
of the thing, attaches anything we please to anything else, and at the
same time our loyalty to the covenant is the only pledge for the
execution of what was agreed on ; as it will not execute itself.

[1] [v. note on § 19, *supra*.]

In all natural processes on the contrary the S to which a result F is supposed to correspond is the actual and appropriate ground G of this consequent F; such as not only demands the result in question but brings it about by itself and unaided by any ordinance of ours; hence the cases $s_1 s_2 s_3$ which we have a right to regard as other quantitative values of the same condition S cannot be without effect, but must in like manner, produce the consequents $f_1 f_2 f_3$ proportional to their own magnitudes and of the same kind with F. Hence arises the possibility of regarding the amount of a natural phenomenon obtained under a condition F as the sum of the individual consequents produced in succession by the successive increments of the condition. But this possibility is at the same time in a certain sense a necessity. We are not here concerned with a relation of dependence, valid irrespective of time, between the ideal content of F and that of G its sufficient reason, but with the genesis of an effect F which did not exist before; so that the condition S in like manner cannot be an eternally subsisting relation, but can only be a fact which did not exist before and has now come into being.

Now, if we chose to assume that S arose all at once with its highest quantitative value, no doubt it would seem that F as the consequence of this cause could not but enter upon its reality all at once; but in fact it would not still have to enter upon its reality, for it would be in existence simultaneously with S; nothing could conceivably have the power to interpose an interval of time, vacant as in that case it would be, between cause and consequence. The same would hold good regressively; if S arose all at once, the cause of *its* reality too must have arisen all at once, and therefore, strictly speaking, have existed contemporaneously with S rather than arisen before it. Thus we find that it is impossible to regard the course of the world as a series of sudden discrete states conditioning each other without completely re-transforming it into a mere system of elements which all have their validity or existence simultaneously; quite unlike reality, the terms of which are successive because mutually exclusive. I shall not prolong this investigation; it was only meant to show that continuity of transition is not a formal predicate of still problematic validity, which we might assign to Becoming after some hesitation as true in fact; its validity is rather an indispensable presupposition without which the reality of Becoming in general is inconceivable.

160. I have now to give a somewhat different form to the ideas with which I began. In the artificial arrangements which we mentioned, the conscious deliberation of the parties to the agreement had

previously determined the result which was to follow from a particular occurrence in the future; and in the same way in all our actions the representation in our minds of an aim that is not yet realised, of a goal that has yet to be reached, may itself be present and effectual among the conditions of the activities which are set in motion to attain our purpose. We should be wrong in transferring this analogy to our present subject-matter, by choosing to regard the altered sum of relations Σ which by itself would be the cause of the quiescence of the element e at the point π, as being at the same time the cause of its seeking and finding this new place. There cannot be an inner state q of any thing such as to be for that thing the *condition* of its *being* in another particular state r. Our reflexion might anticipate with certainty that this state r would contain no reason for further change; but the thing itself could not feel that it was so until the state began, and turned out to be the condition of a more perfect or quite perfect equilibrium.

Thus in our instance; the sum Σ of a thing's relations, if it had always existed, would have corresponded to the place π; but when something new has to arise out of the transition from S to Σ, its action cannot consist in assigning to the thing a new particular place π, as one which would suit the thing better, if it once were there; it can only consist in expelling the thing from the place p where its nature and conditions no longer hold it in equilibrium. But in the real world the negation of an existing state can only be the affirmation of another; besides, there can be no such thing as want of equilibrium in general, but only between specific points in relation, and between them only with a specific degree of vivacity. Therefore, the power of negation exerted by a state which is to act as the condition of a fresh occurrence can only consist in displacing the element in question from its present intelligible relations in a specific direction, which we have still in the first place to conceive as unspatial, and with a specific intensity. The spatial phenomenon corresponding to this process would be a specific velocity with which the element departs from its place p in a specific direction, impelled therefore *a tergo* without a predetermined goal but not attracted *a fronte* by the new place π; this latter cannot act either by retaining or by impelling, till it is reached. So what takes place in the things themselves, and what we might call, of course in quite a different sense from that recognised in mechanics, the *vis viva* of their motion, is this velocity, with which in the intelligible system of realities they leave the place where they were out of equilibrium, or, to our perception, appear to leave a situation in space; what length of space they may traverse, whether with uniform or varying motion, whether in straight lines or in curves, is the result of the existing circumstances; that is, of the new positions into which they are brought by the actual motion which takes place, which positions react on that motion as modifying factors.

161. In this way we have arrived directly at the law of Persistence, the first principle of the doctrines of mechanics, according to which every element maintains its state of rest or motion unaltered as long as it does not come in contact with the modifying influence of external causes. The first part of the law, the persistence of rest, has seldom caused any difficulty; for it can hardly be urged as a serious objection that the nature of an actual element e is quite inaccessible to us and that element may contain inner reasons unknown to us for setting itself in motion. Whatever unconjecturable states the inner being of a thing may experience, still they can only set up a motion which did not exist before by beginning at a particular moment to manifest themselves as reasons for that motion. In that case they presuppose a previous history of a Becoming within the thing; but if there had once been a moment of complete rest, in which all states of things were in equilibrium with each other, and there was no velocity inherited from an antecedent process of Becoming with which they might have made their way through the position of equilibrium, such quiescence could never have given rise to a beginning of change. Our ignorance of the real nature of things only justifies us in assuming as a possibility that such a succession of states remains for a time a movement within the thing, neither conditioned by influences from without, nor capable of altering the relations of the thing to external related points; and that, as a result of this hidden labour, a reason sufficient to alter even those external relations whether to other things or to surrounding space, may be generated as a new factor at one particular moment. But even then the movement in space would not be produced out of a state of rest, but out of a hidden movement which was not of the same kind with it; as is the case with animated bodies which initiate their changes of place by independent impulse. In the first place, however, even these owe the activity within them which generates their resolutions to the stimuli of the outer world; and in the second place their resolutions can only give rise to movement in space by a precontrived connexion of several parts which are accessible to the action of the mind and under its influence move in the directions prescribed to

them by their permanent position in the plan of the organic structure and their situation at the moment in external space.

This analogy is not transferable to a solitary element, to be conceived as setting itself in motion in empty space. In animated beings the element which is charged with the unspatial work within does not set itself in motion, but only other elements with which it is in interaction; and it does so by destroying the equilibrium of the forces operative between them, and leaving the want of equilibrium which results to determine the amount and direction of the motion to be generated. The solitary element has none of these determining reasons; it could not move without taking a definite direction through the point s of empty space to the exclusion of all others; to secure this it would not be enough that the direction $e s$ should be geometrically distinct from any other; the distinction would have to be brought to the cognisance of e's inner nature, that is, s would have to act on e differently from any other point in space. But as an empty point it is in no way distinguished from all the other points; it could only be given pre-eminence before all the others by the presence of a real element occupying it. So even if we admit an abundance of inner life in every thing, still we cannot derive the initiation of a movement in space from that life, but only from external determining conditions.

Still, this is an expression which we shall do well to modify. Whatever attractive or repulsive force we conceive to proceed from s; it cannot determine e to motion by reason of its own starting from s, but only by reason of its arrival at e, or rather through the alteration which it effects in the inner states of e. It is therefore, in fact, this state of inner want of equilibrium which hinders e from remaining at rest; only this state cannot have arisen in a way to determine the line of motion, unless e is conceived as part of a universe which by the configuration of its other parts at any moment helps to determine that of e's inner being.

162. The other part of the law, the continuance of every motion that has once begun, remains a paradox even when we are convinced of it. If we separate the requirements which we may attempt to satisfy; in the first place the certainty of the law, or its validity in point of fact, is vouched for both by the results of experiment and by its place in the system of science. The better we succeed in excluding the resistances we are aware of as interfering with a motion that has been imparted, the longer and more uniformly it continues; we rightly conclude that it would continue unvaryingly for ever, if it

were permanently left to itself without any counteraction. And on the other hand, however a motion that is going on may be modified at every moment by the influence of fresh conditions, still we know that our only way of arriving at the actual process in calculation is to estimate the velocity attained in every moment as continuing, in order to combine it with the effect of the next succeeding force.

If we go on to ask whether this doctrine being certain in point of fact has also any justification as conceivable and rational, we can at least see the futility of the assumptions which prevailed in antiquity, when, under the influence of inappropriate analogies, men held that the gradual slackening of all motion was the behaviour more naturally to be expected. If they had said that all motion is wholly extinguished in the very moment in which the condition that produces it ceases to act, the idea put forward would at least have been an intelligible one in itself; but by treating the motion as becoming gradually weaker they actually admitted the law of persistence for as much of the motion as at any given moment had not disappeared. Still, the more definitely we assume the ordinary ideas of motion, the more remarkable does the law of persistence appear; if motion is nothing but an alteration of external relations by which the inner being of the moving object is in no way affected, and which in no way proceeds from any impulse belonging to that object, why should such an alteration continue when the condition which compelled it has ceased?

We look in vain for more general principles which might decide the question. I said above in the Logic (§ 261) that the law 'Cessante causa cessat effectus' cannot safely be held to mean more than that after the cessation of a cause we do not find the effect which the cause would have had if it had continued; but that it remains doubtful whether the effect which is already produced requires a preserving cause for its continuance. It appeared to me then that every state which had in reality once been produced would continue to exist, if it were neither in contradiction with the nature of the subject to which it occurs, nor with the totality of the conditions under which that subject stands towards other things. But even this formula is useless; for there is still this very question, whether motion which has been generated in a thing not by its own nature, but only by means of external conditions, is to count among the states which are conceivable as going on to infinity without contradicting that nature and those relations. On the other hand, it has been suggested that the reason for the persistence of states of motion in things must in every case lie in the actual nature of the things; I am convinced that no explanation is to be found in this direction; we should only be obliged, after executing some useless circuits, to assert the principle of Persistence about some motion or other within real things, with no more success in deducing it than if we had taken the shorter way of granting its validity at once for motion in space. Instead of a direct demonstration of the law, I believe that nothing more is possible than an indirect treatment, which I subjoin.

163. Let C_1 be the condition which sets in motion an element e with definite velocity and direction so as to traverse the distance dx in the time dt. Let us suppose that the activity and effect of C_1 continue through the duration of dt, but cease when at the end of that interval e has traversed the short distance dx, has thus changed its position, and has for this reason come under the influence of the new condition C_2. This again, if operative during an equal time dt, will make another equal journey dx possible for e, and will cease when e has traversed it. It is plain that as long as we treat dx as a real distance however small, the element e, acted upon by this series of successively annihilated influences, will pass through a finite length of space in the time t.

But our assumptions, as we made them just now, have to be modified. C_1 must cease to act not *when*, but *before*, e has arrived at the extreme point of the first distance dx; by the time e has accomplished the smallest portion of that short distance its position would be changed, and would no longer be that which acted upon it as the motive impulse C_1; if in spite of this we suppose e to traverse the whole distance dx in consequence of the impulse C_1, the only possible reason for its doing so will be the postulated validity of the law of persistence; the motion produced by C_1 will have lasted after C_1 itself had ceased to exist or act. But if we do not regard this law as valid, then not even the smallest portion of the short journey in question will really be achieved; the moment that C_1 so much as threatens to change the place of e, and so transform itself into C_2, the determining force with which it purposed to produce this result must disappear at once, and the matter will never get as far as the entrance into action of the fresh condition C_2 which could maintain the motion; for the motion never begins. If y is a function of x, there may be a finite integral of the formula $y\,dx$ as long as we regard dx as a real magnitude; and the calculation would be more exact as this interval is less for which we take a value of y as constant; but the whole integral becomes o, if we regard dx as vanishing entirely.

In the present case we should apply this common mode of representation as follows ; if y is the velocity generated by C_1, or existing along with some initial value of x, according to the law of Persistence this y will hold good for the whole interval for which the integral is required. The succeeding condition C_2 will be partly satisfied, in respect of what it has in common with C_1, by the motion y which already takes place in consequence of C_1; only that in which C_2 deviates from C_1 is a fresh active condition whose consequence dy, a positive or negative increment of velocity, continues in like manner from that moment through the entire interval of the integration. It is the summation of the initial value y, and of these continuously succeeding increases or decreases, that gives the total of the result obtained between the limits in question.

The tendency of all this is obvious ; of course it cannot tell us how, strictly speaking, it comes to pass that motion when once generated maintains itself ; but still we can see that the law of Persistence is not a marvellous novelty of which it might be questioned whether it would or would not be true of a given natural motion ; in fact its truth is an integral part of our idea of motion. Either there is no such thing as motion, or, if and as there is, it necessarily obeys the law of Persistence, and could not come to pass at all if really and strictly the effect produced had to end with the cause that produced it. For the law holds good not merely as applied to motion, but with this more general significance. No condition can act without having a result which is, speaking generally, a modification of the state of things that contained the stimulus or impulse to action ; and therefore apart from the principle of Persistence no result could ever be reached ; the excitation would begin to be inactive at the moment in which it began to act.

164. If two elements change their distance from one another in space, real motion must in any case have occurred ; but it remains doubtful which of the two moved or whether both did so, and in the latter case the same new position may have been brought about either by opposite motions of the two, or by motions in the same direction but of different amount. This possibility of interpreting what to our perception is the same result by different constructions continues to exist most obviously as long as we look exclusively to the reciprocal relations of two elements without regard to their common environment ; nor does it cease when we consider the latter also ; only in that case the possible constructions will not all seem equally appropriate. We should prefer to regard as in motion the element which is alone in altering its position relatively to many which retain their reciprocal situations; still there is nothing to prevent us from conceiving that one as at rest, and the whole system of the numerous others as moving in the opposite direction. I need not pursue the advantages which we gain in practice from this plasticity of our ideas; but the casuistic difficulties which metaphysic attaches to this Relativity of motion, seem to me to rest on mere misapprehensions.

Let us conceive to begin with a solitary element in a perfectly void world of space; is there any meaning in saying that it moves, and that in a particular direction? Again, in what can its motion consist, seeing that the element cannot by moving alter its relations to related points, as there are none, while we should not even be able to distinguish the direction in which it would move from the other directions in which it would not move? I think we must answer without hesitation: as long as we adhere to ordinary ideas by speaking of real space, and by setting down the traversing of it under whatever condition as a possible occurrence, there is no reason against regarding the motion of this solitary element as one which actually takes place, and none therefore against recognising so-called 'absolute motion' as a reality. If perfectly empty space is wholly devoid of related points for purposes of comparison, even of distinctions between the quarters of the heavens, still this does not plunge the motion itself into any such ambiguity or indefiniteness of nature as to prohibit it from actually occurring; only we lose all possibility of designating what occurs. However little we may be in a position to distinguish intelligibly between the point s which is in the direction of the moving object e and other points which are not, still it would be distinct from all others as long as we regard as real the extension of space which by its definite position towards all other points it helps to constitute. And however little we could distinguish the direction $e\,s$ in which e moves from other directions, before we had a given line in a particular plane which would define the position of $e\,s$ by help of the angle formed between them, still $e\,s$ would be in itself a perfectly definite direction; for such an angle would not be capable of being ever ascertained and determined, unless the position of $e\,s$ were already unambiguously fixed at the moment when we applied our standard of comparison in order to define it.

So the assertion that a motion is real is certainly not dependent for admissibility on the implication of a change of relations in which the real element in motion stands to others like it. Indeed, during every moment for which we conceive a previously attained velocity to continue according to the law of persistence, the moving element moves with precisely the kind of reality which is held in the above case to be of doubtful possibility.   True, in this case we are in a position to assign the direction of the motion, within a world in which it took place, by relations to other realities and to the space which they divide and indicate.   Still all these relations in this case only enter into consideration as interfering or modifying causes ; the persistent velocity of the element, which we must not leave out of our calculation, is in itself, in fact, simply such a motion of a solitary element that takes no account of anything else.   Thus, so far from being a doubtful case, it is truer to say that absolute motion is an occurrence which is really contained in all motion that takes place, only latent under other accretions.   On the other hand, if we intended to acknowledge no motion but what is relative, in what way should we suppose it to take place ?   If we understand by it one which involves a real and assignable change of relative position on the part of the elements, how can this change have arisen unless one or several of the elements in order to approach or to separate from each other had actually traversed the lengths of space which form the interval that distinguishes their new place from their old ?   But suppose we understood by relative motion one which was merely apparent, in which the real distances between pairs of elements underwent no change.   Still it is clear that such an appearance could not itself be produced apart from motion really occurring somewhere, such that the subject to whom the appearance is presented changes its position towards one or more of the elements in question.

165. Our conclusion would naturally be just the same about the other case which is often adduced ; the rotation of a solitary sphere in empty space.   No doubt it would be absolutely undefinable till a given system of co-ordinates should determine directions of axes, with which its axis could be compared.   But there is also no doubt that the specific direction of the rotation is not *made* by these axes which serve to designate it ; the rotation must begin by being thoroughly definite in itself, and different from all others, that it may be capable of being unambiguously reduced to a system of co-ordinates.   All that such a reduction is wanted for, is to make it definable ; but what happens happens, whether we can define it or not ; of course a capacity for being known demands plenty of auxiliary conditions, whose absence no one would conceive as destroying the possibility of the occurrence itself.   Suppose we had the clearest possible system of co-ordinates at our disposal, and saw a sphere in a particular place of that system ; still we should fail to ascertain whether it was turning or not, or in what direction, if it consisted of perfectly similar parts *a* distinguished to our eye neither by colouring nor by variable reflexions of light. At every moment we should observe the similar appearance *a* in the same point of space; we should have no means of distinguishing one instance of the impression from another; are we to infer from this that a sphere of uniform colour cannot turn round in space, but only a chequered one; and even this only with a limited velocity, for fear the different impressions of colour should blend into an undistinguishable mixture to our eyes?

Hence we may be sure that such absolute rotation about an axis is perfectly conceivable; in fact it is not in the least a problematic case, but is continually going on. We have no proof of any action of the heaven of the fixed stars on the motions within our planetary system, nor is it required to explain those motions; both it and the influences of the other planets can never claim to be regarded as more than disturbing causes when we are considering the revolution of the earth and sun round their common centre of gravity; these two bodies therefore actually move as a solitary pair in universal space. And again, the earth, by itself, continues its existing rotation about its axis without help or hindrance in it from its relation to the sun. So in fact, rotation of this kind, the possibility of which is doubted, really occurs, only concealed by accessory circumstances which have no influence on it; indeed the instance of a spinning top which maintains its plane of rotation and opposes resistance to any change of it, presents it strikingly to our senses. The idea of the reality of an infinite empty space and the other of an absolute motion of real elements in space are thus most naturally united and are equally justifiable; nor will it ever be feasible to substitute for this mode of representation another which could form as clear a picture in the mind.

166. As we have surrendered the former of these ideas, we have now to reconcile the latter with the contrary notion which we adopt. Our observations up to this point could not do more than prove that the absolute motion of an element in empty space was conceivable as a process already in action; what still appeared impossible was its beginning and the choice of a direction and velocity out of the infinite number of equally possible ones. This alone would give no decisive argument against an existing space and an actual motion through it; whatever inner development we choose to substitute for this apparent state of facts as the real and true occurrence, the impossibility of a first beginning will always recur. We should have to be satisfied with setting down the fact of motion with its direction and

velocity along with the other original realities which we have to look
on as simply given, and which we cannot deduce from a yet undecided choice between different possibilities. In fact, every permanent
property of things, the degree of every force, and all physical constants whatever, might give rise in infinite recurrence to the same
question; why are they of this specific amount and no other, out of
the innumerable amounts conceivable?

I need only mention in passing once more, that the unavoidable
relativity of all our designations of such constants is not to seduce us
into the mistake of considering the constants themselves as indefinite.
The units to which we refer the measurement of a certain force g,
and in which we express it, are arbitrarily chosen; but after they are
chosen it results from the peculiar and definite intensity of the force
that according to this standard its measurement must be g and cannot be ng. A semicircular movement which goes from right to left
when looked at from the zenith, will go from left to right when
looked at from the nadir of its axis. This does not prove that its
direction is only determined relatively to our position, but just the
reverse; that it is definite in itself independently of that position, and
therefore to suit the observer's different points of view must be expressed by different definitions relating to those points.

Undoubtedly therefore, the real world is full of such constants,
perfectly definite, yet taken by themselves incapable of being designated; they must be set down as definite even while they vary in
value according to a law, under varying conditions; for, to adhere to
the example of the force above mentioned, its intensity under a new
and definite condition will always be measured by a function of g,
and never by the same function of ng. It is, as has been observed
more than once already, only by application of our movable thought,
with its comparisons of different real things, that there can arise
either the idea of countless possibilities, which might equally well
have existed but do not; or the strange habit of looking on what is
real as existent to some extent before it exists, and as then proceeding to acquire complete existence by a selection from among possibilities. Therefore, if we recognise that the first genesis of real
things is altogether incapable of being brought before our minds by
us, though we find their continuance intelligible, we may accept
absolute motion in space and its direction as one of the immemorial
data from which our further considerations must start.

187. But it cannot be denied that one thorny question is left.
We admit all constants which, speaking generally, form the essence

of the thing whose further behaviour is to be accounted for; but here we have on one side an empty space which is absolutely indifferent to all real things and could exist without them, and on the other side a world of real things which, even supposing it to seem to us in need of a spatial extension of its own, is yet expressly conceived as wholly indifferent to the place which it occupies, and therefore just as indifferent. to the change of that place, and incapable of determining by its own resources the direction of any motion to be initiated, although actually engaged in one motion out of infinitely many. Sensuous perception may find no difficulty in such a fundamental incoherence between determinations which nevertheless do cohere together; but thought must pronounce it quite incredible; for the endeavours of thought will always be directed to deriving the causes which determine the destiny of existing things from the nature of the things themselves. To say that motion is the natural state of things is utterly worthless as a philosophical idea; nothing is natural to a thing but to be what it is; states of it may be called matter of fact, but cannot be called natural; they must always have their conditions either in the things or without them. Each particular thing, on the other hand, cannot be in motion merely in general, but its motion must have a certain direction and velocity; further, the whole assumption of original motion is only of use by ascribing different directions and velocities to different elements; but as, at the same time, it persists in regarding the elements as uniform, it is all the less able to conceive such differences as natural states, and is compelled to treat them simply as matter of fact, and indeed as alien to the nature of the thing.

In reality it was this causelessness that was the principal obstacle to the recognition of absolute motion; for what, strictly speaking, does happen if the advancing element *e* traverses one empty space-point after another, without being in itself at all different when it reaches the third from what it was when in the first or second? or, fruitless as the transition is, without so much as receiving an indication of the fact of its fruitless occurrence: finally, without making it possible for even an observer from without, were it only by help of relations to other objects, so much as to give a bare designation of the supposed proceeding? And are we to suppose that a process so unreal as this, a becoming which brings nothing to pass, must of necessity last for ever when once stimulated to action, though to begin with incapable of originating without external stimulus? These inconceivabilities have at all times led to some rebellion against the view adopted by mechanics (though it yields so clear a mental picture and is so indispensable in practice), which makes the moving element merely the substratum of the motion, without any peculiar nature which is affected by the motion or generates it by being affected. It is objected that motion cannot consist in the mere change of external relations, but must in every moment be a true inner state of the moving body in which it is other than it would be in a moment of rest or of different movement. Then can the view which concedes to space no more than a phenomenal validity offer anything satisfactory by way of a resolution of this doubt?

168. Let us suppose a real element e to be in inner states which we will sum up in the expression p. Then the question for us could not be whether e_p would produce a motion in space, but only whether e_p could form the ground of an apparent motion of e within space for a consciousness which should possess the perception of such space. We will begin by making the same assumption as we made in the discussion of time[1]; that the consciousness in question is an absolutely immediate knowledge of everything, including therefore e_p; and is not based on the acquisition of impressions by means of any effect produced by e_p on the knowing subject; and therefore does not compel us to attribute to this subject any specific and assignable relation to e_p.

Then, I think, we may consistently conclude as follows. Such a consciousness has no more ground for ascribing a particular spot in the space of which it has a mental picture, or motion in a particular direction, to the e_p of which it is aware, than e_p has power in an actual empty space to prefer one place to another as its abode, or one direction to others for its motion which has to be initiated. If we want to bring before ourselves in sensuous form what appears the reasonable result under such imaginary conditions,—we can only think of a musical note, to which we do no doubt ascribe reality in space, but localise it most imperfectly, and then only in respect of its origin: or we must think of a succession of notes, which we do not exactly take to sound outside space, but which still remains a purely intensive succession, and has definite direction only in the realm of sound, and not in space.

I should not adduce such utterly fictitious circumstances, were they not about on a par with what is usually put forward by popular accounts of the Kantian view; a ready-made innate perception of space, without any definite relations between the subject which has it, and the objects which that subject has to apprehend under it. But

---

[1] [Cp. p. 256 sup.]

in reality we find the consciousness in question invariably attached to a definite individual being ϵ, and in place of immediate knowledge we find a cognition which is always confined to the operations of e on ϵ. Besides this postulate, however, something more is required for the genesis of phenomena of motion in the experience of ϵ. Whatever the inner state p within ϵ may be, and in whatever way it may alter into q and its effect π on ϵ into κ, still, for an ϵ that is simple and undifferentiated in itself all this could only be the ground for a perception of successive contents, not for their localisation in space and for their apparent motion. More is required than even a plurality of elements, ϵ_p, ϵ_q, ϵ_r, in different states of excitation, operating simultaneously on a simple ϵ. No doubt, the felt differences of their action might furnish ϵ, supposing it able and obliged to apprehend them by spatial perception, with a clue to the determination of the *relative* positions which their images would have to occupy in space. And alterations of their action would then lead to the perception of the relative motions by which these images changed their apparent places as compared with each other. But the whole of the collective mental picture which had thus arisen, whether at rest or in motion, would still be without any definite situation relatively to the subject which perceived it. The complete homogeneousness of this latter would make it analogous to a uniform sphere, so that it could turn round within the multiplicity which it pictured to itself without experiencing, in doing so, any alteration in the actions to which it is subjected, or any, therefore, in its own perceptions. To make one arrangement of phenomena $a\,b\,c$ distinguishable from another arrangement $c\,b\,a$ or a downward motion to the right from its counterpart in an upward motion to the left, it is essential that the directions in question should be unmistakably distinguished in the space-image for ϵ itself by a qualitative mark; then ϵ will be able to refer every action or modification of an element to that direction to which it belongs according to the qualitative nature of the impression made or of the modification of that impression.

The result of the argument comes to this, after the insertion of some intermediate ideas which I reserve for the psychology. It is true that a simple atom, endowed with a perception of space, might find occasion in the qualitative differences of the impressions received from innumerable others to project a spatial picture of phenomena with a definite configuration of its own. But for this same atom there would be no meaning in the question what place or direction in absolute space such images or their motions occupied or pursued.

What could be meant by such an expression in general would not become intelligible to it till it had ceased to be an isolated atom endowed with knowledge, and had come into permanent union with a plurality of other elements, we may say at once, with an Organism ; such that its systematic fabric, though still to be conceived as itself unspatial, should supply polar contrasts between the qualitatively definite impressions conducted from its different limbs to the conscious centre. The directions along which consciousness distributes these impressions as they reach it, in its picture of space, and in which it disposes such images as appear to it of its own bodily organism, would alone furnish consciousness with a primary and unambiguous system of co-ordinates, to which further all impressions would have to be reduced which might arise from variable intercourse with other elements. *ε* the subject of perception may then gain further experiences in this intercourse, such as prove to it that permanent relations exist between the other elements towards the totality of which *ε* can give itself and its body varying positions ; and then the inducement arises to look in the spatially presented picture of the outer world for a fresh system of co-ordinates belonging to that world, to which both its permanent relations and *ε*'s varying positions shall be most readily reducible.

But it will again be essential to any such fresh system that it should be defined by a qualitative distinction between the perceptions which are assigned to the opposite extremities of one of its axes[1] ; though on the other hand what place this whole system with its inner articulation holds in absolute space, or in what direction of absolute space this or that of its axes extends, are questions which on our view would cease to have any assignable meaning at all. For this is just what does not exist, an absolute space in which it is possible for the subject of spatial perception with all the objects of its perception, to be contained over again, and occupy a place here or there. Space only exists within such subjects, as a mental image for them ; and is so articulated for them by the qualitative difference of their impressions, that they are able to assign the appearances of other elements their definite places in it ; and finally, it is the thorough coherence of all reality which brings about that each of these subjects also presents itself in the space pictured by every other in a station appropriate to the totality of its relations with the rest of what the world contains; and thus it happens that each of them can regard the space which is

[1] [This alludes to the distinction of ' up ' and ' down ' furnished by the feeling of resistance to the force of gravity. Cp. § 287.]

in its own perception as a stage common to all, on which it can itself
meet with other percipient subjects than itself, and can be in relations
which agree with theirs, to yet another set of subjects.

109. But it is still necessary to return expressly to the two cases
given above, in order to insist on the points in them which remain
obscure. We saw that they present no special difficulties on the
common view ; if we have once decided to accept empty space as a
real extension, and motion as an actual passage through it, then
rectilinear progress and rotation of a solitary element might be
accepted into the bargain as processes no less real although undefinable. But we should now have to substitute for both of them an
internal condition of ϵ, say p, whose action π on an ϵ endowed with
perception produces in this latter the spectacle of a motion of ϵ through
the space mentally represented by ϵ. Now according to the common
view the absolute motion of ϵ, whether progressive or rotatory, though
it really took place, yet was undefinable. The reason was that the
observing consciousness which had to define it was treated only as an
omnipresent immediate knowledge, possessing itself no peculiar relation
with its object which helped to define its perception ; therefore the designation of the actual occurrence would have been effected in this case
by co-ordinates independent of the observer ; and as none such were
found in empty space the problem of designating this occurrence remained insoluble, though its reality was not thereby made less real.

For us the case is different. What we want to explain is not a real
movement outside us, but the semblance of one, which does not take
place outside, within us ; therefore for us the presence of the observing
subject ϵ for whom the semblance is supposed to be forthcoming, and
the definite relation of ϵ to the external efficient cause of this semblance,
is not merely the condition of a possible designation and definition of
the apparent motion, but is at the same time the condition of its
occurrence, as apparent. So we too, within the phenomenal world
which we represent to our minds, may accept the progress or rotation
of a solitary ϵ for a real occurrence, if we do not forget to include
ourselves in the conception as the observer ϵ, in whose mind alone
there can be a semblance at all. For then there must in any case
be a reaction and a varying one between ϵ and ϵ as elements in one
and the same world, and it is the way in which the action of ϵ on us
changes from π to κ while ϵ is itself undergoing an inner modification,
that will define the direction of the apparent motion in question with
reference to some system of co-ordinates with which we must imagine
the space-perceiving ϵ to be equipped from the first if its universal

perception is to admit of any method of application to particular things.

170. Still I feel that these doctrines are inadequate, as strongly as I am persuaded that they are correct; they leave in obscurity a particular point on which I will not pretend to see more clearly than others. It concerns that transition of *e* from one inner state to another which in acting on us produces for us the semblance of a motion of *e*. It must of course be conceived as going on at times when it does not act on us, or before it begins to act on us; and at those times it can be nothing but an inner unspatial occurrence which has a capacity of appearing at some later time as motion in space by means of that action upon us which it is for the moment without. Here we are obstructed by an inconvenience of our doctrine which I regret, but cannot remove; we have no life-like idea of inner states of things. We are forced to assume them in order to give a possibility of fulfilling certain postulates of cognition which were discussed above; but we cannot portray them; and any-one who absolutely scorns to conceive them as even analogous to the mental states which we experience in ourselves, has no possible image or illustration of the constitution by help of which they accomplish this fulfilment of essential requirements.

This lack of pictorial realisation would not in itself be a hindrance to a metaphysical enquiry; but it becomes one in this particular case where we are dealing with the conceivability of the motions in question. When the element *e* traverses an apparent path in our perception it is true that the beginning of the series of inner states, whose successive action on us causes this phenomenon, must be looked for not in *e* itself, but in the influence of other elements; but still the undeniable validity of the law of persistence compels us to the assumption that an impulse to motion when it has once arisen in *e* becomes to our perception independently of any further influences the cause · of an apparent change of place of the sense-image, with uniform continuance. The same assumption is forced on us by another instance, that of two similar elements *e* which unceasingly traverse the same circle, being at the opposite extremities of its diameter.

We can easily employ the ordinary ideas of mechanics to help out our view so far as to assume an inner reaction between the two elements, which, if left to itself, would shorten the distance between their sense-images in our perception; then there would still remain to be explained the rectilinear tangential motion, which, continuing in consequence of the Law of Persistence, would counteract this attraction to the amount needed to form the phenomenal circle.

Now what inner constitution can we conceive e to possess, capable of producing in our eyes the phenomenon of this inertia of motion? Considered as a quiescent state it could never condition anything but a permanent station of e in our space; considered as a process it still ought not to change e, into e, in such a way that the new momentary state q should remove the reason for the continuance of the same process which took place during e,; we should have to suppose an event that never ceases occurring, like a river that flows on ever the same without stopping, or an unresting endeavour, a process which the result that it generates neither hinders nor prohibits from continuing to produce it afresh. This conception appears extraordinary enough, and justifies a mistrust which objects to admitting it before it is proved by an example to signify something that does happen, and not to be a mere creation of the brain.

It is certainly my belief, though I will not attempt a more definite proof, that mental life would present instances of such a self-perpetuating process, which would correspond in their own way to the idea, extraordinary as it is though not foreign to mechanics, of a *state* of motion. Perhaps there may even be someone who cares to devote himself to pursuing these thoughts further; after we have been so long occupied with the unattainable purpose of reducing all true occurrence to mere change of external relations between substrata which are in themselves unmoved, even fashion might require a transition to an attempt at a comprehensive system of mechanics of inner states; then we should perhaps find out what species are admitted as possible or excluded as impossible by this conception of a *state* as such, which has hitherto been as a rule rather carelessly handled. Till then, our notions on the subject have not the clearness that might be desired, and the law of persistence remains a paradox for us as for others; I will only add that it presents no more enigmas on our view than on the common one. The fact of such an eternal continuance of one and the same process is actually admitted by mechanics; the strangeness of the fact is what it ignores by help of the convenient expression which I have quoted, '*State* of motion.'

171. I may expect to be met with the question whether it would not be more advisable to abstain from such fruitless considerations; it is not, however, merely the peculiarity of the presuppositions that we happen to have made which occasions them. Poisson, in § 112 of his 'Mechanics,' in speaking of uniform motion according to the law of persistence, observes; 'the space traversed in a unit of time

is only the measure of velocity, not the velocity itself; the velocity of a material point which is in motion, is something which resides in that point, moves it, and distinguishes it from a material point which is at rest;' and he adds that it is incapable of detailed explanation. I am better pleased that the illustrious teacher should have expressed himself somewhat cavalierly on a difficult problem, the solution of which was not demanded by his immediate purpose, than if he had philosophised about it out of season. He, however, is not open to the charge of taking a mere formula of measurement furnished by our comparing cognition for a reality in things; on the contrary, he justly censures the common notion as overlooking a reality to which that formula should only serve as measure. Velocity and acceleration are not merely the first and second differential quotients of space and time; in that case they would only have a real value in as far as a length of space was actually traversed; but it is not only within an infinitely short distance, but in every indivisible moment that the moving body is distinguished from one not moving; although if the time is zero, that which distinguishes them has no opportunity to make itself cognisable by the body describing a path in space and by the ratio of that interval to the time expended.

It is impossible to deny this while we speak of the law of persistence. If an element in motion, that *passes through* a point, were even in the unextended moment of passing precisely like another which merely *is* in the point, its condition of rest would according to that law last for ever. Therefore, we shall not indeed conclude with Zeno that the flying arrow is always at rest, because it is at rest in *every* point of its course. But we shall maintain that it would have to remain at rest for ever if it were at rest in a single point, and that so it would never be able to reach the other places in which, according to Zeno's sophism (which rather forgets itself at this point), the same state of rest is to be assigned to it. Now if that in which this essence of motion consists cannot exist in an indivisible moment as velocity, i.e. as a relation of space and time, but nevertheless must exist with full reality in such a moment, then of course nothing remains but to regard it as an inner state or impulse of the moving object which is in existence prior to its result. We may admit too that this impulse moves the element; for however it may itself have arisen by the action of external forces, still Poisson and we were only speaking of the impulse which has arisen, in as far as it is for the future the cause of the persistence of the motion.

172. The parallelogram of motions teaches us the result of the

meeting of two impulses in the same movable material point. Its validity is so certain that all proofs which only aim at establishing its certainty have merely logical interest; we should here be exclusively concerned with any which might adduce at the same time the meaning of the doctrine, or the *ratio legis* which finds in this proposition its mathematical expression as applicable to facts.

If a subject S has a predicate p attributed to it under a condition π this same S as determined by π could possess no other predicate q; for every condition can be the ground of one consequent only and of no other. Thus, the two propositions S_π is p, and S_κ is q, each of which may be correct in itself, speak of two different cases or two different subjects; mere logical consideration gives no determining principle to decide for what predicate ground would be given by the coexistence of the two conditions π and κ in the same case or in the same subject. The real world is constantly presenting this problem; different conditions may seize upon an element, which they can determine, not merely in succession, but at once; and as long as no special presuppositions are made no one of them can be postponed or preferred to the others. Just as little can the conflict of their claims remain undecided; in every case a result must be generated which is determined by the two conditions together.

I thought this characteristic of the real world worth a few words of express notice; it is generally presupposed as self-evident and attention turned at once to determining the form of such a result. If we are to attempt this in an absolutely general way, we shall first have to reflect on the possibility that the conditioning force of the two may depend on their priority in time, and consequently there may be a different result if κ follows π and if π follows κ. In the case of motion this doubt is solved by the law of persistence. The element moved by the condition π is at every moment in the exact state of motion into which it was thrown at the moment in which the motion was first imparted. Therefore at whatever moment the second condition κ begins to act all the relations are just the same as if π was only beginning to exert its influence simultaneously with κ, and so the order of the two conditions in time is indifferent. But even so it remains doubtful whether κ will endeavour to give an element e acted on at the same time by the condition π the same new movement q which it would have imparted to it in the absence of π. If we conceived p as the motion produced first by π alone, then the motion resulting from the two conditions might possibly be not merely $p + q$ or $p\,q$, but also $(p + q)(1 \pm \delta)$ or $p\,q\,(1 \pm \delta)$; if, first, q had been produced alone by κ, the addition of π would turn it into $q\,p\,(1 \pm \epsilon)$ or $(p+q)$ $(1 \pm \epsilon)$. It is obviously indifferent which of the two formulæ we choose; the only function of the mathematical symbol is to designate p and q as absolutely equal in rank; the result which is produced is strictly speaking neither sum nor product. Now as the order in time of the conditions is indifferent, $p\,q\,(1 \pm \delta)$ must $= p\,q\,(1 \pm \epsilon)$; and this equation is satisfied by either of two assumptions; that $\delta = \epsilon$, or that both $= 0$. I do not think it possible to decide on general grounds for one or other of these assumptions with reference to the joint action of any two conceivable conditions however constituted; on the contrary, I am convinced that the first has its sphere of application as well as the other; therefore though it is a familiar fact that the second holds good for motions and their combinations, I can only regard it, in its place in my treatment of the subject, as a fact of the real world, such as is easily interpreted when established on other evidence, but such as in default of that confirmation could not be reliably proved *a priori.* The meaning of this fact then is, that n simultaneous motions produce in the element e in a unit of time the same change of place which they would have produced in n units of time if they had acted on e successively, each beginning at the place which e had already reached. It is unnecessary to observe how the final place of e and also, as the same relations hold good for every infinitely small portion of time, the path of e as well, determine themselves by this principle in accordance with the parallelogram of forces.

This behaviour of things is akin in significance to the law of persistence; just as by the latter a motion once in existence is never lost if left to itself, so too in its composition with others none of it is lost, in so far as the collective result completely includes the result of each separate motion. Only, the process by which this collective consequence is attained must be single at every moment and cannot contain the multiplicity of impulses as a persistent multiplicity; it is the resultant, which blends them. The expression $p+q$ would correspond to the former idea by indicating the two motions which may be allowed to succeed one another with a view to obtaining the same result; the other, $p\,q$, would express the latter, the process by which this result is reached; namely that the motion in the direction p would be continuously displaced parallel to itself through the condition q.

173. In declining the problem of a deduction of the law of the parallelogram I expressly said that I only did so in its place in my discussion. But if we make the ordinary assumptions of mechanics I believe that the restriction of it to mere empirical validity is quite

baseless. I find it maintained that all attempts to prove it as a necessary truth of the understanding have to meet the argument that there is nothing in our reason to compel us to assume precisely this arrangement to exist in nature. There would be, it is said, no contradiction to the nature of our reason in such an assumption as that the physical or chemical quality of the material points and the mode of generation of the forces brought into play had an influence on the amount and direction of the resultant. For instance, forces of electric origin might influence degree and direction of the resultant differently from forces of gravitation, or attractive forces differently from repulsive ; it is admitted that this is not the case, but alleged that it is only experience that tells us so. As against this argument I must remind my readers that the general science of mechanics treats of forces only in as far as they are causes of perfectly homogeneous motions, distinguished by nothing but direction, velocity, and intensity, and not with reference to other and secret properties. The law of the parallelogram applies directly to none but the above motions, and to them only as already imparted and so brought under the uniform law of persistence ; and this application excludes all reference to the history of what precedes their origin. In the same way the movable elements are taken to be simply and solely substrata of motion, and perfectly indifferent to it. That component, with respect to which they are purely homogeneous masses possessing a quantitatively measurable influence on the course of their motions only by the resistance of inertia, is conceived as standing out separately to begin with from the rest of their qualitative nature.

Granting these postulates our reason has no longer a number of possible cases before it ; on the contrary, it is certain that two motions which are nothing but changes of place, and have no force behind them which can influence their persistence, can produce no more than their sum if they are similar, or their difference if they are opposed. This determines the maximum and minimum of the change, because no increase or diminution of what exists can take place without a reason. But supposing that there are other relations between two motions besides complete agreement and complete opposition, it is equally certain that if the nature of the case admits of both impulses being obeyed at once both will have to be satisfied as far as it admits ; for again, nothing can be subtracted from their complete satisfaction unless the new phenomenon of subtraction has a compelling cause that hinders the complete continuance of what already exists. Now it is the nature of space which in virtue of the infinite variety of directions possible in it admits of these relations of imperfect opposition between motions. And this same nature of space, by permitting the different directions to be combined, and compensated by each other, makes possible the complete and simultaneous fulfilment of the different impulses; and therefore the determination of the result in accordance with the law of the parallelogram is of course a necessity and there is no alternative which can be treated as equally possible. This was the proper occasion to notice the objection just refuted; for as long as the question was how the inner movements of things modify each other it was possible for the total result of two simultaneous impulses to be an increase or diminution of the phenomenon in question dependent on the qualitative peculiarities of the impulse itself. But when it comes to be decided that their results in the e which is acted on are nothing but two homogeneous motions, and when these motions come to be regarded as already produced or as communicated to e, then the further composition of the motions can only result according to a simple law that regards what they are at the moment and not the utterly extinct history of their past.

CHAPTER V.

The theoretical construction of Materiality.

174. THE elements of Real Existence have hitherto been spoken of only in so far as regards the positions occupied by them in Space and the changes in those positions; as regards the form and nature of that which takes up and changes its positions, we have been silent. This latter question, which at the point we have now reached we shall be called on to consider, is usually stated as the theoretical construction of Matter. If I were to give this name to the following investigations, it could only be with the reservation that I understand the philosophical problem which is commonly designated by it in a changed sense. For this Matter, the construction of which is required, is not a ready-made fact open to observation. Real Existence—as known to us in Space—consists merely of an indefinite number of individual objects variously distinguished by inherent differences in their sensible qualities. At the same time, however, we learn by observation and comparison of these objects to perceive a number of common properties in which they all, to a greater or less extent, participate. They are all alike extended in space; all alike show a certain tendency to maintain their positions against any attempt to change them; they all oppose a certain *vis inertiae* to any efforts to move them. These common properties of things, which are consistent also with the most manifold differences, may be classed together under the generic name Materiality, and Matter would then be a general term standing for anything which participated, to whatever extent, in the above-mentioned modes of behaviour. The problem of philosophy would be to determine what is the subject of which these are the attributes, and under what conditions there arise in their successive grades the forms of existence and of action which we comprehend under the name of 'Materiality.'

A general consideration of these questions must have regard to two possible modes of answering them. Conceivably the Real Existence which appears to us under forms of action so homogeneous, may be not merely of like, but of quite identical nature throughout, and may owe the differences which characterise it to subsequent accessory conditions. But it is equally conceivable, that Beings originally distinct, and such as cannot be comprehended in the totality of their nature under any one notion, should yet be bound by the plan of the world, in which they are all included, to express their own inmost and heterogeneous Being, where they come into mutual relations, in a language of common currency, i. e. by means of the properties of matter.

175. I shall not now attempt to determine, whether the present age with its more extended knowledge of nature has discovered grounds decisively favouring the first of these suppositions—what is certain is, that the ancients, who first propounded this view, proceeded on no such sufficient grounds. The conception of an attribute admitting of being applied to things differing from each other, they hastily transformed into the conception of a real identical subject underlying the varieties of phenomena. This example has unfortunately been very generally followed by Philosophy in subsequent times, and the days are still quite recent when the most strenuous attempts were made to construct this universal substratum, though even if it had been shown to exist, it would have been most difficult, if not altogether impossible, to deduce from it the different material bodies to the explanation of which it was supposed to be necessary. In any case, this universal matter could not have been adequately determined by reference to those predicates which constitute its materiality. For, all of them, extension, reaction, *vis inertiae*, denote merely the manner or mode in which a thing behaves or is related. They do not in any way touch the nature of that to which these changes of behaviour are attributed.

There are two ways in which it may be attempted to get the better of this difficulty. As we are under no obligation to lay claim to universal knowledge, so it may simply be granted, that Matter is a real determinate thing, but known to us and intelligible only in respect of its behaviour. This is roughly the point of view which is adopted by Physical Science. Science distinguishes that which is extended and operative in space from the empty environment in which it appears. But it leaves the original nature of this substratum undefined, or ascribes to it only such general characteristics as are forced upon it by the analysis of individual objects. By so doing, Science gives up the attempt to construct a theory of a universal matter, preferring rather to examine into the nature of phenomena singly, whilst assuming the existence of a common basis underlying them. On the other method, if we attempt to deduce the general properties of matter from the nature of the real thing of which they are predicates, we are met by a well-known difficulty. We convinced ourselves, when treating of ontology, that to look for the essence of a thing in a fixed quality and then to represent the modes of its activity as consequences derivative from this, was a method which could never be successful[1]. We saw, that all those forms of insight which seemed to explain the inner nature of things were only possible because they were nothing but forms of vision, appearances such as a consciousness may present to itself. What lay at the bottom of such perceptions, in external reality, always converted itself into some kind of activity or process or mode of relation. And however strong may be the impulse to attribute these living processes to some subject, we had to give up the attempt to explain the marvellous fact of active being, by representing its activity as the mere predicate of an inactive subject. Similarly, in the present case, it would be labour mis-spent to attempt to describe the reality underlying the forms of material existence previous to and independent of these its manifestations. There does however still remain something to be done, viz. to determine the place which this inaccessible substratum occupies in the sum-total of existence. At any rate we must be clear as to whether we mean to regard it as something absolutely original and specific, standing in no connexion with other forms of reality, or as itself, no less than its properties, an intelligible part of the order of the universe. The attempt to explain the origin of matter mechanically is now regarded as impossible; no theory of a universal matter can show how the existence of matter first became possible and then actual. All that can be done is to indicate the *manner* of its existence and its place in the order of the world. Not until the nature of matter had been thus explained, and so could be taken for granted, could the attempt be renewed to derive individual phenomena by mechanical laws from the universal fact of matter.

176. There has never been a dearth of such attempts; I shall content myself with a brief mention of only a few; confining myself to those which stand in the closest relation to existing opinions on the same subject. According to Descartes, extension and consciousness constituted together the two ultimate facts of perception, both being equally clear and neither admitting of being merged in the other. Having made this discovery, Descartes proceeded with a light heart to treat also the *res extensa* and the *res cogitans* as equally simple and clear. He considered that these were the two original elements

[1] [Cp. Bk. I. Chap. 2, § 21.]

of the world, and he maintained that they had no further community of nature than such as followed from their having both sprung from the will of the creator, and being involved in a relation of cause and effect, which the same will had established. Doubtless, an advance was made upon this view by Spinoza, in so far as he conceived of conscious life and material existence not merely as springing from the arbitrary will of the creator, but as two parallel lines of development, into which, by reason of its two essential attributes, the one absolute substance separated itself. At any rate, it was established that the material world does not proceed from any principle peculiar to itself, and of undemonstrable origin: the Reality underlying the forms and relations of matter in space is the same as the Reality, which in the intelligible world assumes the form of Thought.

But I cannot convince myself that Spinoza got further than this point towards a solution of the questions now before us. Though insisting on the necessary concatenation of all things, even to the extent of denying every kind of freedom, he hindered the development of his view, by introducing barren logical conceptions of relation, the metaphysical value of which remained obscure. A logical expression may often be found for the content of a conception by enumer. ing a number of attributes co-ordinated in it. All that this really means is that every such determination a is imposed upon the single object in question by the given condition p, with the same immediate necessity with which in another case the determination b would follow upon the occurrence of q. But we cannot tell in what consists the unity of a substance, which apart from all such conditions exhibits two original disparate sets of attributes, leaving it open as to whether these are eternal forms of Being (*essentia*), and as such help to constitute the nature of the substance, or whether we are to understand by them merely two modes in which the nature of this substance is apprehended by us. The fact that in respect to the infinite substance every influence of external conditions must be denied, makes it all the more necessary that the inner relations which are contained in its essential unity, issuing as they do in such very different modes of manifestation, should be explained and harmonised. The striking peculiarity of the circumstance that Thought and Extension should be the attributes thus colligated, is not explained away, it is only hidden from view by the suggestion that besides these attributes, there are an infinite number of others, which though inaccessible to our knowledge are yet co-ordinated together in the nature of the absolute after the same incomprehensible fashion.

Again, every individual existence in the material world may be logically subsumed under the universal attribute which is called by the not very appropriate name extension, as species or subspecies ; but, in the merely formal conception of absolute substance, there is nothing whatever to determine why out of the infinitude of possible modifications of the absolute substance which are logically conceivable, one should exist in reality and another should not—or, supposing it to be held that in the infinite unexplored totality of existence all these numberless possibilities as a matter of fact *are* realised, there must still be some reason why the events within the limits of our own experience take place in the order in which they do and not in another. Those two attributes of the infinite substance would, if left to themselves, be able to develope merely the system of all possible consequences derivable from them ; but such is not the reality which we find before us ; in order to arrive at that we need either a plurality of underived existences, or a simple plan capable of being the reason why of the possible consequences of those principles some occur often, others but rarely, and all in such infinitely various combinations.

Once more, it is true that no modification of the one attribute can be derived out of a modification of the other, and therefore thought cannot be derived from extension nor extension from thought. But the logical impossibility of deriving one from the other analytically cannot invalidate the possibility of their synthetic combination in actual reality, except on a view which treats logical subordination as if it were the same with dependence in fact, and confuses a condition with a cause. The necessary admission that in Being there are elements which cohere and mutually affect each other, though in thought they are incommensurable, cannot be replaced by the wearisome repetition of the assertion, ‘ ordo et connexio rerum idem est atque ordo et connexio idearum.’ Whatever reference this proposition may be supposed to have, whether to the parallelism of the forms of Being in the totality of the world, or to the combination of physical and psychical functions in the life of each individual, as long as consciousness and extension have admittedly no common term, there can be no common term between the order and connexion of their respective modifications. Their alleged identity can only be understood in the restricted sense that always and in every case the modification b of the attribute B corresponds with the modification a of the attribute A, and that the change of a into a is followed always by a corresponding change of b into β. But there is no proof that the correspondence which is exhibited as a matter of fact between $a - a$

and $b - \beta$ rests on any identity of nature; or, in other words, that the transition between two modifications of the one attribute is or expresses or repeats the same thing in a different form as the corresponding transition in the other. I cannot, therefore, discover that Spinoza has advanced the explanation of the material world in its relation to the spiritual. Instead of a metaphysical theory, what he gives is scarcely more than a logical classification. According to this, material and spiritual existences may be ranked under two disparate categories, which, both as real determinations in the nature of the absolute, and in all that is produced from it, are, not indeed by any inner necessity, but always as a matter of fact, combined. It is quite possible that we may not be able to make any advance worth speaking of beyond this point; but, in that case, we must admit that we have arrived at a result which is worth almost nothing, and we shall not feel bound to make any profession of enthusiasm on account of such a trifling addition to our knowledge.

177. I shall touch only briefly on the kindred speculations which our own idealist philosophy has developed more recently. Schelling contented himself at first, as Spinoza had done, with the recognition of that Law of Polarity, which as a fact constrains the absolute to develope itself under the twofold form of Ideality and Reality. He interested himself more, however, in showing the constant presence of these two elements in every phenomenon, and explained the manifold differences of things as arising from the preponderance of one or other of them. But it soon became evident (as would have appeared even more clearly if his demonstration had been successful) that he intended to regard this duality not as a mere fact, but as a necessary process of differentiation involved in the original nature of the Absolute. At a later period, he was dominated, as was Hegel, by the thought of a development within which the material world appears as an anticipation of the higher life of the Spirit. Of this development Hegel believed himself to have discovered the law.

It would be impossible, without going to extreme length, to give a representation of the governing purpose of Hegel's account, which should be at once faithful to the original, and at the same time adapted to our present habits of thought. I shall confine myself, therefore, to attempting to show that he has confused two classes of questions which ought to be kept distinct. After satisfying oneself that the purpose of the world is the realisation of some one all-comprehensive idea, and after being further assured that the arrangement of the forms of existence and activity in a fixed system is required as a means to this realisation, one may proceed to ask, what is the place of matter in such a system? what necessary and peculiar function is served by it? It would then be natural to speak first of matter in its most universal form, *i.e.* materiality as such; and we might hope to find that the same inner process of development, following which the original idea of matter breaks itself up into certain definite postulates of existence, necessitated by the correspondence of the idea with the whole sphere of reality, would be followed in like manner by the concrete forms which different objects assume in filling in the common outline, and that these would be similarly developed. No one now believes in the pleasant dream that this project is realisable, still less that it has been realised. Still, there is nothing unintelligible in the notion itself. What troubles us is the obscurity of the connexion between this project and the second of the problems I alluded to, that of showing how the postulates dictated by the Idea are satisfied both in existence as a whole, and in the complex course of actual events in particular. As regards the former point, it may be sufficient to bear in mind that the self-developing idea is no mere system of conceivable possibilities of thought, but itself living reality. The same reflexion cannot, however, as often it is wrongly made to do, serve the place of a system of mechanics, determining in reference to each concrete existence in Space and Time why precisely here and now this rather than some other manifestation of the idea should necessarily be realised.

178. More in accordance with the scientific views at present held is the teaching of Kant. I can remember how a few decades ago the student used to hear it said that of all Kant's epoch-making works the deepest were those which treated of the Metaphysical basis of Natural Science. While admitting the worth of what Kant has written on this subject, I cannot value it quite so highly. I lament, in the first place, the gap which separates the results of these speculations from those of the Critique of the Reason. The ideal nature of space which is asserted in the Critique is here left almost out of account; the construction of matter is attempted exclusively from the ordinary point of view, according to which there is a real extension, and there must be activities adapted to fill it. I lament no less what has previously been observed by Hegel, viz. that there should remain such uncertainty as to the subject to which the activities thus manifesting themselves in Space, and so constituting matter, are to be attributed. That this subject is what moves in Space, and that it is the reality which underlies our sensations, these seem to be the only determinations of it

which are not derived from what the properties of matter show themselves to be by their subsequent effects.   Who or what this is that is
thus movable or real remains unexplained.   Taking into consideration the fact that Kant used to speak of things in themselves in the
plural, it seems probable that his thoughts on this subject did not pass
beyond the conception of an indefinite multiplicity of real elements,
an obvious hypothesis, which was likely to recommend itself to him
for the purposes of Physical Science.   This view is confirmed by his
mode of deriving the differences of individual existences from combinations of the two [1] primary forces in varying degrees of intensity,
which is his invariable explanation of matter as a phenomenon.   Now
these differences of combination would have nothing to stand upon if
they are not based on specific differences of nature in the real elements
which they combine.   Although, therefore, it is not explicitly laid down
that the Real elements are originally distinct, still this interpretation is
quite as little excluded, and it may be admitted that what Kant is
endeavouring to explain is not a universal matter, but rather the
universal form of materiality, together with the special manifestations
which are developed within this form in consequence of the characteristic nature of the Reality which the form contains.   But, supposing
this to be admitted, we should still be at a loss to explain how this
real existence is related to Space, in which it thus makes its appearance.   If we refer back to the Critique of the Reason, we find one
thing settled, but only in the negative.   True Being can neither be
itself extended, nor can the relations in which it is expressed be other
than purely intelligible ones.   The problem would then have been to
show how the elements of Real existence are able to present themselves to our consciousness [2]—in which alone space is contained—in
such a way that they not merely take up definite positions, but also
have the appearance of being extended in Space.   Kant never really
handled this question.   The forces of attraction and repulsion which he
mentions can only be understood on the supposition of certain definite
points from which they are put in operation by the ultimate elements.
Moreover, if Space which is continuous is to be continuously filled with
matter, differing indeed in degrees of density, but still such that no
smallest particle of it can be absolutely driven out of Space even by the
greatest pressure, and if matter is to an unlimited extent divisible into
parts which still remain matter, there seems to be nothing left for our
imagination but to conceive of extension in Space and impenetrability
as original and fixed characteristics of the real substratum, which

[1] [I. e. attraction and repulsion.]   [2] ['Anschauung.']

thus becomes the basis of further enquiry. But in that case, what we should have would be neither a universal matter nor the universal form of materiality. The latter would be merely assumed as the common characteristic in real elements otherwise diverse, in order that it might serve as a basis for investigation into the relations subsisting between different material existences. This result would not be very unlike that which is soon reached by the ordinary reflexion upon Nature. Different kinds of unknown elements are . assumed, which owing to causes also unknown we come upon, each of them in numerous specimens, at different points in Space. At these different points each fills a certain volume with its presence; their presence is manifested by the changes of position which they originate, and by the resistance which they offer to any attempts coming from without to remove them from their occupancy or to lessen its extent. In other words, we think that there are many different kinds of matter which are distinguished for us by the different coefficients which we are compelled to assign in each of them to the action of certain forces or inherent tendencies common to them all.

179. The application of this conception of force in order to explain the fundamental qualities of matter has always been regarded as the most valuable advance of Kant's Philosophy of Nature, though to some it has seemed to go further than experience would warrant. Kant himself does not appear to me to have allowed the motive clearly enough to emerge which led him to this view, though there can be no doubt as to what it was, and we may trace it thus. He mentions [1] Lambert's account of Solidity as a necessary property of all material existence. According to Lambert, it follows from the very conception of Reality, or, in other words, it is a consequence of the Law of Contradiction, that the mere fact of the presence of a thing in Space makes it impossible that any other thing should occupy the same position at the same time. Against this it was contended by Kant that the Law of Contradiction could not by itself keep back any part of matter from approaching and making its way into a position already occupied by some other part. This objection is not quite fair. We should not expect the physical impossibility referred to to be produced *by* the Principle of Contradiction, but only *in accordance with* that principle and *by* the fact of solidity which for practical purposes, we assume as an attribute of Real Existence. And

[1] [Kant. Metaphysische Anfangsgründe der Naturwiss. Dynamik. Lehrsatz I. Anmerkung.]

why should we not make this assumption if there is nothing at variance with it in experience? It is no sufficient reason against doing so to urge, as Kant does in the course of his 'Proof' of this 'Precept No. 1' of his 'Dynamic,' that to make way into a position is a motion; and that in order for there to be a decrease or cessation of motion there must be a motion proceeding from an opposite quarter, or rather a something which can produce such a motion, a motive force. For the view of atomism according to which the smallest particles of matter are possessed of solidity, though it would admit that motion makes its way up to the surface of a body, would not admit that it makes its way into the body; yet, according to this view, the effects of the impact communicated would not vanish without producing an effect at the surface of the solid matter, but would be distributed from one atom to another, or to several atoms, and so become imperceptible. Whatever difficulties may attend the explanation of the phenomena by this method, at any rate a closer investigation than has been entered on by Kant would have been required in order to exhibit them.

Again, what Kant adds in his note is not to me convincing. He admits that in constructing a conception it is allowable to assume any datum to start with, e.g. solidity, without attempting to explain what the datum itself is. This, however, he says, gives us no right to affirm that the hypothesis is altogether incapable of being explained by mathematics. It seems to him that such a view would only hinder us in the attempt to penetrate to the first principles of science. But supposing we were willing to go so far with Kant as to assume the force of expansion, to which he gives precedence, would this be more than a datum, which could be used certainly to explain subsequent manifestations, but was itself taken for granted and would not admit of being deduced from the nature of real existence as such? The point at which a man will declare himself satisfied in this matter really depends in each case on his individual taste. There could be no real necessity to follow Kant in assuming something more than solidity as a fact pure and simple, unless it could be shown that solidity itself is either impossible or inadequate. Now the question whether it is impossible must for the present be left out of account; inadequate, however, it certainly is. The fact that no visible body is of unvarying extension, but all are susceptible of compression or expansion, would, it is true, apart from Kant's assumption of a continuous *plenum* in space, form no immediate disproof of the solidity in question, though this obviously implies the allegation of unvarying volume. The atomic theory, postulating empty spaces between its solid elements, would have a different explanation for the varying size of substances. But all the phenomena of elasticity, in which bodies resume their former shapes so soon as the external agencies which determined them to change have ceased to operate, prove beyond question that there must lie in the very nature of real exist-ence conditions capable of producing states of Being which as yet are not. The form and extension, consequently, which an object of sensible perception assumes, cannot attach to it as an original and fixed property, but are rather a varying state of its existence, determined by inner conditions inherent in its Being. Sometimes, the object is permitted to appear in its true form, sometimes it is hindered from doing so; in the latter case, however, i. e. where the inner states of Being are prevented from giving themselves expression, they make known their existence by the resistance which they offer to the adverse influences. These inner determinations may be spoken of as *forces*, in order to distinguish them from *properties*. It will then be seen not to be enough to ascribe solidity, as a property, though it were only to the smallest particles of matter. The atoms themselves must have certain moving forces attaching to them, in order to make the ever-changing volume even of composite bodies intelligible.

Thus we may say provisionally that Kant regarded as fundamental in this problem of Science that principle which we cannot dis-pense with even though we prefer the other principle; but which may very well help to explain that other principle. This solid matter was not a fact open to observation; it was not so even as applied to the smallest particles; it was an hypothesis. Hence, it could be denied, and every occupation of space not merely by large visible bodies, but by their smallest elements, could be regarded as a perpetually changing state produced by the force of expansion, according as its action was free or impeded. Stated in a few words the case stands thus. If every material existence, remaining always indivisible, occupied the same space at one time as at another, solidity might be predicated of it as an original quality which it must not be attempted further to ex-plain. But, now, inasmuch as extension, though a *character indelebilis*, is not a *character invariabilis* of matter, the extension which a thing has at any moment is the result of conditions which though present at that moment may vary at other moments; one of these conditions lies in matter itself, and offers a resistance, though not an insuperable one, to those which come from without.

. 180. I wish to dwell for a moment longer on the difference to

which I have referred between a fixed quality and a force. We have
been long convinced that what we ordinarily call properties of things
are really only modes which they assume, or manifestations which
become known to us as the result of their interaction. Things do
not have colour except as seen by us, and at the moment when in
combination with waves of light they stimulate the eye. They are
not hard, except in relation to the hand which attempts to move or
pierce them. As a matter of fact, then, we should be at a loss to
point to an indubitable instance of what we mean by a quality of a
thing. All we can say is, that we are clear ourselves as to what
we mean. By a quality is meant that which a thing is for itself and
independently of any of its relations to other things. Hence, in order
to exist, a quality neither requires these other things, nor is interfered
with by them. A force, on the other hand, is not, like a quality,
something belonging to things as such. In order, therefore, for a
thing to be what it is, we do not attribute to it any force of being ;
though we do speak of its having a force of self-conservation, in
opposition to certain conditions which we assume to be capable of
changing it. Our conception of a force, therefore, involves the thought
that the character of a thing is neither unchanging, nor yet on the
other hand determinable to an unlimited extent from without. Rather,
it implies that when the two things meet, they both undergo a real
change, the change of the one depending on the nature of the other,
but each at the same time by its own nature forbidding a change
without limits or one which would amount to a surrender of its essential Being. If qualities attach to things in their isolation, forces can
only belong to them in consideration of their relation to each other ;
they are, in fact, conditions which enable one thing to affect another
and to place itself to it in different relations. It is in this sense that
Kant speaks of the forces which fill space ; they belong to the separate
parts of matter, and are brought into activity by these parts in their
mutual relations ; their appropriate effects they either succeed in producing, or else show to be present by the resistance which they offer
to other forces tending to hinder them. Here, however, it may be
objected that Kant did not confine himself to the exposition of
this process, but that taking this for granted as a universally presupposed fact, he imported into the discussion considerations of quite
a different order, attaching to the term 'Force,' which he selected.
I do not believe that Kant himself is liable to the charge here made
against him ; but the popular view of nature which was suggested by

his doctrines, has given rise to a number of false opinions, and these therefore we shall now proceed to examine more at length.

181. It is no doubt most useful to be able to express the import of an intricate relation between several connected points, by means of a single word ; at the same time, there is danger in doing this. After the word has been called into existence, not only are we able to combine it with other words, but we are led to suppose that every such grammatical combination has something real correspond· ing with it in fact. Thus, we speak first of all of force, and then of the force of matter. The use of the genitive in this instance, implying as it does that matter is possessed of force, or, that force is exercised by matter, has suggested these interminable questions concerning the nature of force as such, and its relation to matter of which it is a function. Such questions cannot be easily answered at once, when stated in this form. To understand, however, the *applications* of which this conception of force admits, we have only to observe the ordinary usage of Physical Science. Physics makes no mention of Force in itself, but only of its effects, i. e. of the changes to which it gives rise, or which it hinders. It is moreover against the Law of Persistence that an element should of itself modify its own states ; the impulse to change must come from some other element. Thus, an element a is not possessed of a force p until a second element b is presented to it on which it may take effect. The force is really produced in a by the relation to b ; and it changes to q or r if either the nature of the second element or the relation of a to it is changed. Now, observation shows that there is nothing impossible in the attempt to determine the nature of the elements, the relations in which they may stand to each other, and the changes which they undergo in consequence of these relations. We can understand how, when elements containing specific amounts of generic properties enter into specific forms of some general relation, there are general effects which follow and vary proportionally according to definite laws. The proposition, a is possessed of the force p, when all that it implies is fully stated, in the first instance merely conveys the assurance that whenever a is brought into a specific relation m with a given element b, changes of state will be experienced both by a and by b which will go together to form the new occurrence, of fixed character and amount, π. Having arrived at this result we may then go on to speak of this fixed determinate force in another way, as if, i. e. it were present in a in an ineffective and indeterminate form, its definite effect being supposed to depend on subsequent conditioning circumstances,

e. g. the nature of the elements *b* or *c* which come into contact with *a*, the peculiarity of the relation *m* or *n* into which *a* is brought, the presence or absence of some third circumstance. To all these causes the actual realisation of the result *π* or *x* might be ascribed. Even this mode of statement, however, expresses no more than a presumption as to what will necessarily happen in a given supposed case. It follows in accordance with the general law which connects the changes of things with one another, that the circumstances being such as they are, no other result could have happened. Each of the elements, in virtue of its own nature, contributes to this result, and it is an allowable mode of statement first of all to represent them as containing severally and individually all the required conditions, and then to rectify the error of such an assumption by adding that the force potentially inherent in each element cannot become active until the element enters into some specially determined relation. As a matter of fact, it is this special relation which gives rise to the force. If we desire a definition of force, we may say that it is that quantitatively and qualitatively determined result, which may or must ensue, whenever any one element enters into a specific relation with any other. It is only for convenience of speech that this future result, which under given conditions we are justified in expecting, is antedated as a property already present though inoperative in the element. This being understood, there can be no harm in thus speaking of a force as being asleep and awaiting the moment of its awakenment, according as the conditions, which together with the specific nature of the element constitute all that is necessary to produce the result, are present or absent. We shall perhaps make the matter clearer, if we adduce other instances besides those of physical forces with which we are more immediately concerned. Thus, it is the same conception of force which we have in view, when we speak of the powers of the mind, the revenue-yielding power of a country, or the purchasing power of money. In this last case, no one seriously believes that the current coin contains some latent property which gives it its value. The possibility of obtaining a given quantity of goods in exchange for so much money depends on highly complex relations which men enter into for purposes of traffic; and the value of the money changes not owing to any change in the substance of the metal, but to a change in some one of the conditions by which the value of the money is for the time being determined. There would be no power of purchase in money if there were no market in which to exchange it. Similarly we are quite justified in speaking of the

Power of Judgment as a property of mind. When we make an assertion in regard to any given matter before us, which is what properly
constitutes a judgment, it is certainly our intellectual nature that is
called into exercise ; at the same time, however, it would be nonsense
to speak of a power of judgment, which belonged to us before we
came to make use of it, or one which was constantly being exercised
without reference to any distinct object-matter. It is impossible to
say more than that we are constituted by nature in such a way that
the mind, when it is acted upon by impressions from without, not only
receives the impressions singly, but reacts upon them in that way of
comparison of their different contents which we call judgment. It is
only at the moment when it is exercised that the Power of Judgment
is living and present, and this applies not only to the reality of the
activity, but also to its nature and content ; these likewise being
dependent on the conditions which bring them into existence for the
time being. We may say the same as regards the conception of force
which obtains in Mechanics. Thus when we speak of centrifugal force,
we do not mean that this force is possessed by Bodies as such, when they
are at rest. We at once see that we are speaking of effects which may
or must take place when bodies are rotating or being swung round.
If we distinguish from these forces certain others, such viz. as the
attraction which bodies exercise upon each other, and call the latter
primary forces inherent in the bodies as such, all that we mean is that
the conditions under which such forces arise are extremely simple and
always fulfilled. In order for two elements to be drawn to each other
by the force of attraction, all that is required is that they should exist
at the same time in the same world of space. This one condition,
however, is indispensable ; it would have no meaning to say that an
element gravitated, if there were no second element to determine the
direction of its motion.

We shall not, therefore, attempt to determine what actual relation
subsists between forces and the bodies which are their substrata, believing as we do that the problem itself results from a misunderstanding. No such relation exists in the sense that a force can in any
way be separated from the body which we call its substratum. Its
name ' force ' is only a substantive-name employed to express a proposition, the sense of which is, that certain consequences follow upon
certain conditions. What it signifies is neither a thing, nor any existing property of a thing, nor again is it a means of which a thing could
avail itself in order to produce any given result. It merely affirms the
certainty that a given result will happen in a given case, supposing all

the necessary conditions to be complied with. Nor can we ourselves attach any meaning to those hastily-conceived maxims, which are popularly held to express the truth on this subject, such e. g. as that there can be no force without matter; and, no matter without force. These equally stale propositions merely add a small grain of truth to the old error in a more perverted form. It is rather true that there is no force inherent in any matter, and no matter which by itself has or brings with it any kind of force. Every force attaches to some specific relation between at least two elements. On the other hand no opinion is here expressed with regard to the question as to whether it is possible for two elements thus to be brought in relation without some force being engendered. It is dangerous to attempt to lay down propositions by the way with regard to matters of fact, merely for the sake of making a verbal antithesis.

182. If these considerations are regarded as conclusive, the term force will be understood, not indeed in the sense in which it is sometimes used, viz. as a Law according to which things take place, but as an assertion in regard to each single case to which the term is applied that we have in that case an instance of the operation of the Law. Thus understood, the term will not suggest any meta-physical explanation as to why the particular facts *must* fall under the general Law.

It is this sense which Physical science is content to adopt when making use of the term. For the practical aim of science, that of connecting events in such a way as will enable us on the ·basis of present facts to predict the Future or unriddle the Past, it is found amply sufficient to know the general law of the succession of pheno-mena and by inserting the special modifications of its conditions which occasion prescribes to determine the nature of the result. Science can afford to be indifferent as to the inner connexion by which results are made to follow antecedents. It cannot be main-tained that this was all that Kant intended to be understood by his conception of force. He everywhere speaks as if he meant to ex-plain extension not as a simple consequence of the existence of matter, but rather as due to the action of a force. This is a very different conception of force from that according to which it is regarded simply as the connexion of phenomena in accordance with Law. Clearly he means by Force something which is active in the strict sense of the word, something which, he believes, will produce real changes of state; whereas, the counter-theory, confining itself within narrower limits, asserts only that they follow each other in orderly succession. The popular view of nature which based itself on Kant's doctrine imported into the idea of physical force all those associations which are suggested by reflexion on our own conscious activity. In order for this doctrine not to seem to be at variance with the observed facts of the outer world, it had to be toned down, and, in spite of the manifold contradictions which the idea involved, the activity was regarded as Will or Impulse unconscious of itself. These latter-day developments of Kant's view I shall for the present leave to take care of themselves. It will, however, be understood after what I have urged in the ontological portion of my work as regards the relation of cause and effect, that this view which has been made to bear so heavily on Kant, is one in which I fully agree with him—I agree with him in the general recognition of an inner process and activity, in virtue of which things are able to be that which, according to the frequent expression of Physicists, it alone belongs to them of right to be, viz. interconnected points serving as the basis of ever-varying combinations, centres from which forces proceed and to which they return, points of intersection at which different converging processes meet and cross each other in fixed succession. I do not regret that Kant should have refused to put this view on one side. I regret rather that he should not have brought us to closer quarters with it. The general position for which I have already contended does not require to be further elaborated in reference to this special case of Physical causation. An element a cannot produce the effect p merely because there is a general law L, which prescribes that when a stands in the relation m to b, the result p shall follow. No doubt this result *does* follow in the given case, i.e. we who are the spectators see and know that it does so. But, in order for the change itself to take place, in order for a to give birth to an activity under these new conditions which it did not previously produce, it must undergo an experience through being placed in the relation m which otherwise it would not have undergone, and, similarly, the effect p could never be brought home to b, merely because the relation m existed between b and a. The existence of the relation m must have been *felt* by b before it could have been acted on. Hence, the results which arise in each case are not consequences of mere *relations* which subsist between a and b. These relations, as we call them, are really inner states of Being, which things experience as the result of their mutual activity. It is not to be expected that this theory of an unceasing activity of the inner life of things will be of much real assistance in the explanation of each separate fact of nature. It is a supposition, however, which it is necessary for us to entertain if we are to cease to regard the world from a point of view, which however useful it may be for practical purposes is full of inconceivabilities, the view, viz. that the elements of existence are without individuality and without life, endowed with reality merely because a network of relations is established between them by the agency of general laws. The usefulness of this latter point of view, if considered merely as a half statement of the truth, I shall not dispute, whilst at the same time I shall point out how far it is applicable and justifiable, and when and where it is necessary to recur to what actually takes place in the nature of things.

183. Out of the multitude of opinions which offer themselves for consideration at this point I shall make mention, first, of Kant's view, according to which there are two forces necessary to every material existence, the force of attraction, by which things are made to cohere, and the force of repulsion by which they are expanded; the two together forming a standing element in the countless attempts at explanation which have been made since Kant's time. I must confess myself that I do not feel much interest in these two forces. When the point is raised as to how it can be that a given matter has definitely fixed limits of extension, it is easy to see that there must be some reason why it is what it is—neither more nor less—i. e. there must be an attraction of the parts, which if it were allowed to work alone would reduce the extension to nil, and there must also be repulsion, which similarly, if it were the only principle at work, would make the extension infinite. This is simply a logical analysis which might be applied to the conception of any real existence which has a definite magnitude in space. The enquiry does not become metaphysical until it deals with two further questions; how, that is, these two mutually opposed forces are possible, both attaching as they do to the same subject; and what that is which produces and maintains them in such varying proportions as are required in order to give rise to the manifold differences of material things in point of extension?

The first of these two questions has been made a subject of investigation by Physics. It was considered that to ascribe to matter two equally original opposed forces would involve a contradiction in terms. The attempt was therefore made to assign the two forces to different subjects. The mutual attraction of the parts proceeded from the ponderable elements, the repulsion was regarded as confined to particles of imponderable ether; and thirdly, an interchange of activities between the two classes of elements was admitted, in order to explain those varying states of equilibrium between attraction and repulsion which the facts required. Whether this last result was secured by the hypotheses is for our present purpose indifferent. It may be admitted that the reasoning is logically sound, though the conclusion is only necessary, if, in compliance with the usage of language, both forces are conceived as original and essential attributes of the subjects to which they attach. How the whole matter may be regarded from a different point of view, for which the course of my argument will already have prepared the way, I shall now proceed gradually to unfold, ignoring provisionally arguments derived from the alleged ideality of space. Even if we adopted the ordinary view of the nature of space, it would not really become any less difficult to explain, why the mutual relation between two elements, belonging to the same world, should be one of absolute repulsion, when this fact would seem rather to show that the world to which they belonged was *not* the same; nor would it be less wonderful that two other elements, both of them, similarly, supposed to belong to one and the same world of extended matter, should be drawn towards each other by such an absolute force of attraction, as that if there were no counteracting principle, the whole possibility of their extension would be annihilated. Once grant that the world is a single whole, and not a mere confused aggregate of existences, and it will follow that its component elements cannot be governed by any abstract principles of attraction or repulsion, driving them continually out of or into one another, but must aim at the conservation of the whole order, which, in accordance with the intention of the whole, assigns to each one of them its place at each moment of time. The force which proceeds from the collective mass of the elements, is one which determines the position of those elements and which, while it seems to reside in each individual element, really sets itself against any deviation from the law imposed on all. It sets limits to the nearness or remoteness of objects as regards each other, appearing in the one case as the force of repulsion, in the other, as that of attraction; in both cases acting as a corrective wherever there is a tendency in the object to oppose the requirements of the whole. I wish to see the order of our thoughts on this subject reversed. We are accustomed to regard the position of a thing as the result of certain forces acting upon it. The first consideration, as I think, on the contrary, is precisely the position which a thing occupies, as deter-

mined by its nature and character, in the world-system, and the first
and only function which a thing as an individual has to perform,
seems to me to be to retain this position; while attraction and repulsion we may represent to ourselves as the two elements into which
this self-conservation of things admits logically of being analysed.
In reality however what happens is that the self-conservation assumes
one or other of these forms according as the needs of the moment
give occasion to it. We must postpone the consideration of the
question, as to what takes place in the inner nature of things when
the place in which they find themselves at any given moment is out of
harmony with the place marked out for them. As a phenomenon in
space, the tendency to return to an equilibrium must necessarily appear
in its simplest form, either as the approximation or as the separation of
two elements. Hence it is possible to refer all physical processes to
motive forces consisting of attraction or repulsion. But it is not the case
that on all other occasions things are empty of content, and that
these forces attach to them merely for the time being. Rather, like
the gestures of living beings, the forces are merely the outward expression of what is going on within.

184. Thus far, no doubt, the statement of our views has conveyed
the impression that we regarded the world like a picture having fixed
outlines, within which every single point invariably occupies the same
position and clings to it with equal tenacity. Such a picture would
be little in accordance with the facts. We have long known that the
world is never at rest and that the picture which it presents is for
ever changing. Yet, the whole case is not stated even when we have
grasped this truth. Admitting that the picture of reality is what it is
at any given moment in virtue of its essential connexion with the
arrangement that prevailed the moment before and that which is to
prevail the moment after, the forces emanating from the different
points of space must still derive their power to act on each separate
occasion from the law which pervades the whole. The connexion
between the whole and the part is peculiar to each case, and is very
different from a mere instance of the operation of law in general,
such as is known to us by observation and makes it possible to us
to apprehend the process of the world as the result of innumerable
individual forces working by invariable rules. I have, however,
already[1] endeavoured to show that this plan or idea cannot be made
real in this off-hand way of itself and without means; rather indeed
that it presupposes uniformity of action on the part of the elements,

[1] [E. g. § 67.]

so that under like conditions like consequences flow from them, quite independently of the place which each occupies in the universal plan. Hence, even assuming that the world is ceaselessly in a state of flux, our view that the permanent tendency of each thing is to maintain the place which belongs to it in the system of the universe, and that this is what gives to it its force, does not exclude the opposite or physical view according to which the course of events in the world is explained as due to varying combinations of constant forces. I may add that the supposition of a number of forces attaching to the same elements at the same time, but acting in different directions, does not seem to me to be liable to any of the objections which are commonly urged against it. No doubt, it would be unintelligible as applied to two elements working in isolation, but it is not so as applied to elements between which a connexion has been established owing to their belonging to one and the same world. We may learn to comprehend this by the experience of our own lives. Our actions are conditioned by many different systems of motives, which operate on us at the same time. The satisfaction of our physical wants may e. g. be inconsistent with the social good. What family-affection requires of us may conflict with our duty as citizens, and within this last sphere we find ourselves parts of many different institutions whose claims it is not always easy to harmonise. A like interpretation must be given of the world in which we live. When we speak of a systematic connexion between things, we do not mean a single uniform classification in which we could find any given member by following out one principle of division. Rather, there are many cross-purposes at work, each of which requires that the elements should be distributed exclusively with reference to its own satisfaction. Each element may be stationed at the intersecting point of several different tendencies which unite and divide the world. As long, therefore, as two elements are considered as belonging to such a world, there is no reason why their mutual activities should not be regarded as the result of a plurality of forces acting simultaneously, and differing entirely in the effects they produce in response to each change in the circumstances of the environment; owing to the different points of view under which they bring the same set of circumstances, and to the consequent variety of the reactions set up.

186. There still remains to be considered the question as to whether it is allowable to speak of forces which take effect from a distance, or whether those are not right who regard the possibility of a thing's acting where it is not as inconceivable. I cannot help

adding to the two conflicting views which are held on this question, a third one of my own. It seems to me that *motion* can *only* be an effect of forces acting at a distance; to speak of action when the elements are in close contact, I regard as a contradiction. Let us suppose two spherical bodies of equal diameter and density to be placed so as exactly to contain each other. If, then, the nature of the materials of which the bodies are composed is such as to admit of their reciprocal action, and if we are to disregard all possibility of effects taking place at a distance, it will follow that every point *a* of the one body will produce an effect on the point *b* of the other body, with which it coincides. Now, I do not dispute that the two elements may be affected in a very real way by reason of this coincidence at the same geometrical point. But, whether the effects thus produced are such as to intensify or such as to diminish the condition in which the elements find themselves, i.e. whether they tend to attraction or repulsion, in no case can these inner occurrences result in motion. *a* and *b* being already stationed at the same point of space cannot by any attraction be brought nearer; nor could any force of mutual repulsion, however actively manifested in other ways, avail to part them asunder, there being no reason why the initial movement tending to separate them should take any one direction rather than another.

Nor need we confine ourselves to bodies perfectly coincident in extension. No matter what form the two bodies assume, they would never be able to affect each other's motions, if there were no distance intervening between them; for those parts of the two bodies which were coincident would admit only of being affected *internally* by their mutual action, and thus there would be no external motion. It makes no difference as regards this conclusion, that effects are spoken of as taking place between *contiguous* bodies, and that the ambiguity to which this mathematical conception so easily lends itself, is made to yield a perplexing solution of a difficulty which is one of fact. If we confine ourselves to the case in which the two bodies are spheres, their volumes can only meet at one point. Now, we must be sure that what we have in view is a real contact of the bodies in question, and we must banish from our minds all thought of there being *any* distance, even an infinitesimal one, intervening between them. As long as we have any such idea we have in principle admitted the action of force at a distance, though without any reason restricting the distance to an infinitesimally small one; a conception which, besides other difficulties, it is, to say the least, not easy to explain on physical principles. It is equally inadmissible to substitute for a *point* of contact an infinitely small *surface*, or, supposing the contact to be between flat surfaces, to imagine that the layers which are in contact and which thus produce the effect, can have any, even the smallest conceivable degree of thickness. It must be taken as settled that the bodies which are in contact have their boundaries common or coincident, in the first case, in a point without extension, in the second, in a surface without thickness. Whatever way we may try to turn these ideas, the fact will always remain, that real elements which occupy the same position in Space will exercise no effect as regards the production of motion, and such effect as does take place will spring only from those parts of the bodies which are really separated from each other by intervals of Space. As for a contact which does not involve either separation or coincidence at the same point in Space, the idea is intelligible enough as applied to the whole volume of each of the two bodies brought into contact, but it has no meaning as applied to a possible interaction of single points such as we have been here considering.

This same observation holds good as regards the attempt to substitute, instead of forces operating between different elements, a reflexive power of expansion or contraction, in virtue of which a thing assigns to itself a greater or less space of its own accord. If the 'thing' here spoken of is understood as a material existence extended and divisible, this power of self-extension belonging to the whole must in every case be capable of being finally referred to the reciprocal repulsion of the parts, these being already distinguished in Space. If, on the other hand, the thing is held to be endowed with this power in consequence of a real metaphysical unity prior to its multiplication in Space, we shall then have to face another enquiry, which is for the most part overlooked in these attempts to construct a theory of matter, viz. this, How did this reality first get form and extension in Space—that form and extension which are always presupposed, in order that forces of the kind mentioned above may be furnished with points to which to attach themselves? This question we propose to consider in the next chapter.

186. All that the above demonstration proves is that mere contact of elements cannot produce motion. If, however, it should be found to be equally inconceivable that effects should take place at a distance, we shall be compelled to deny that motion is a result of force in any shape whatever, and our task will then be limited to the attempt to conceive of physical effects as taking place owing to the supply of

motion already in existence being perpetuated. But it soon appears that
the expression, *communication* or *distribution of motion*, though enabling
us to picture to ourselves results which are constantly passing before
our eyes, does not give any tenable conception of the process to which
the results in question are due. Take, e. g. the familiar instance
of the effects of impact on inelastic bodies. Suppose b to be a body
in motion and a a body at rest, then, when b strikes against a, we say
that it communicates to it a certain part of its own motion, and this,
no doubt, is an extremely convenient way of signalising the new fact
which has taken place, in consequence of the two bodies having been
brought together. We cannot, however, seriously suppose that the
motion produced the result by changing its place. If we may repeat
what has before been said [1], it is for ever impossible to conceive that a
state q, by which a real thing b is affected, should loose itself from b,
and pass over to a; yet this is such a case; before the motion could
transfer itself from the limits of b to a, it would have to traverse, no
matter in how short a time, a certain space intermediate between the
two, and during this time it would be a state which was the state of
nothing. The absurdity of this notion is here even further increased
by the fact that it is only by a free use of language that we are able to
speak of motion as a *state* at all. Motion, in fact, is not a quality
permanently attaching to anything; it is an occurrence merely, or a
change which the thing moved undergoes. Hence, the very conception of a motion, which is itself set in motion in order to pass from
one thing to another, is *ipso facto* impossible. But what should we
have gained, supposing that this inconceivability were a fact? If the
motion has passed over to a, it is now where a is, but that would not
make it a state of a, nor would it explain why it should ever move a.
Inasmuch as it was possible for the motion to become detached from
b, either wholly or in part, why should it not continue on its course
according to the same law of Persistence which it followed whilst on
the way from b to a? Why should it not leave a at rest, and again
become a motion belonging to no one as before, and so on *ad infinitum?* It results, therefore, that this theory fails to give any reason for
the motion of the body which receives the impulse, and it gives only
an obscure reason for the decreased motion of the body from which
the impulse proceeds. Of course, it will be argued that both these
facts are due to the impenetrable nature of bodies, which makes it
impossible for one of them to find a passage for itself through the
space occupied by the other. But this impossibility taken by

[1] [§ 56.]

itself rather suggests a dilemma than furnishes us with a solution of it.

If two bodies cannot both occupy the same position in Space, and if nevertheless it is this at which one of them aims, the question arises as to how these two conflicting propositions are to be reconciled. How they are reconciled as a matter of fact we see before us ; we see motion originated in the one case, and a corresponding decrease of motion in the other. But we cannot suppose that this happy solution comes to pass of its own accord because it is an ingenious idea ; it must rather be the necessary consequence of what the bodies are in themselves, and of what they pass through at the time. If, further, we bear in mind that in order adequately to estimate the result, account must be taken of the mass of the two bodies, we shall be led back to the conclusion that this impenetrability, which the communication of motion requires, is an effect produced by the conflicting tendencies of various forces, which thus give rise to motions in opposite directions, so that bodies at rest are supplied with motion which before they were without, whilst the bodies set in motion lose some of their velocity owing to the resistance of the bodies at rest. But it is impossible to represent such a repulsion as arising when the bodies are in contact, and not before. For, if at the point of contact there is no inter-penetration of the two surfaces, the contact instead of being a real one becomes a mere geometrical relation ; it can have no influence on the bodies themselves, but only on the limits by which they are bounded. If, however, we suppose that the bodies *do* interpenetrate each other at the point of contact, it will follow from our previous conclusions that the forces proceeding from the two bodies can only affect each other's motions at those points which are still separated by an interval of space. Nor can it be said that the motion q, which is communicated to a body at rest a by a body in motion b, determines what would otherwise be undetermined, viz. the direction of the two bodies at the moment of their divergence. For, from the mere fact that the mutual repulsion takes place at the moment that the body b, whilst tending in the direction q, comes into immediate contact with a, it could only be argued, in opposition to all experience, that b would pass through a in its former direction q with accelerated speed, whilst a would begin to move in the direction $- q$. It seems to me, therefore, that under these circumstances we cannot but conclude that even the communication of motion is an effect dependent on the action of moving forces, and that, in this case as in all others, forces can only produce motion when the bodies are removed from each other, while, contrariwise, they are powerless to produce it when the bodies are in contact.

187. All this reasoning would be to no purpose, if there was really any insuperable difficulty in conceiving of forces as taking effect at a distance. But I must say for myself that it quite passes my comprehension to understand on what grounds it can be maintained to be the most self-evident of facts that a thing can only act where it *is*. What, we may ask, is the meaning of the assertion, *a is* at the point *a*? Can there ever be any other evidence or manifestation of a thing's Being, than by means of the effects which are transmitted from *a* to the point *p*, where we ourselves are? Of course, it will be instantly objected : ' No doubt, the effects of *a* and the directions which these follow in the course of their transmission to us, are the only sources of the *knowledge* which justifies us in concluding that *a* is at the point *a* ; the fact itself, however, is independent of the means by which we come to know it.' But what conception can be formed of this fact itself, if we abstract all the effects which the given form of existence *a* emits from the point *a*, where it is stationed? Is the existence of *a* in general a conception which has anything definite corresponding with it? and how can the limitation of *a* to the point *a* be understood, if it does not give rise to any effects at that point distinguishing that point from all other similar points of Space, where *a* is not present? It is an illusion to believe that the mere *being* at a certain place can give a thing any determinate character, and that it acquires *subsequently* to this the capacity to produce the effects which seem to be diffused around that point. We ought rather to say, on the contrary : Because, in the disposition and systematic arrangement of the world as a whole, and in the world of Space which is its counterpart, *a* is a meeting point for relations of the most various kinds, and acts upon the other elements as these relations prescribe, for this reason and for no other, it has its fixed place amongst them ; or more correctly—it is this which justifies us in making use of the common forms of speech, *a* is at the point *a* and acts from thence.

This, however, will form the subject of further investigations. Putting this question as to the relation between real existence and Space for the present aside, we shall make use of a very simple idea to expose the fallacy of the doctrine here referred to. Let us suppose that at the commencement of their existence things were stationed each at some one point of Space, e.g. *a* and *β* : what reason would there be why the interval *aβ* between them should prevent them from mutually affecting each other? ' It is obvious and self-evident that it would do so;' it will be replied,—'the body set in motion does not feel the impulse to move, until the impelling body reaches it. There can be no sense of vision until the nerves have been touched by the moving particles of the ether. That which is incapable of transmission has no effect, and is for us as if it had no existence.' These instances, however, may be met by others. The stone falls without requiring first to be impelled; an electric repulsion takes place to all appearance quite independently of any connecting medium. If anyone wishes to refer these phenomena to the communication of motion already in existence, he may do so; but he will be appealing not to observed facts, but to his own hypotheses; he will be employing without any just reason the particular form which one class of effects assumes, as if it were the universal form which must necessarily be assumed by all other effects. And yet even these hypotheses, which aim at the avoidance of all distant effects in the case of large bodies, cannot help interposing Spaces between the infinitesimal particles of the media which are held to explain the transmission of the impulse. There could be no presumption in favour of the above interpretation unless it could be shown that contact in Space was as obviously a condition favourable to the action of force, as separation in Space is maintained to be unfavourable to it. But this is not true with regard to contact in Space. For, it cannot be concluded that anything must of necessity happen from the mere fact that two elements touch at the same limit, or are stationed at the same point of Space; nothing can come of the contact of the elements if they are not fitted by Nature mutually to affect each other, and when this condition is wanting, spatial contact cannot produce it. As for the assertion that elements which have this capacity to affect each other, require contact in Space in order to make its exercise possible, it rests on that arbitrary selection of instances mentioned above; with those in whom it has become a cherished prejudice it is ineradicable, but it is not in itself necessary, nor capable of being shown by the evidence of undoubted facts to hold good universally. We ourselves, it is true, are not endowed with any capacity for producing effects at a distance. The objects on which we attempt to bring our activity to bear, we, no doubt, set in motion by means of a continuous succession of intermediate effects, which serve to bring us and them together. But this is not enough to make us conclude that two elements, between which there is an interval of Space, belong, as it were, to two different worlds separated by a gulf which nothing can bridge over. We are compelled, in order to understand their subsequent effects, to conceive of them both as subject to the same laws ; a fact which we are accustomed to consider as self-evident, without enquiring into the presuppositions which it involves. This fact obliges us to regard, without exception, all things throughout Space as interconnected parts of one world, and as united together by a bond of sympathy to which separation in Space acts as no hindrance. It is only because the elements of the world are not all of the same kind, and, instead of being simply co-ordinated, are related in the most various ways, that this unfailing sympathetic *rapport*, by means of which all things act on each other at a distance, is not in all cases equally apparent, but differs in degree of intensity, being in some cases widely diffused, in others contracting itself within narrow and scarcely perceptible limits.

CHAPTER VI.

The Simple Elements of Matter.

THE confused notions which the different theoretical constructions showed to exist in regard to the true nature of Matter, led us in the first place to examine into the conception of the forces, the operation of which gives rise to the changing qualities of material things. There remains now to be considered the question as to the form in which the real thing, from which these forces themselves emanate, takes up its position in Space. The subject to which we shall be introduced by this question is the antithesis between atomism and the theory of a continuous extension in Space.

188. What appears to be the evidence of immediate perception on this point must not be misrepresented at starting by a slovenly mode of statement. Of a single continuously extended Matter it tells us nothing; all that it presents to us is a vast variety of different material objects which for the most part are separated from each other by clearly defined limits and are but rarely blended and confused together. This multiplicity of things is all that can be affirmed at starting— many, however, even of these things the naked eye at once perceives to be composed of parts existing side by side, but differing in kind. Others, which appear to be extended in Space with unbroken continuity, are seen by means of the microscope to fall asunder into a distinguishable variety of divergent elements. It is not proved by this, but it is made probable, that the apparent continuity of the rest merely conceals a juxtaposition of discrete elements. But, what *is* proved for everyone who has eyes to see is, that substances composed of atoms may produce on the senses the impression of perfect continuity of extension. The frequently-urged objection, that a combination of discrete parts would never account for the coherent surface and the solid interior structure of material bodies, does not really require any *metaphysical* refutation. The sharp edge of a knife, when placed beneath a microscope, appears to be notched like a saw,

and the surface, which feels quite smooth, becomes a region of mountains. Spots of colour again, even if seen only from a short distance, take the form of a continuous line. These recognised facts are a sufficient proof that the nature of our sensible organs makes consciousness of what intervenes between successive vivid impressions impossible for us, when the intervals are either empty of all content, or such that they only faintly affect us. Though, therefore, the appearance of continuous extension, no doubt, may correspond with a real fact, it arises none the less certainly and inevitably from a sufficiently close approximation of discrete parts. Now, what induces us to adopt this last hypothesis in explanation of the whole is this, that even substances which seem to be continuous admit of being divided, to an apparently unlimited extent. For, as the parts which spring from this division retain unimpaired the same material qualities which belonged to the undivided whole, it would seem that they cannot owe their origin simply to the division of this whole; but that they existed before it, and formed it by their combination. Later on, I shall give reasons for suspecting the soundness of this conclusion; but, at first sight, it is convincing enough, and in all ages it has given rise to attempts to exhibit the parts of Matter as elements whose metaphysical unity of nature expressed itself in terms of Space as indivisibility.

I shall offer some remarks—not intended to be historically exhaustive—on the forms of Atomism which thus arose. Two points I shall mention here in advance. First, let it be remembered that this hypothesis of a multitude of interconnected points admitting of changeable and precisely determinable relations and interactions, is the only practical means of satisfactorily explaining the extremely complex phenomena for which an explanation is sought; and that in contrast with this explanation, the bare general supposition of the uniformity of Matter, not less than the special one of its continuity in Space, has never led to any fruitful solution of the facts given in experience. To prove this would be only to repeat what has been so clearly and convincingly stated by Fechner (cp. his 'Doctrine of Atoms'). Taking it then for granted that the real world of nature is presented to us primarily under the form of an infinite number of discrete centres of activity, I shall confine myself merely to a metaphysical investigation into the nature of these centres. This is a question which Physics is not practically called upon to decide, nor is her certainty about it at all equal to the ingenuity with which she avails herself of the advantages which the hypothesis offers to her. Again, I am entirely at one

with Fechner in regard to his second conclusion. I believe with him
that the atomic view of the Physical world is peculiarly adapted to satisfy the æsthetic needs of the mind. For what we long to see exhibited
everywhere and in the smallest particulars, is precisely this, organization, symmetric and harmonious relations, order visible throughout the
whole, and a clear view of the possible transitions from one definite
form into another. The demonstration of this point I likewise do not
repeat. I wish only to say that I have never been able to comprehend the reason of that tendency, which for a long time past our
German Philosophy has shown, to look down upon atomic theories as
of an inferior and superficial character; whilst the theory of a continuous matter was opposed to them as quite incontrovertibly a truth
of a higher kind. If there were proofs at hand to establish the necessity of this latter conclusion, they should have been set forth in a more
convincing form than they have yet received. There is, however,
really nothing to admire in the theory of continuity, either when considered in itself, or in regard to the results which have been derived from
it. It seems as if a mystical power of attraction had been given to it
merely owing to the mathematical difficulties in which the whole conception is involved.

189. The following are the chief characteristics of general interest
which distinguished the atomism of antiquity, as represented by
Lucretius. Theoretic knowledge of the changes of things would be
impossible for us, if we were restricted to observation of the co-exist-
ence of qualities, and the modes of their succession; there being no
fixed standard, by which to estimate their relationship, opposition, and
quantitative difference. We cannot be in a position to deduce from
such conditions any conclusion of real value, unless we are able to
exhibit the states which succeed each other as comparable forms of a
homogeneous existence and occurrence, or unless, at any rate, we
can show how effects disparate in themselves can yet be annexed to
comparable relations of comparable elements. The conviction that
this was what had to be shown, led by steps of reasoning which can
easily be supplied to the attempt to refer the varieties of sensible
phenomena to differences of shape, size, combination, and motion, in
certain absolutely homogeneous and unchangeable elements. The
working out of the theory in detail was extremely defective and rudi-
mentary. It was not so much that it was left unexplained how the
sensible appearances which attach to these mathematical groupings
can arise out of them, but the impossible assertion was made that the
sensible qualities *are* nothing but these very mathematical determina-

tions themselves. Setting aside, however, these imperfections, the general conception of Atomism is one of the few Philosophical Speculations of antiquity which have hands and feet belonging to them, and which, therefore, live on and lead to ever fresh results, whilst other theories, with perhaps more head, find a place now only in the History of Ideas. The hard and fast line of distinction that was drawn between the equality of the several parts of Being, as opposed to the inequality of their relations, excluded all original differences from the ultimate elements themselves; these latter, however, if they had been so completely equal, could never have served as a basis for the manifold appearances which spring out of them; they had, therefore, at any rate to be assumed to differ in size and shape.

But this admission was no sooner made than it was seen to be inconsistent with the uniform oneness of all existing things. Hence, these differences were held to obtain merely as facts, which in the order of nature as it now exists cannot be reversed, but which are not in themselves original, having come into Being only at the commencement of the present age of the world's history. At any rate, I think I have shown that Lucretius distinguishes between the multiform atoms, which are the unchanging causes of the *present* order of phenomena in the world, and those infinitesimal and essentially uniform particles, from the combination of which the atoms are themselves ultimately formed. He supposes that there are different ages of the world, during each of which the combination of the atoms for the time being is dissolved by the stream of change. It is only the *combination* of the atoms which is dissolved; the atoms themselves do not change, but are combined afresh. At the *end*, however, of each age the atoms likewise are reduced back to their homogeneous first elements, and these latter being again united so as to form new atoms, are what constitute the material substances out of which are met the demands for the phenomena of the next succeeding age. We see here a recognition of the metaphysical difficulty mentioned above, though not a solution of it; it still remains that the form which the atoms are to assume is determined by an arbitrary cause.

The further elaboration of the system presents little that can interest us. The common nature of what is real, which was declared to be the true substantive existence contained in all the countless atoms, might, one would have thought, have suggested the hypothesis of an inner relation existing between them, and from this might have been developed the conception of forces by which they mutually affect each other; forces, which would assume different modes of operation, according as the ultimate component particles of the atoms were differently combined. But no use was made of this thought. The communication of motion by impact remained as the sole form in which things affect each other ; and the resistance which they oppose to the falling asunder of their parts was no less inadequately explained than the invincible tendency of the ultimate elements to combine in the form of an atom.

190. Passing over the various forms which Atomism assumed after it had been revived by Physical Science, I shall mention only the last of them. As long as extension and shape were ascribed to the atoms, no matter whether all were supposed to be the same in these respects, or, some to be different from others, it could not but appear that a question was being solved in reference to the larger bodies by the assumption of the smaller ones which was left unsolved as regarded those smaller ones. It was impossible to go on for ever deriving each atom from atoms still smaller ; some point of space must at last be reached which is continuously filled by the Real thing. But here a doubt suggested itself. How can the continuous substratum be indivisible, if the space which it occupies is infinitely divisible ? That a portion of space should be held intact against all attempts to encroach upon it, would seem to be conceivable only as the combined effect of activities proceeding from points external to each other, and prescribing to each its fixed position in relation to the rest. Such active points, however, would inevitably come again to be regarded as so many discrete elements, from which the whole is formed only by aggregation. It seems to me that the regression into infinity which would thus result, could not be escaped from by any appeal to the metaphysical unity of the essence which forms the real content of an atom, and which preserves it from the division of its appearance in Space. This distinction between the real essence and its appearance in Space would be a meaningless rhetorical phrase if it did not suggest questions far deeper than any of those with which Atomism is concerned and quite indifferent to it.

Atomism considers extended and tangible matter as reality pure and simple, not as a mode in which Reality manifests itself, and which requires a process of intermediation to connect it with Reality. Now it is most difficult for many reasons to apply to this extended Real thing the conception of unity. I do not mean to maintain that the question is at once decided by the fact that in order for a form of matter to remain unaffected by all external forces, it would have to be credited with a simply unlimited power of resistance, such as would be very little in harmony with the first principles of our knowledge of mechanics. I do not say this; for in the last resort there would be nothing to prevent us from conceiving of the atoms as elastic; and then each atom would really undergo a change of form proportioned to the force acting upon it; only that there would be an accompanying reaction, sufficient to restore to the atom its original outline, and preserve it from disintegration. No doubt, in a sense it is true that the atom would require to have an unlimited power of cohesion in order to admit of this process. But there is nothing in it inconsistent with what we know about mechanics in other respects. The force inherent in an atom would not be indifferent to all external influences; rather, it would react with a degree of intensity precisely corresponding with the original stimulus.

But another requisition must be complied with if the metaphysical unity of an extended real thing is to make itself felt as an actual fact and not be a mere name. Essential unity of nature cannot contain parts, which are affected by experiences peculiar to themselves, and not shared by the rest. Every impression by which the point a of any such unity A is affected, must at once be a state or impression of the whole A, without any process of intercommunication being required, to transmit the impression from a to b, or to the other points contained in the volume. At all events, if the parts of A are so different that what each experiences has to be transmitted to the rest, I fail to see in what would consist its essential unity, or how, since a system of discrete elements would necessarily proceed in precisely the same way, there can be any difference between the two. Before proceeding further, I must guard these statements against a possible misunderstanding. I cannot find that there is anything incompatible between the essential unity of A and the existence at the same time of different modes of its Being $a \beta \gamma$, which are necessitated by different influences acting upon A simultaneously: I only wish to maintain that both a and β are equally states of the whole A, and therefore that they are neither of them produced by influences which merely affect themselves, but are both modified by the fact of their contemporaneous existence in the same essential unity. Let us suppose a and β to be motive stimuli affecting two points a and b in the same atom. The result would not be two separate movements of these two points, which at some later period merged in a common result; but in the point a, which was the part affected by a, the whole Real thing would be present in the same complete fulness as in the point b, which is affected by β. The immediate effects of both impulses would be felt equally at both points, and the resultant ρ would be but one motion which would at once lay hold of the whole extended substance. Further, since every change requires for its occurrence a certain space of time, and according to the law of Persistence, leaves a trace of itself behind, it is quite intelligible that a primary stimulus a should not till after some interval show itself as the condition of the next stimulus β; and that a new impression of the kind a should make itself felt in modifying the states connected with it before it modifies those that are connected with β. When this happens, we are accustomed to say: 'only one *side* of the whole Being of the thing was affected; the other remained untouched.' But by the use of this figure derived from Space, we express most inappropriately our better and truer meaning. At each moment, the whole essential Being is both acting and being acted upon; only it belongs to the nature of this indivisible unity that the several activities which external conditions elicit from it should, as they succeed each other, exhibit the most various degrees of mutual dependence, and should be some more and some less closely associated together.

Let us apply these legitimate ideas to the case before us. What we should be entitled to say would be, not that the atom A responds so immediately to the stimulus a by producing the result a that there is absolutely *no* intervening interval of time, but rather that the reaction in it does always follow upon the stimulus, at however infinitesimally small an interval of time; so that what takes place here too is that A is first affected on its receptive side, and only *afterwards* and in consequence of this on its side of reaction. This imagined splitting up of the substance into parts has nothing in common with the false notion of there being *in fact* any such separation between them, as would be the case, if we meant that an impression a produced upon an atom is confined to a point a, from which point it does not pass on to the remaining points b and c, until after some lapse of time. In such a case, there would, as I have before remarked, be nothing left to distinguish the pretended unity of this A from the communication of effects which takes place in every assemblage of discrete and independent elements when brought into active contact. If we are serious in supposing this unity to exist, we must assert that every motion communicated to a point a in an atom, *is* also literally a motion of the point a^1 at the other end of a diameter of the atom. The motion, consequently, would have to be transmitted all along the intervening line $a\,a^1$ in absolutely no time at all; and the ordinary rule according to which the intensity of a force varies with its distance, would have in this case to be suspended ; the effect produced upon the remoter point a^1 must be as strong as that produced upon a. These consequences which, as it appears to me, are inevitable, cannot be reconciled with the ordinary principles of Mechanics. But if they are to be avoided, either the unity of the atom or its extension must be given up.

191. Physical theories in favour of the latter of these two alternatives have assumed a variety of forms. Though they have not been expressly based on the above-mentioned arguments, which have led me to infer that extension is not a predicate of a simple or single substance, but the appearance assumed by many different elements when combined, they have originated in a general feeling that the very thing which it was intended to explain in composite bodies by means of the atoms, could not be consistently assumed as already existing in the atoms. The extension of the simple elements was not a fact given in experience ; nor was there any necessity for assuming it. All that was required was, certain points in space, from which forces of attraction and repulsion could operate with a certain intensity. The unextended atoms, as the vehicles of these forces, served quite as well to explain phenomena, as they would have done on the almost inconceivable hypothesis of their extension. Hence, since all that was needed was a working hypothesis, it became the custom for Physicists to describe the atoms simply as centres, to and from which Forces and Operations are transmitted, leaving it unexplained how these real points are distinguished from the empty points of space which they fill. This omission may easily be supplied. A real thing could never by being extended in space produce an effect which it was not in virtue of its nature capable of producing when in relation with the other thing in question. At most, the space which it occupies could only prescribe the sphere of operation, within which capacities due not to extension but to the inherent nature of what the thing is, are exercised. If, further, it is impossible to conceive of motion as produced under conditions of actual contact and if distance is necessary to the operation of force, actual reality becomes independent of extension in space, and the elements, though they have indeed positions in space, are without either volume or shape.

This point of view grew up not merely as a conclusion arrived at by Physics ; it is an ancient possession of Philosophy. Herbart refers back to Leibnitz ; for myself, I prefer his own definite exposition to the doctrines of his forerunner, which can only be arrived at by a somewhat dubious interpretation. Herbart's ontology starts from the assumption of countless simple substances without parts or extension, which form the elements of the world. His construction of matter could, therefore, only lead to Atomism : and, he tells us quite clearly what are the original subjects from which the activities formative of matter proceed, and as to which we found Kant's explanation unsatisfactory. Herbart distinguishes his own theory from the theories of the Physicists, by calling it 'Qualitative Atomism.' He gives it this name, not only to show that his simple substances owing to their qualitative differences are endowed with distinct concrete natures, and not merely substantiated abstractions of a single homogeneous reality ; he uses the term in a far more important signification than this to imply that from the inner experiences to which these differences of nature give rise, all these Forces and Laws of relation are derived, which the common modes of speaking and thinking in Physical Science represent, without any further attempt at explanation, as predicates inherently attaching to the ultimate elements. Being, as I am, quite at one with Herbart in regard to this general conception, I regret that owing to a certain ontological doctrine, which I do not myself share with him, he should have been deprived of the fruits of these conclusions in constructing his theory of matter.

The entire independence which he ascribed to each of the essential elements prevented him from holding the doctrine of a pervading connexion, in virtue of which the states by which one is affected become the immediate condition for what is experienced by the rest. Another of his assumptions, the origin of which I am ignorant of, led him to regard contact in space as the only cause capable of disturbing the mutual indifference of the elements and forcing them into active relationship. As, on this view, it was impossible for the essential elements to act on each other from a distance, Herbart became involved in the hopeless attempt to show how points unextended, though real, are brought into contact in order that they may act upon each other, but yet not absolutely into contact, in order that their combined effects may endow a multiplicity with an extension which attaches to no single one of its component parts. It is a view which requires to be changed only in a single point, though this no doubt is a vital one. The simple elements of reality, on which the constitution of the world primarily depends, must be regarded as conditioned, not independent, and therefore as in unceasing relation to each other. By making forces which act at a distance emanate from the simple elements, elements not empty but of a definite internal character, we can frame an intelligible picture of the forms of matter, as systems of real unextended points, limited in space, and endowed with forces of cohesion and resistance in very various degrees.

192. Now, at this point we might stop, if it were not for another assumption which these theories commonly contain, that viz. of an actually extended space, in which the real elements take up their positions. The contrary conviction, in support of which I have contended, compels me to introduce some further modifications into the view which I have first stated in order to arrive gradually at the idea which I wish ultimately to establish. I continue for the present to make the assumption of an indefinite number of individual existences; an assumption from which the explanation of the variety of phenomena must always make its first start. Not much need be added to what has been already said as to the general relation of these existences to space. These simple elements, having as such no connexion with space, stand to each other in a vast variety of relations, which only for our modes of apprehension assume the forms of position and distance in space. It is for Psychology to supplement the suggestions which have been already made by telling us how this mode of apprehension is originated; here, we are only concerned with the ideas which we must form of the nature of Real existence, in order to make intelligible the particular mode in which it presents itself to our subjective consciousness.

In the first place, then, to repeat what I have already mentioned, it is requisite that we should reverse one of our ordinary ways of thinking. When a certain element a is in a certain position a, we think of this fact as if it was something in itself, as if it was in virtue of this that the element had the power to produce effects on other things in certain definite ways. But, according to all the results at which we have so far arrived, we ought contrariwise to say :—That the element a 'is in the position a,' can only mean for it, that it has received so many and such impressions from all the other elements which belong to the same world to which it belongs, that, if we regard the whole mass of existing facts of that world under the form of space no place except a corresponds to that which is assigned to a in the universal order. Hence, the position which an element occupies must always be regarded by us as the result of the forces that determine it, and in so determining it, are in a state of equilibrium. This conclusion the Science of Mechanics only half admits. It admits, no doubt, that during every moment that an element remains at rest, the forces working upon it must be in a state of equilibrium. But the conception still remains possible that an element might occupy a position in space without any action of force whatever, and that forces arising subsequently might find it there and act upon it.

Further, I have abundantly shown that by this systematised arrangement of unextended points, which I believe to be what constitutes the world as a whole, I understand not the order of a rigid classification, but an order which, incessant as is the movement of things, and manifold and various as are the forms which the sum of conditions at each moment assumes, maintains throughout a continuous and unchanging purpose. The position, therefore, which an element assumes, when it appears in space, does not simply indicate the place which it occupies from all eternity in a classification of the world's contents, but, rather, the place which, at that moment, was the only point at which the changing conditions to which it is subject came to a changeable equilibrium. It would be too simple an explanation of what takes place, to suppose that when two elements a and b make their appearance at two points of space in close proximity, a and β, they have been accredited to these positions owing to the special sympathy of their natures or the intimacy of their interaction. Rather they may have been quite indifferent to each other, and yet have been forced into this juxtaposition, simply because the demands made by all the rest of the elements and their motions can find no better satisfaction than in the momentary proximity of these two elements, though it may not answer to any vital connexion between the elements themselves. Reflexion upon this constant motion of the world will cause us to modify our previous view, or, at all events, to define it more accurately. The position a of an element a, though always no doubt it expresses the balance of the several forces for the time being affecting a, may also at the same time be the expression of an unavoidable want of equilibrium between the present state of a, and that state to which its nature gives it a claim in the totality of existence; an expression, therefore, of a discordant Tension, which remains until, in the course of events, the causes which occasioned it again disappear.

I make these remarks, in order to give an idea of the complex kinds of relations which here present themselves, and in order to remove the impression that there is any correspondence between the appearance of the world in space at any given moment and an intelligible order of things, in which the position of each element would correspond with the conception which permanently represents its nature. But I hasten to add that this reference to a disproportion of

states, in the above-mentioned sense, must not be mixed up with any
secondary associations of ' that which ought not to be,' ' that which is
out of place,' or ' which contradicts the purpose of the whole.'
Whether anything of this kind ever happens, whether, i. e. there is
anything in the world's course which can be compared with discords
in a musical progression, we shall not here enquire.   The disproportion of which we have been speaking is primarily nothing but the
impulse to a change of state which is suggested by the course of
events, and which tends to or accomplishes the transition to positions according as it is impeded or unimpeded.   Turning now from
these general considerations, we will apply ourselves to the solution
of certain special questions, which acquire from our present point of
view either the whole of their significance or a different significance
from that which is commonly assigned to them.

193. Let us start as before from the supposition of a given plurality
of active elements; remembering at the same time how frequently
it happens, as has been proved by experiments, that apparently
different properties are really only the result of different combinations of a single homogeneous substance.   The question will, then,
obviously be, must we, in order to explain the facts, assume the existence of a multiplicity of originally distinct materials? or, shall we
explain even the characteristic differences between the chemical Elements as mere modifications of a single homogeneous matter?   The
eagerness which is now shown in favour of the attempt to explain
away these differences seems to me to be based to some extent on a
false principle of method.   For practical purposes Science is, of
course, always interested in reducing the number of independent
principles upon which to base its explanations, and in making calculable the course of events by subordinating the complex derivative
premises to a few primary ones.   But not less certain is it that
Science cannot desire any more complete unity than actually exists,
and until the point is decided by experience, a unity which remains
still unknown must not be presupposed as certainly existing except
in cases in which without it a contradiction would be introduced
into the nature of the subject-matter.

Now, our idea of Nature implies three things.   (1) A system of
universal laws, which determine the sequence of cause and consequence.   (2) A multitude of concrete points to which these laws may
attach and so find their application.   (3) Lastly, a purpose to realise
which these actual existences are combined together.   Every theory of
Science admits the two first of these postulates ; the last is, no doubt,

the subject of conflicting opinions. But wherever the thought of a purpose in nature is cherished, it stands to reason that there can be but one, and that all seemingly independent tendencies must be really subordinate to this unity and appear as moments in its Being. Not less necessary is the unity of the supreme laws which govern the connexion of events. These consist not so much in the rules to which various forces variously conform, as in the universal truths of mathematics, to which any self-consistent world, even though it were quite otherwise constituted than the existing one, would always have to submit throughout its whole extent alike. It is impossible to conceive an order of nature, unless it can be determined according to the same rules of measurement in every instance what results may be deduced from the presence of active elements in given proportions, and from their reciprocal interactions in calculable degrees of intensity. On the other hand, the actual existence, which has to furnish these laws with cases in which they will apply, has to fulfil no requirement but the primary one of being manifold. Nor is there the slightest reason why a theory which takes no exception to the doctrine of an original plurality of homogeneous atoms should regard with suspicion the hypotheses of original differences of quality. No further likeness of nature need be attributed to the atoms than such as is required to enable them to combine together in the same order of things. It must be possible in so far as they affect each other by way of interaction, to exhibit their natures as combining in definite degrees of intensity certain universal modes of activity. But there appears to me to be no necessity for regarding the group of specific coefficients which these general modes of action are found to take in any particular element, as attached to a substance of like nature throughout, or, more strictly, as attached to what is merely the substantiated abstraction of reality. The group may equally well be regarded as the expression of a specific quality, so far as such expression is allowed by the mutual intercourse of the various forms of matter.

Practically the importance of the difference between these two views would consist in this, that the latter would altogether exclude the possibility of one chemical element passing into another, whilst, according to the former view, this would be at any rate conceivable. It would indeed be more than conceivable. It would rather be inexplicable that throughout the endless process of combination, dissolution and transformation to which the parts of matter are subject, no element should ever lose its identity or merge its own individuality

in that of some other. If the essential character of each element
depends merely on a peculiar arrangement of homogeneous particles,
it may be conjectured that the same course of events which gave
birth to one of the forms thus composed, might again produce the
conditions which would lead to its being either dissolved or transformed into some other shape. But if it was meant that it could
be shown that there are certain forms of combination which having
once originated can never by any possible conjunction of forces
be dissolved, it would still be open to ask, Why at any rate there
is not, through a further composition of the simpler structures, a
constant increase in the number of these irrevocable combinations?
Finally, if it is to be regarded as an eternal fact that these combinations are all alike indestructible and at the same time incapable
of further development, it would be difficult to say in what would
consist the difference between this view and that which assumes an
original difference between the elements. As regards the practical
explanation of nature there would be no difference between the two
ideas; it would be a difference merely of theoretical view. The
probability of all reality being homogeneous in essence, unless confirmed by future experience, could only be maintained upon considerations of a different and more indirect kind.

194. To this class of considerations belong the views commonly
held in regard to the mass of matter, its constancy, and its influence
in determining the character of different kinds of effects. It is now
quite superfluous to recur to what was once a mistake of frequent
occurrence in philosophy, by pointing out that the idea of mass is
not exclusively associated with that of weight and heaviness; but, that,
as applied to the reciprocal action of any two material bodies, the
term expresses the intensity of the force which each contributes to
the common result. Let us suppose that we have formed two bodies
from m and μ numbers of units of the same matter, and have observed their behaviour to a third body c in regard to a certain effect of
the kind p. If having observed this, we then find that two other
bodies, both demonstrably formed from the same material, behave
in the same way towards c as the two first in respect of the same
effect p, we rightly conclude that they also contain m and μ number of
units of the same matter. Suppose, however, these two latter bodies
exhibited divergent properties, so that their consisting of the same
matter was open to doubt, and yet that, in regard to p, they were
affected towards c precisely as those two substances had been which
we had ourselves formed from a demonstrably common matter, it

would no doubt be a natural and obvious conjecture that their behaviour was also due to the presence of m and μ units of a homogeneous substance, though this likeness was hidden in their case by secondary differences of quality. At the same time, this conjecture would go beyond the facts. All that the facts teach is that in respect of the effect p, the two bodies in question are *equivalent to* m and μ numbers of the before-mentioned matter; not that they actually *consist of* them. There is nothing to prevent them from being in all other respects different in original quality from each other, and from c, and yet being capable of a special interaction of the kind p between them and c, in which their contribution to the common result admits of a numerical expression m and μ, as identical or comparable with that of the two bodies first considered. If now, assuming them to have the above specific quality, we proceed to consider their interaction with a fresh body d resulting in a different kind of effect q, we shall not be justified in assuming that the proportion in which they contribute to this result is the same, viz. $m : \mu$, as that in which they contributed to produce p. Rather, it is conceivable that in their new relation to d, bringing into play as it would new forces, they would be like where they had before been unlike, and unlike where they had been like ; or, in a word, that in regard to the effect q, they would assume the quantities m_1 and μ_1, different from the previous quantities m and μ. In point of fact, at any rate at first sight, this is how the several effects p, q, r, produced by the reciprocal action of the bodies in question, are related, and it is never certain that a which in regard to p is greater in quantity than b, will still remain so in regard to q.

Whether these differences can be intelligibly explained on the hypothesis of a homogeneous matter, as secondary effects due to different modes of combination, must here be left undecided. Owing to the extreme variety of the phenomena to be taken into account, such a conclusion could only be established, if at all, in the distant future. On the other hand, it is of course always possible to express each of the new quantitative determinations that arise, e.g. m_1 and μ_1, by means of the old ones, i.e. by $k\,m$ and $\kappa\,\mu$, and so by assigning for each kind of special effect a specific coefficient to bring the fiction of a homogeneous mass into harmony with the given facts. In a metaphysical point of view this would decide nothing. The possible qualitative difference between the parts of matter is as little made to disappear by this reduction, as corn and meat cease to be two different things after their value has been expressed in the common term of money. The doctrine then which I am maintaining is not open to any general objection on these grounds, though it cannot be applied to explain the particular facts. It, at any rate, does not oblige us to think of the different elements as differing without any principle. Belonging as they would to one and the same world, their qualities would be mutually related members of a single interconnected system, within which they would be combined in different directions, in different senses, and with various degrees of intimacy. Stationed at the meeting-point of many opposing tendencies, an element might on one of its sides display a greater degree of force than its neighbour, on another an equal degree, whilst on a third side its force might be less; and, if we knew the purpose of the whole system, which we do not know, we should be able to deduce from the mass which an element exhibited in the production of the effect p, the specific coefficients which belong to it for the actions q, r, &c., and to exhibit those coefficients as a series of mutually dependent functions.

195. I have made these observations, still proceeding on the assumption that a plurality of individual elements is what forms the ultimate constituents of the world. We shall see, however, that they have equal force, if viewed in connexion with a result established by our ontological investigations, according to which these multitudinous elements are but modifications of one and the same Being[1]. To hold this latter opinion, seems at first to be equivalent to repeating the very view against which we have been contending. It appears as if we could have no real interest in establishing the fact of difference amongst the elements, if this is not to be regarded as ultimate and irremovable. But the doctrine here maintained, is essentially distinguished from the doctrine of Physics. I understand by this absolute Being, not a Real existence infinite in quantity and of like nature throughout, which has no other inherent capacity than that of falling into countless homogeneous parts, and which only is in a secondary sense, by means of the various possible combinations of those parts, the ground of a diversity existing in the content of the world. I conceive it rather as a living idea, the import of which, inaccessible in its essence to any quantitative measurement, is no mere homogeneous aggregate of ideas, but a self-articulated whole of variously interwoven parts; each one of these parts, as well as the several elements which compose it, acquiring a determinate quantity according to its value and position in the whole.

Let us give an illustration. If this idea could be expressed in terms of our thought, it could only be so by means of a number of propo-

---

[1] [Cp. Chapters vi. and vii. of Book I.]

sitions which would be towards one another in those extremely various modes of dependence in which the different parts of a scientific system are connected together. But these principles would be meaningless, if they were not again composed of words—words of which the meanings while different and unchangeably fixed, are still not immeasurably different, but so precisely determined in relation to each other that they admit of being joined together in very various syntactical combinations, to serve as vehicles by which the Idea is articulated into its parts. With these words I compare the elementary materials of nature. In themselves they are nothing; they are merely forms of a common principle underlying the world, a principle, however, which maintains them as constantly uniform activities, so that in every case in which they occur and enter into mutual relations they observe the same laws of behaviour. But, although thus involved from all eternity in a network of relations, they still remain different as regards each other, and incapable of being referred to mere division and re-combination of a uniform substratum. The mathematical mode of regarding the question which favours this latter view, and which has very extensive rights in the treatment of nature, is still not the only way of conceiving its unity, nor does it penetrate to the ultimate ground of things. Merely, within the limits of our observation, this mathematical connexion of things, secondary though it is, presents itself first. That whole world of quantitative and numerical determinations is itself based on an order of things, the synthetic connexions of which we could never have arrived at by any logical analysis. We have called this order 'systematic,' and now we may replace this imperfect expression by another, that of the 'æsthetic' unity of purpose in the world, which, as in some work of art, combines with convincing justice things which in their isolation would seem incoherent and scarcely to stand in any relation to one another at all. Or, lastly, we might prefer to use the term 'dialectical unity,' in memory of a late phase of our German Philosophy, which was thoroughly alive to the truth of this doctrine, but failed, as it seems to me, because it believed itself able to apply to details of fact principles which can only in a rough way prescribe a general direction to our thoughts.

196. This transformation of our views introduces us to a further question, which to Physical Atomism appears to be no question at all. It is assumed that a countless number of individual atoms fill the world. Now, be they the same or be they different in kind, whence comes their plurality? If they are regarded as starting-points to be

assumed, beyond which we cannot go in thought, no doubt their dispersion throughout space can also be included in the number of facts to be taken for granted, which we must recognise without attempting to explain.   But, to us, who have conceived every qualitatively distinct element as one of a connected series of acts emanating from the supreme principle of the universe, it is necessarily perplexing to find that the instances in which each element occurs are scattered over a countless number of different points in space.   Nor is this an enigma merely from our point of view.   We can, no doubt, by an act of thought easily represent to ourselves the same content *a* a thousand times over, and we can distinguish the thousand creations of our imagination, by localising them at different points of space, or by enumerating them according to the different moments of time when they first suggested themselves to us.   But how, strictly speaking, can it be conceived that, in actual fact, the same *a* occurs several times over?   Must not the mere fact that there are several, make it necessary that *a* should be in one case something different from what it is in another, though it ought in every case to be the same?   What constitutes the objective difference between them, which makes a truth of fact of the logical assumption that they are so many like instances of a general notion or a common nature?   We remember what a stumbling-block this question was to Leibnitz.   It seemed to him to be impossible that two things should actually occur, unless their duality was based on a difference of nature between them.   He would not even allow that two leaves of a tree could be exactly alike.   This difficulty scarcely attracts any attention now.   I must say, however, that it seems to me to have been somewhat too hastily passed over by those who have followed in the footsteps of Kant.   What Thought could not achieve, was held to be made possible by spatial perception.   It was in and through space that it was clearly shown how things could be at once like and manifold; they might differ in position, but be perfectly identical in the nature which occupies the position.

Certainly, this is clear enough; but I cannot see in this clearness a solution of the difficulty.   All that it does is to bring the problem itself vividly before us; but for this phenomenon, indeed, the difficulty would scarcely have been suggested.   Now, if science admitted to an unlimited extent the possibility of things acting upon each other at a distance, it might no doubt be granted that one atom is never subject to precisely the same sum of external influences as another.   And, if it were further granted that the atoms experience changes of inner

state corresponding with these external influences, it would follow
that an atom *a* would be in some way different at each moment of
its existence from a second and otherwise similar atom, since its
internal states at any moment are not an extraneous appendage to
its nature, but an actual constituent of what it is at that moment.
But this mode of statement would still involve the latent supposition,
that though the states by which an atom is affected change, yet
through all this change the atom itself remains as a constant
quantity, which would have maintained itself in its position even if
there had been no forces acting on it, and which only becomes
distinguished from other atoms of the same kind as itself owing to
external influences which *might* not have been operative. Thus we
are brought back to the question, What does it really mean that an
element occupies a point in space? and how can it be that in virtue
of its position it is distinguished from other elements, seeing that all
points of space, both in themselves and in their effects on the
elements, are precisely alike? I have tried to give an answer to this
question. Its very terms are meaningless from the point of view
which regards space as something actually existing by reference to
which things are determined. Things do not first find themselves in
certain positions, and then become enabled to take effect, but it is
the kind and degree of the effects which they already exercise upon
each other that makes them occupy those positions for the perceptive consciousness, which seem to us to be those which originally
belong to them.

This answer, however, does not at once remove our present difficulty. In order to find a reason why these qualitatively-distinguished
elements should assume the form of a scattered multitude of individual atoms, it seems as if we should be compelled to suppose
that in the intelligible world, which reflects itself in space, that action
or thought which we designated as the nature of an element, must
repeat itself as often as its phenomenon. Is such a repetition less
unaccountable than the easy hypothesis of a plurality of atoms with
which Physics is content? I feel myself able to answer in the
affirmative. It is merely owing to the effect of constant association
with the forms of space, that when we come to represent to ourselves
these 'repeated' actions, we conceive them as falling into a number
of disconnected groups separated from one another by empty intervals
just as the parts of space are separated by their lines of distance.
There is really no such relation between them. Just as in our own
inner experience the self-same principle or the same conception recurs

in the most various connexions, and exercises a limiting or determining influence of many different kinds on any other of our thoughts
with which it happens to be associated, just so the idea, which determines the qualitative nature of an element of matter, serves in the
order of the universe as a point of intersection for the different
tendencies which make that universe into a connected whole; connected, as I must again insist, not merely as a rigidly classified system,
but as an eternally progressive history. We are, therefore, not called
upon, nor are we interested to maintain that there are distinct special
existences corresponding in number with the functions which the
same idea must fulfil when thus associated with others in these
various combinations. The number of scattered atoms is merely the
number of the separate appearances which an element assumes to
our perception of space owing to the manifold relations in which it is
involved with other elements.

197. The extremely paradoxical nature of this conclusion shall not
prevent me from mentioning also a certain corollary which follows
from it. We have now arrived at a point of view from which the
atomic theory can no longer satisfy us, not even after that transformation of its fundamental idea, to which it seemed not to be disinclined.
So long as the unextended points, from and to which forces proceed,
points which have indeed positions in space but no volume, were conceived as having, not less than the extended atoms had previously
been conceived as having, an obstinately indestructible nature, there
could of course be no mention of a further division into parts; since
that which was to be divided was, as its very name implied, indivisible.
This mode of representation no longer holds good. If the single real
idea which determines the nature of a qualitative element necessarily
manifests itself under a number of distinct forms, and if there is no
limit to the multiplication of the relations which it may assume towards
other ideas, why should it be specially attached to just those points in
space where it happens to be active at any given moment? Why should
not the positions which it may occupy also admit of being multiplied
indefinitely, seeing that none of the manifestations of the element
have any other claim to a separate existence except such as depends
on the mandate of the whole order which assigns to them this and no
-other position?

Not that I have any desire to return to the notion of a continuously
extended, infinitely divisible matter, nor to that other notion, according to which the real atom, at least in its spatial phenomenon, is
quite continuous in the sense that it is equally present at every point

within its own narrow volume. I would rather not in any way depart from the results of Atomism most recently arrived at, according to which an atom is conceived as developing its activity from a geometric point. But I can see no reason for regarding the amount of force which is thus diffused—a force which is now no longer in any sense an indestructible metaphysical unity—as eternally attached to this one point. Rather, it would admit of partition in space, just as it is itself only a partial manifestation of a single identical function of the whole. In proportion as new combinations of phenomena were required to exist by the course of the world, each centre of activities would have the power of breaking itself up into several centres, which would then assume different positions in space according as the new conditions to which they were subject prescribed. These conditions may be very different. Their effect may be not merely to compel the new centres of activity to combine with atoms belonging to other elements, but also to cause an increase in the volume of any particular atom by forcing its constituent elements to expand and fall asunder. There would thus come to be differences in the density of the atoms. Owing to this constant process of inner dissolution, new points of departure for effects would be multiplied, and there would arise the appearances which were formerly believed to be only capable of being explained on the hypothesis of a continuous and real extension in space, and which are only accounted for at the cost of a permanent improbability by those who believe, with ordinary atomism, that all things are ultimately analysable into real existence and empty space. In this way we should be brought to the idea of an infinite dynamic divisibility of unextended atoms, an idea which, it is to be hoped, will seem less frightful than the barbarous name by which, in order to distinguish it from the traditional theories, I believe that it may most briefly be described. It will no doubt have been taken for granted that the degree of intensity, or, to put it shortly, the mass of each of the parts will be diminished, while the sum of these masses remains the same. I have nothing to say against this addition, but the principle on which it is made will require further discussion in our next chapter.

CHAPTER VII.

The Laws of the Activities of Things.

Of the inner movements of things we know nothing. Still less do we know what are the constant modes of co-operation which the order of the Universe requires them to assume. Hence, experience alone can discover to us the motive forces into which the course of natural events can be analysed and the law according to which each of these several forces may be conceived as taking effect. But a sufficiently careful and comprehensive observation has long since established certain general results, which deserve, by way of supplement at any rate, to receive an interpretation in connexion with metaphysical views, and which suggest the question whether they are really nothing but the expression of what has been observed to take place, and not rather of necessities of thought to which experience has directed our attention only subsequently? I shall attempt to investigate this point, though well-knowing beforehand that my labours are not likely to produce any considerable result. They will serve merely to draw attention to the ambiguity of those speculations, philosophical no less than scientific, which will never cease to be directed to this unpromising subject.

198. In the first place, it is universally admitted that the intensity of the effects which a force produces at a distance, is dependent on the interval between the elements between which it operates. And to this conclusion the doctrine which is here maintained must also lead, though it remains to be seen later by what steps. If the positions of things in space are merely expressions of the forces which are already acting upon them, *a fortiori* every impulse to further activity will depend upon these interactions between the elements and on the distances in which those interactions manifest themselves. This merely general characterisation is not, however, enough to determine precisely the nature of the connexion between forces and the distances of elements. But the other assumption,

which is asserted with almost as much assurance as the last, viz. that the intensity of the effect is in an inverse ratio to the distance, has nothing to recommend it if we exclude the familiar instances furnished by experience, except the inadmissible idea that space acts as an obstacle which cannot be overcome except by a partial sacrifice of force.

Other preconceived notions combine with this one to produce an impression that this decrease of force is a fixed law, holding good in all cases in which forces act at a distance. That a force, emanating from a certain starting-point, diffuses itself through space, is not merely our mode of expressing the fact that its effects differ in degree at different distances. Unfortunately, we believe ourselves to be describing not only a fact but an actual process by which the necessity of this difference is explained. As the force is transmitted to larger and larger spherical shells it seems as if its tenuity must increase in the same ratio as the area which it occupies, the ratio of the square of the distance from its starting-point. ˙This coincidence of a simple geometric relation with a general law which we see illustrated in the effects produced at a distance by gravitation and by electric and magnetic agencies, is too tempting not to invite often-repeated attempts to establish the closest connexion between them. None of the assumptions, however, which are required as links in the connexion can be admitted. A force cannot be supposed to proceed from a point c, without at once being regarded as an independent fluid medium. That its tenuity should increase with its increasing extension, would no doubt not be altogether inconceivable. But still we should have to discover to what the motion of the fluid was due. This could only ultimately arise from a new force, a force of repulsion, exerted upon the fluid by the thing present at c. We should have also to show what becomes of the force thus diffused, if it meets with no object on which to take effect, and further from what source the constant supply of force at c is derived. These questions cannot be evaded by supposing that the force does not *diffuse* itself around c, but is, as it were, a permanent atmosphere already diffused around it. To deny the fact of the movement of radiation, would be to take away the only justification for the principle that the density decreases with increase of distance, whilst it would contribute nothing towards the explanation of the effect eventually produced. Let us, then, suppose that a given force whilst proceeding from c, meets in the point p with an object which it is to act on. How is this action possible? and how can the force impart to the body

motion in any particular direction? All that could be concluded
from the arrival of the force at p, would be that it was now present at
that point—not, that a body situated at that point must, owing to the
action of the force, be set in motion. But, even granting that it were
thus set in motion, what direction could the motion take? The
motion as such could stand in no relation to the point c; for, if the
activity of the force is made to depend on this process of its diffusion,
it follows that it only acts at p just as far as it is there; it makes no
difference whether it was there always or whence it came there.
Supposing it then to coincide exactly at p with some real element,
it could not impart any motion to that element, for there would be
no reason why it should prefer one direction to another[1]. If, on the
other hand, we suppose that at the first moment of its beginning to
exercise its activity, the force *is* separated from the element by ever
so small an interval, we are making action at a slight distance serve
as an explanation of action at any and every distance, though we
cannot bring the former under any definite law, and must therefore
fail in the attempt to deduce the law of the latter.

Even if these difficulties could be got rid of, it would still remain a
question whether the resulting motion will take the direction $c\,p$, or the
opposite one $p\,c$. For this process of radiation would be just the same
for an attractive and for a repulsive force. Each smallest particle of
the fluid would, in such a case, still have to exert attraction or repulsion
upon whatever it might meet with at the point to which it had come,
as a property peculiar to itself and not admitting of further explanation. But, if that were so, there would be no longer any occasion for
confining these effects to the parts of the force which come *before* p
in the line $c\,p$; the parts on the other side of p, which lay in the
course of this line when produced, or which had come there, would
exercise an influence on the element at p, of the same kind though
in a contrary direction. The ensuing motion would then be the
result of these different impulses; at any rate it could not correspond
with the simple law which it was hoped could be deduced from it.
Finally, the attempt may be made to get rid of these difficulties, by
supposing that the radiating force imparts its own motion to the
element which it lights upon, and determines by its own direction
that which the element in its turn is to take. Putting aside, however,
that this is a transition from one idea, that of a force acting at a
distance, to another, the idea of communication of already existing
motion, all that would be explained by this method would be the

[1] [Cp. § 185.]

centrifugal effects of repulsion; every case of attraction would require a centripetal pressure, such as has, indeed, often been assumed, but has not hitherto had any intelligible explanation given of it.

199. On these grounds I not only hold that these attempted deductions have failed to establish their own special conclusions, but the spirit in which they have been undertaken seems to me to be inconsistent with itself. Of course, many of the occurrences which take place in the world are of a compound character, and arise from mechanical combination of others. It is possible that gravitation and other similar phenomena which seem to us to be the expressions of the simplest primary forces, may really be compound results produced by forces still more simple: An elaboration of experience so advanced as to show this to be the case would really have succeeded in furnishing us with a genetic theory of the Law of Gravitation. If, on the other hand, these and all effects of a like kind are regarded as the expression of simple and primary forces, we must not attempt, as is done by these theories, to give a mechanical explanation of their origin, by referring them to a diffusion and attenuation of force. The only proof that can be expected of these elementary processes and their Laws is the speculative one, that they have a necessary place in the rational order of things. The *Ratio legis* might be given, but not the machinery by which it is carried into effect. The treatment, then, of this problem belongs, without doubt, to Philosophy, nor do I complain that there should have been such innumerable attempts to solve it, though unfortunately I know of none that has been successful. I do not therefore continue my own investigation with any hope of arriving at a result that can be final, but merely in order to bring out more clearly some of the distinctive features of my general view.

200. Owing to the doctrine which I have already expounded in regard to the nature of forces, I do not feel touched by an objection which Physicists have urged against the absolute validity of the Law of Gravitation, an objection which, if it held good, would render untenable the whole of this doctrine which speculation so obstinately attempts to deduce *a priori*; where the distance = o, the attracting force must according to this law, it is said, be infinite. I will not now stop to enquire whether this result is altogether inadmissible. It would be open for those who maintain the ordinary hypothesis of a continuously extended matter to urge that contact takes place only between points, lines, or surfaces without thickness, and consequently that the masses whose distance = o, must in every case themselves = o

also. If the hypothesis of unextended atoms conceived as points be preferred, we should certainly have to ascribe to them an infinite power of resisting separation, in case they had once got united in the same point by attraction. But all that would be necessary would be to take care that such a case never arose. It would be easy indeed so to alter the formula of the law, that in case of all observable distances, even the smallest, it should correspond as nearly as possible with the results of observation, while in the case of vanishing distances it should still not imply the infinity of the force of attraction. But, I think we can achieve the same end, without introducing a modification of the law such as would be purely arbitrary and incapable of ever being proved. All the several forces which Physics is led by experience to assume, stand in our view merely for the various components into which the single power of interaction inherent in the nature of things admits of being analysed. It is not therefore at all surprising that a law which expresses with perfect precision the operation of one out of this number of components should nevertheless yield infinite degrees of intensity or other inapplicable values if the component is supposed to continue its operation isolated and uncontrolled. These cases of isolated action are precisely those which are never met with; they express merely what would occur under certain imagined conditions, but what under existing conditions, never does occur. Hence it is not necessary to modify the formula of the well-known law of gravitation, considered as simply claiming to indicate the variations of the attractive action; in this sense the formula may be perfectly precise; only the limiting case never occurs for which alone it would yield such problematical values. In proportion as the elements which are attracted approach each other more nearly the tendency to repulsion will be found to grow even more rapidly, and if any one of the proposed modifications of the law could be shown to hold good in actual experience it would not be a more correct expression of the attraction taken by itself so much as of the total effect in which attraction and repulsion are already united. Moreover, it is easy to see that without this supposition this partial law expressing mere attraction would yield results which would be not so much inconceivable as merely inapplicable in our view of nature. Let us suppose two elements a and b between which there is attraction but never repulsion, to approach the point c from opposite sides. They would then at the moment of meeting have not only an infinite attraction g, but also infinite though opposite velocities $\pm v$. Now as the velocity last reached, v, has arisen by the summation of all the accelerations which have been increasing infinitely up to the value *g*, we cannot but regard the infinite quantity *v* as greater than the infinite quantity *g*. Hence, if there were no repulsion, *g* could not prevent the two elements from passing with opposite velocities through each other's midst and thus distance would be restored between them and the amount of their attraction would become finite again.

201. A special objection to the received views has been urged by Herbart. He will not himself admit the operation of forces at a distance ; but for those who do admit it, he holds that the only legitimate assumption is this, that the intensity of each force is diminished in proportion as it is satisfied by the attainment of its result. That a repulsion, therefore, should decrease with the distance which it has produced, requires no explanation. On the other hand, the force of attraction, which becomes always more intense in proportion as it has drawn its object nearer, remains a paradox for him. This objection is plausible enough if the object is to explain the observed effects of the law by reference to its inner meaning; but I cannot think that the particular psychological analogy, to which it owes its conclusiveness, will admit of this general application. I entirely agree with Herbart that there are inner processes in things, from which the forces moving them are derived, and I will concede to him that in both the cases which are here brought together—psychical endeavour and physical motion—the impulse to what is done lies in a difference between a thing's actual state of being and some other state, which, if it could be realised, would be more in correspondence with its nature. But I dispute the conclusion which is so hastily drawn from these premisses. Herbart shows himself in this matter to be influenced by his main conception, according to which each changing state of a thing is a disturbance of its original nature, so that the only manifestation of activity which can fairly be attributed to real existence, is that of self-conservation or recurrence to the *status quo ante.* In that case, no doubt, supposing *M* to be this permanently fixed aim, and *q* the state which is a departure from it, the result to be achieved at each instant would correspond to this difference *M* − *q*. Strictly speaking, it does not at once follow from this, that the intensity of the force which exerts itself to recover the former state, must vary directly as the amount of divergence represented by *q*, and inversely as the result already obtained. All that would be measured by *M* − *q* is the extent of what is required in order to attain the given end. But there would be nothing to prevent the force from continuing to operate with unvarying intensity until this difference had been made to disappear, just as a labourer in filling up a pit does not at first work more rapidly and afterwards with less energy, because the space to be heaped in was at the beginning larger and has since become smaller, but he works throughout at the same pace.

Even, however, admitting this assumption, insufficiently proved as it is, I doubt the relevancy of the analogy which would make the occurrence of a physical effect correspond with the satisfaction of a psychical impulse. If, taking an imaginary case, we compare a supposed quantity M, of which we have an idea, with a smaller given quantity q, no doubt we know that $M-q$ expresses the amount which must be added to q, in order to make it $= M$. In this case, M, though not present before us in external reality, was as adequately represented by its idea as q was, and the estimation of the difference between them was thus made possible. If, however, we experienced a state q, and this were merely a manner or mode of our consciousness, some form perhaps of feeling, this feeling would not be able by itself to produce in our minds the idea of the absent M. Knowledge of the character and extent of the difference between q and M could only arise if we had had a real experience of M as well, and it were to enter into our consciousness in the form of a feeling or the remembrance of a feeling similar in kind to q. Although, therefore, the disturbed state of our feelings may depend upon the difference $M-q$, yet this difference only exists primarily for the comparing mind of an onlooker; it is not a real element in the experience of the being who is affected by the state q, at any rate not unless there is some remembrance of the state M. It cannot therefore be the obvious standard by which such a being, with a sort of preference for what is reasonable and just, determines the intensity of the effort which it has to make. However far, then, we may go in assimilating the inner states of things to processes of mind, so long as we do not believe that the physical operations of things are regulated like acts of our own by rules drawn from experience, as long as we believe rather that there is a necessity imposed on them to come to pass as they do, the difference between an actual and a better state of things cannot be the determining reason by which Physical effects are regulated.

On the other hand, there is no reason, we at least can see none, why the order of the universe which prescribes to all things their nature and mode of working, should not have attached to q a blind and unpurposed activity, which was as a matter of fact measured by the difference $M-q$, though the individual thing which was affected by the difference was itself unconscious of it. But, as there is nothing to hinder this supposition, so there is nothing to make it necessary; it remains a possible though but an arbitrary assumption that the course of things is nothing but a constant effort to attain to an equilibrium and to reproduce a state M, which can only be effected by getting rid of their present state. There is equally nothing to prevent us from admitting the claims, though not the exclusive claims, of the opposite view, according to which the attainment of a state q means a change in the condition of things, which tends to reproduce itself in a more emphatic and intensified form. That other theory, the watchword of which is 'disturbance,' has thought only of pain; and then it seems quite natural that the self-conservative activity directed to the removal of pain, should decrease in proportion as it succeeds. It has taken no account of pleasure, which just as naturally creates a stimulus to the intensification of the state which was desired and is pleasant. For it is not true, except in those cases in which the source of enjoyment lies partly in the body, that pleasure is dulled with satisfaction. The body, no doubt, is forbidden by the habits of its action from contributing to the intensification of feeling, and interrupts it by weariness and satiety. It will not, however, be maintained that the pursuit of knowledge and its results, or the aspiration after beauty and goodness, is lessened by approximation to the ideal.

But we will leave these analogies, which decide nothing. The general conclusion to which we come is this: there is a blind tendency in each thing, owing to its place in the all-embracing order of the world, whenever it is in any given state q, to produce an effect. The character and extent of this effect are not regulated by any law inherent in the nature of substance or force, and binding things without regard to the purport of this universal order. It is this order and this alone, which, in accordance with its own aims, connects reason and consequent, and it is as able to determine that the force of reaction should increase with the attainment of results, as that it should diminish in proportion as they are attained.

202. It is easy to see the consequences that follow from this conviction. As we do not know the idea which is endeavouring to realise itself in the world, it is from experience only, as I have before remarked, that we can derive our knowledge of the recurrent operations of things according to general laws. We cannot, therefore, take it amiss that Physics, following the lead of observation, should

assign to the different forces, the assumption of which is found to be necessary, laws of action of the most various kinds. These Laws Physics regards merely as expressions of the facts, without attempting any metaphysical interpretation of them, and every idea of this kind, serving to clear up a group of interconnected phenomena, and enabling us to infer the future from the present, deserves respect, as an enlargement of our knowledge. Philosophy is altogether in the wrong, when she depreciates results obtained in this way, merely because they do not penetrate to the ultimate truth; but she is certainly within her right, when, starting from her own point of view she attempts to supply the interpretation which is still lacking to those results. Whatever may be thought of space and of existence in space, if once the intensity of interaction between two elements is made to depend on the distance separating them, and just so far as it is made to depend *only* on this, it seems to be impossible that different forces could be determined by this cause to act in different ways; the same distance, it would seem, could only make itself felt by the elements and determine all their reciprocal effects in the same way. It is this which has prepossessed philosophers in favour of the view that the different modes of action which Physics assumes, when it makes different forces dependent on different powers of the distance, cannot have a primary right of existence, but that there must be one fundamental law for the relation of action to distance, and the deviations from it which experience compels us to admit must be due merely to the complexity of the circumstances. By an easily understood transition, this fundamental law then came to be identified with the Law of Gravitation, this being a Law which is obeyed by many familiar effects, differing from each other and occurring under apparently very simple circumstances. I cannot myself share this prepossession, except with great reservations. It is necessary first of all that a certain assumption from which all such attempted explanations start, should be clearly stated. That assumption is made by thinkers, by whom perhaps in their ultimate essence all things are mysteriously merged in the unity of an infinite substance and a single creative plan, when they afterwards leave out of sight the continuous operation of this single principle, and explain the whole course of the world merely from the permanent qualities and the changing relations of individual existences, and the consequences which, by common logic, seem to follow from these two premisses.

203. Upon this assumption we are not justified, according to all that has preceded, in regarding the interval of distance itself as that

which determines the amount of force exercised between two elements, *a* and *b*. This is due only to the inner states of the elements which correspond to the distances between them. Every mode of treating the question must admit so much as this. Even if we adopt the ordinary view of space as objectively existing, the distance of things will still only be distance *between* them ; the distance and its measure is, therefore, a reality *prima facie* only for an observer who is able to represent to himself the space which must be traversed in order to pass from *a* to *b*. If *a* and *b* are to be guided by it in what they do, it must be possible for them and not for an observer only to take note that the distance between them is in one case *d* and in another case *δ* ; thus in order to act they must first be acted upon by the very same condition which is to regulate their activity. This would lead us—supposing the merely phenomenal character of space to be assumed—to the conclusion, that every actual distance between *a* and *b* is nothing but the manifestation in space of the sum of the effects which they experience at the moment from one another and from the whole, and which are also the cause of their effect upon us. The universal order is, however, neither according to our view, nor according to the ordinary view, a rigidly classified system, such that each element persistently occupies the place which corresponds with its conception. Such a system no doubt exists, but its parts which are in a constant state of chaotic flux, are every moment falling into relative positions which do not correspond to the permanent affinities of their natures.

We know what this means in terms of ordinary spatial perception ; it is not the elements which are by their nature most fitted for active intercourse which are always the nearest neighbours ; the action of some third or fourth element may separate those which are coherent, or bring together those which are indifferent. It is indeed impossible to give a picture of what things undergo or experience in their inner nature when they enter into those changing intelligible relations with which we are familiar in spatial perception as distances of greater or less extent. As objects of such perception, i. e. as distances, these relations seem to us obviously to imply a greater or less amount of estrangement or of sympathy in the things, and upon this the degree of their reciprocal action is naturally supposed to depend. Yet our previous investigations have shown that we cannot account for the manner and degree in which things act upon each other from the mere fact of their being outside one another ; it is only from what they get or experience from this fact, or from the way in which it *connects* them, that we can do so.   We cannot, therefore, say that the distance between things itself exercises an influence on the intensity of their force ; it is merely the mode in which the greater or less 'degree of their metaphysical affinity is manifested to us, varying as this does with the different combinations into which they are brought by the course of events.   Throughout this process the things remain what they are, and continue to act upon each other conformably to their natures.   At the same time, the different degrees to which they are temporarily displaced from their position in the system, cannot but have some influence on their behaviour; a change in the closeness of their metaphysical affinity involves a corresponding change in the amount i of the intensity p with which they stimulate those mutual actions for which their nature has fitted them.   If these very abstract considerations have so far inspired any confidence, we now stand before a conclusion which seems certain, and before an alternative which we are quite unable to decide.   No reason can be anywhere discovered why this metaphysical affinity should correspond to any but the first power of the distance, which is the distance itself.   On the other hand, after what has been said above, it seems quite as possible that the effect of this affinity should vary directly as that it should vary inversely, with the distance.   The two formulæ—

$$i = pd \text{ and } i = \frac{p}{d}$$

would be the only formulæ in which this point of view could result, and they would be of equal validity.

204. In making use of these expressions, I wish it to be borne in mind that as I understand them they have not the same meaning as any of those quantities to which the ordinary mechanical view of Physics leads us.   We do not use i to designate any kind or degree of outward performance, but merely the intensity of the stimulus with which, in virtue of the relation at the time being subsisting between two elements, one of them excites and is excited by the other to any or all of those possible forms of reciprocal activity which spring from their affinity, their difference, or generally their respective places in the system.   As regards what follows from the stimulus a fresh and specific determination is required to decide whether it is to be attraction or repulsion, and yet another to decide its amount.   The first requirement, however, may perhaps be assumed to be already fulfilled by the coefficient p.   For though we have hitherto spoken of this merely as a quantity, it is dependent on the nature of the interacting elements, and therefore, strictly speaking, could only be a concrete number.

On the other hand, as regards the amount of the initial motion, I can see no reason why it should not be considered as simply proportionate to the stimulus i, which is its motive cause. I shall not, therefore, make any further comparison between the formula $i = pd$, which would indicate a sort of metaphysical elasticity, and what we meet with under the same name, though generally under highly complicated conditions, in the sphere of Physics. As regards the second formula, I do not see how the desired result, viz. dependence upon the square of the distance, could be shown necessarily to follow from it.

I will however mention the assumption which would have to be made in order to bring this law into ultimate harmony with the other or metaphysical view. I cannot esteem as of any value for such a purpose the appeal to the reciprocity of all effects, which some distinguished authorities have introduced into the discussion. If $i = \dfrac{p}{d}$ is the intensity with which one element is attracted by another and at the same time tends of itself towards it, I cannot see any reason for supposing the result to equal the product of the two activities; like every other resultant, it would be the *sum* of them ; it is only the intensity of the effect, not the function of the distance upon which it depends, that would be affected. Perhaps, however, it will be urged, that the effect of a force depends not merely on what the force intends to do, but also on how much it is able to do, i.e. in the case before us, not merely on the amount of mutual excitation, but also on the conditions which promote or check the satisfaction of the demands. To put the matter shortly and clearly ; the distance d between a and b indicates a degree of estrangement between them, and their willingness to act upon each other is therefore inversely proportional to that distance. But, the weaker will is not only weaker, but has opposed to it the greater obstacle, in the shape of the greater distance which weakened it. The active force, therefore, which can be exerted, is found by multiplying the effort by the reciprocal of the resistance to be overcome ; and accordingly is inversely proportional to the square of the distance. Such a mode of expression could indeed only serve to indicate briefly the essence of the idea ; in real truth, the distance d even if it were a really extended space between a and b, could not be regarded as an obstacle to the effect. According to the view which we have maintained, it could not actively condition the incipient process, except in so far as it was represented within the elements a and b by means of that hidden state of excitation which we found ourselves obliged to assume in all cases. But perhaps it is precisely to this inner state of things that an argument of this kind may seem to be most rigorously applicable. It may perhaps seem incredible that when two elements *a* and *b* are separated by the distances *d* or *δ*, they should in both cases alike, though not excited to action to the same degree, yet in spite of this difference aim at producing one and the same effect. It may be thought that the *object* of their effort as well as its integrity would vary, and vary proportionately to the degree of their excitation. The amount of the actual external result would then be found by multiplying *i* into a quantity proportional to *i*, and would thus vary inversely with the square of the distance.

I shall not attempt to decide whether there is anything of value in this suggestion : I wish only to point out that this new way of characterising an intended result, as one which increases in proportion to the stimulus, is just one which cannot be decisively established, if nothing is assumed but individual elements with their natures and the relations subsisting between them. There is no universal Metaphysic of Mechanics, capable of showing that every time any two existences combine in a relation, they *must* have so combined. Whenever any such relation occurs, it is a matter of fact, which, from a Metaphysical point of view, can only be regarded as an effect of the all-embracing *M*, i.e. the idea of the whole. It must be this idea which is present and active in all individual elements, assigning to each its mode of manifestation in relation to the rest, which otherwise would not flow necessarily from the mere conception and the nature of the elements. But, as we do not know the content of this idea, we cannot affirm positively that it imposes a necessity on things to assume these forms and no others ; and hence, the whole attempt to establish the existence of a single, original, and only legitimate law for the operations of all forces is entirely fruitless.

205. Nothing is left to us, but to accept with thanks the empirical rules which enable Physics to express, in conformity with observation, the effects actually produced by the several forces on each occasion of their activity. Philosophy should not turn away from assumptions, unless they are inherently absurd, and those made by Physics are seldom that. Thus, no one has ever attempted to explain an increase or diminution in the intensity of force as depending on mere Time ; where observation seemed to confirm such a view, the Time was in every instance occupied by actual occurrences, each of which contained in itself the efficient cause of that which was to follow ; these processes, then, and not the mere lapse of Time, must have determined the varying intensity of forces. On the other hand, there is no reason on philosophical grounds to deny that the amount of force which results from the interaction of two elements, depends to some extent also on their motions. For according to our view motion is not merely a change of external relations, which takes no effect on the things themselves ; as those relations depend on inward states of the things, so the rapidity with which they change them is also an inward experience, and one which at every moment may help to determine their subsequent behaviour. Besides the degree of intensity which a force would have, corresponding with the distance at the moment between the two elements from which it] proceeds, there would thus be a positive or negative increase of the force, dependent on the rapidity with which the elements travel through the space which they at present occupy. But it is not expedient to continue the discussion on this point; for while the hypothesis has been employed by Physicists only with extreme reserve, in regard to the interaction of electric currents,—a case in which it seemed to be required,—there would be no limits to its application when treated, as we should have to treat it, as a general principle. Once admitted, the dependence of force upon velocity of motion, and upon its successive accelerations, would apparently have to be regarded as a universal characteristic of physical action.

206. Connected with this question is the other one : Do forces, in order to take effect, require Time ? Stated in this form, indeed, as it occasionally is, the question is ambiguous. It is a universally admitted truth that, every effect, in its final result, is formed by the successive and continuous addition of infinitesimal parts which go on accumulating from zero up to the final amount. In this sense succession, in other words, expenditure of Time, is a characteristic of every effect, and this is what distinguishes an effect from a mere consequence, which holds good simultaneously with its condition. Vain, however, would it be—as we saw in our investigation of Time— to seek to go further than this, and to discover the inscrutable process by means of which succession of events in Time comes to pass at all. The question we are considering was proposed on the assumption of the diffusion of force in Space. Supposing it were possible to instance a moment of Time in which a previously non-existent force came into Being, would all the various effects which it was calculated to produce in different places, both near and remote, be at once realised ? Or, would a certain interval of Time be required, just as it

is in the case of Light, which transmits itself to different objects rapidly, but not instantaneously, and must first come into contact with them before it can be reflected by them.

It is not necessary to embellish the question by introducing conditions which make any decision impossible. There is no need to imagine either the sudden appearance out of nothing of some new body in the world, or the disappearance of one already existing, and then to enquire, whether the addition of gravity, as in the first case (the new body being likewise supposed subject to the law of gravitation), or the subtraction of gravity, as in the second, would make itself felt by distant stars immediately, or not till after a measurable interval? The action of force in its beginnings may he illustrated by examples nearer to hand. Each smallest increase in the velocity of two elements, which are working upon each other—whether by attraction or repulsion—at a distance, by the very fact that the elements are brought nearer to or are parted from each other, brings about an increase of attraction or repulsion, in other words, a new force, though no new vehicle of it. Similarly, the electrical actions of bodies, depending as they do upon a condition which is not always present, furnish an example of a beginning of force in Time. It makes no difference that this condition itself does not come into existence at once and with a permanent intensity, but only by degrees ; at any rate, a moment can be assigned for every one of its degrees before which it did not exist, and from which its effect must begin. Having regard to such cases the question that has been raised can only be answered in the negative ; there could be no possibility of an affirmative answer, except on that supposition of a diffusion of force which we found to be impossible [1]. But even on that supposition, it is the passage through space which, strictly speaking, would have to be regarded as the first work of the diffused force ; the work done upon its arrival at the distant object would be only second and subsequent, for its presence as force would not be felt by the object until it had come into the necessary contact with it. It must not however be supposed that after the force has come into Being, a blank space of time *t* is required to pass before the motion begins to be transmitted ; nor again, that after the force has reached its object, and so secured its control over it, it should require a similar space of Time *t*, in order to take effect. If this space of Time *t* were really blank, everything would remain at the end of it as it was at the beginning, and the effect might just as well be expected to

[1] [Cp. § 198.]

occur at the end of some other space of Time $= n\,t$; if, on the other hand, any positive change in the phenomena takes place during this time, this change is a link, by means of which C, the imperfectly realised condition of the result F, is completed and perfected: that part of C, however, which was already present has, at the moment of its coming to be, immediately produced that corresponding part of F which it was adequate to produce.

207. It will be objected that real events are, as I stated above, not related to each other in the same way as conditions to their consequences, because the result in the former case always follows the cause which produces it; but for this succession, events would be transformed into a system of cotemporaneous parts, which would differ only in the different degrees of their dependence upon the first of the series: C and F, therefore, though it is true there could be no blank interval of Time between them, would always come into contact in the order $C\,F$, not in the order $F\,C$. This true remark again suggests an enigma, the insolubility of which we have already admitted. For succession in time could never arise from these contacts which occupy no time, however often repeated, between members which follow out of one another; we should still have merely a systematic order if C and F did not each fill a certain extent of time of its own; if they did, then, it seems, F would have to wait till C had completed its interval of time. But even this is not a way out of the difficulty. Suppose C and F both to consist of a series of parts following each other in unbroken succession, e. g. $c_1,\ c_2,\ c_3, f_1, f_2, f_3$. Are we then to suppose that the occurrence of F is conditional on the completion of the group C? that it cannot, i.e. commence, until c_3 is reached, and that nothing of the nature of F takes place until this term is realised? There are facts enough which seem to confirm this view, and indicate that the result F is attached to a specific determination of C. A closer examination will, nevertheless, not fail to show that the force C, all the time that it seemed to be increasing in amount without producing any effect, was really already occupied with the removal of hindrances which stood in the way of the occurrence of anything of the nature of F. When, at last, the amount c_3 is reached, this removal is completed, and from this point its first positive and visible effect commences, though not absolutely its first effect. As regards this effect, again, we do not believe that a finite amount of it, f_1, arises *suddenly* so soon as C is ended. Rather, each smallest addition which is made to B, involves a correspondingly small addition to F; but between these two occurrences there is no blank interval of Time; f_n corresponds to c_n immediately. But the assumption we have made as regards C itself involves the same difficulty. If C remains unchanged during the whole space of Time t, which it is supposed to fill, there is no better reason why F should follow at the close of that time than at its commencement. If however we assume, as was assumed, that C traverses the series $c_1 c_2 c_3$, then, as the order of the series is supposed to be fixed, each term must be the condition of the succeeding one, and as in the previous case, if they are to form a succession in time, two adjacent terms can neither have any blank interval of time between them, nor can they be simultaneous.

The conclusion to which this points is clear. The whole nature of Becoming is unknown to us, and we cannot reconstruct the origin of it in theory. In this quite general sense, it is true to say that every operative condition and every force draws its consequences and its effects *after* it. But in order to do this, it is not so much the case that they need a lapse of time, as that they *are* this lapse of time itself; only because they are themselves in a process of becoming can they convey that same process to their consequences. But there is no measurable interval of Time between the condition c_n and its true and immediate result f_n; there is nothing but the enigmatical fact of their contact, a fact which cannot be ignored any more than it can be explained.

If we now leave these general considerations, and return to the subject which first suggested them, that of forces acting at a distance, it must follow from the doctrine which has been stated, that, at the same moment that the force which is active in the element p passes from c_1 into c_2, there will be a similar transition in the element q, no matter how remote it may be, from f_2 to f_3, provided that c and f are causally connected through that inner sympathetic affinity upon which all action depends. Moreover, just as c_2 in the element p can only change into c_3 continuously, that is, by passing through all the intermediate values, in exactly the same way in the element q, f^2 will pass by succession into f^3. But the idea that a lapse of Time is required in order that p should transmit its force to q at all, is barred, among other considerations, by that of the reciprocal action of the two elements, which is universally admitted to be a necessary assumption. No force could be diffused from p towards q, nor could any force even originate in p, unless it were awakened and solicited in p by q; on the other hand, q could not produce this excitation, unless it was invited by p. No action, therefore, could ever take place between p and q, if it were required that a force should first proceed from p to q; for the only thing which could excite this force to set out from p would be the stimulus of another force starting from q; and this stimulus it would never have, because q would be waiting for an invitation from p. This connexion of mutual affinity between the elements, the source of their action upon each other, does not at one time or another *come into* existence through a diffusion of forces in space; it always *exists*, thus rendering it possible that changes of state experienced by one element should involve corresponding changes in another.

208. Owing to the boundless complexity of the manifold conditions which meet in the course of nature, we cannot expect to be able to explain every event directly from the joint action of the forces which combine to produce it. Hence, the desire has often been felt to discover certain customary rules by which, at any rate, the course of the natural world is regulated. It was hoped that in cases where knowledge of the special connexions between things is wanting, we might thus be enabled to establish equations expressive of general conditions with which the results, however unknown may be the manner in which they are brought about, must certainly correspond. Experience itself also leads us to the same ideas, whether, as some believe, it is from this source that they are derived exclusively, or that they are preconceptions which experience merely confirms, and which, as it then seems, we must have arrived at independently.

Opinions are divided between these two alternatives. The Realistic view inclines to treat general principles of this kind either as designations of mere matters of fact, which might have occurred differently, or else their universality is explained by what is called their self-evident truth, though its opposite is not regarded as strictly inconceivable. On the other hand, the Idealist view, which is that which we here adopt, can recognise no supreme law except the one unchanging purpose underlying the multiplicity of phenomena, and seeking for its realisation in them. At the same time, the Idealist, being unable to express the nature of this purpose, or the laws to which it requires that things should conform, cannot regard these universal principles, in so far as they are borne out by experience, as more than habits of nature on a great scale, valid within the circle of our observation, but not infallible as regards the far larger sphere of reality which lies beyond the limits of Time and Space to which our investigations are confined. Hence, instead of establishing any positive truths, the duty which lies before me is the less grateful one of calling in question the unlimited validity of principles, the limited validity of which is one of the most important and unfailing aids to scientific enquiry.

209. One of the simplest of these truths appears to be the invariability and the conservation of mass. Though not especially, or, at any rate, not invariably confirmed by the appearances of every-day life, this doctrine receives such universal support from the systematic view of science, that it would be superfluous to adduce any detailed arguments for its certainty. But now that it has been fully established, I cannot see in it any necessity of thought the late discovery of which need cause surprise. It may indeed be self-evident for a theory which regards the world as composed of individual and mutually independent atoms. Out of the absolute void, which would be all that would lie between these atoms, obviously no new real existence could arise; the principle that out of nothing comes nothing, would hold good absolutely. But this point of view we have been compelled to abandon. In order to conceive reciprocal action, without which no course of nature is intelligible, we were led to regard the individual elements, not as self-conditioned, but as depending for the beginning, continuance, and end of their existence on the determination of the one Being, from which their nature and capacities of action are derived. Now, it is certainly a tempting conclusion, but it is no necessity of thought, to go on to suppose that this one Being at least is a sum of reality which cannot be increased or diminished, and which changes only the forms of its manifestation. And we ourselves inclined above to this idea, when we admitted it to be natural that each individual qualitatively-distinguished element, i. e. each activity of the one existence, when conformably with the plan of the world it splits itself up into various elements, should have a diminished intensity in each of the parts so arising.

But all this world of quantitative determinations has no significance outside that complexity of things and processes which the one and only true reality creates to express itself. It is only their meaning and function, and the value which they thus acquire, that give to the individual elements and forces the particular magnitudes they possess and exhibit in comparison with others. But what lies beneath them all, is not a quantity which is eternally bound to the same limits, and can only represent the same sum in different ways, however variously divided. On the contrary, there is no reason why, if it is required by the Idea which has to be realised, one period of the world should not need the efficient elements to be more, and another less, and why in the former case each part of the whole should not also exert itself with a greater degree of force on the rest. The history of Nature would then resemble a musical melody of varying strength of tone, the swellings and varyings of which do not spring from nothing, nor yet from one another, but each in its place results from the requirement of the whole. I do not mean to affirm that this actually *is* what takes place in Nature. Quite conceivably it may be part of the hidden purpose of the supreme Idea, that all its requirements should depend for their realisation on a fixed sum of real elements, and that the production of variety should be restricted to different adaptations of the same material. Still less ought we to be surprised if the course of Nature, so far as we can observe it, shows this to be practically the case. For, as far as we can see with clearness, we find Nature moving in a cycle, which makes it certain that forms once in existence will maintain themselves in existence. The only phenomena which suggest a progress wholly new, a progress which would go nearest to proving that the materials as well as the results are changed, are those which come from an antiquity so remote as to preclude exact investigation. It would, therefore, be mere folly to call in question the principle of the conservation of mass, so long as we confine our view to the world of accessible facts, and to what we may call the retail dealings of the physical elements in it. But it is the business of Philosophy to be constantly reminding us how limited is that section of the universe which is open to our observation, and that the whole which comprehends it is a reality, though not one which we can make an object of positive knowledge.

210. Similarly, the attempt has been made to conceive of *the sum of motions* in the world as a constant quantity. The general state of knowledge at the time when this idea was first entertained, did not admit of its being substantiated or even rendered probable by evidence derived from experience. For, as long as the effects which things exercise upon each other were explained as due merely to communicated motion, the conclusion could not be evaded that contrary velocities of elements tending in opposite directions would neutralise each other either wholly or in part, and consequently that motion disappeared from the world without any compensation. And ordinary experience seemed to confirm this conclusion by an abundance of examples, which no one knew how to explain in any other way. On the other hand, it was seen that living Beings were centres from which fresh motions were initiated at every instant, which could not but be taken for really new beginnings. So that neither was there anything in experience which was inconsistent with the indefinite

multiplication of motions. Nor, finally, did experience suggest at all that this increase and diminution must balance each other, so as to maintain a constant sum of motion. Such a conception originates in an hypothesis as to the general character of the course of nature. Such an hypothesis was furnished by the idea of a system, having no object but the maintenance of itself, and furnished with fixed resources to this end : one of these means was the sum of motion, as it once for all exists, which in the economy of nature might not be spent, but only differently dispensed.

Recent physical speculations tend to revert to this same idea. So many apparently fixed qualities and conditions of things have been already demonstrated to be a ceaseless process, that it may be doubted whether there is such a thing as Rest at all, except in the indivisible moments of reversal in the minute oscillations with which all things are vibrating. Philosophy can have no motive for objecting to the assumption of such eternal motion as a matter of fact ; it is a mere prejudice to infer, that because from our point of view an element must be first supposed at rest in order that the results of varying motions which condition it may be understood, this quiescence must have been prior in reality, and that the impulse to motion is an addition for which it has to wait. At the same time, it is only as a fact, and not in any other light, that we can regard this perpetual motion. It implies, not merely that motions already in existence may be communicated, but also that fresh motions must be produced in cases where two motions are opposite, and their communication could only result in the neutralisation of both. This elasticity of things, without which it would be impossible for them to counteract the self-annihilation of motion, is only conceivable if there are inner states of their being capable of developing the forces from which motions spring. It is possible, though not probable, that effects produced at a distance—against which there exists an unfounded prejudice—are conveyed in this way by means of motions transmitted from point to point of some connecting medium. But, even in that case, not only the conception of force, but also in a special sense that of force producing effects at a distance, is still indispensable, in order to explain each one of those countless communications of motion, the sum of which is usually held to compose the effects of force at measurable distances. If, however, force alone gives a sufficient reason for expecting that the motion will be replaced, which mere communication would permit to be lost in its antagonism, it cannot be supposed that force itself is the constant quantity which is in request ; its intensity varies with the distance, though this is itself determined by force.   The constant element in the course of Nature can only be an inner connexion between the circumstances which give rise to the operation of forces, a general law governing all combinations and connected successions of effects.   It was thus that that most comprehensive principle, the one which dominates our whole estimate of physical processes, that of the Conservation of Force, first suggested itself, in respect to which I proceed now, though only so far as the connexion of my views requires, to offer the following considerations.

211. The simple principle, that out of nothing comes nothing, requires to be more precisely defined by the addition that even from something no result can follow, so long as that something, the event B, is only the condition or occasion of what is to take place, and remains just the same after the consequence F has been produced as before.   On the contrary, B must be sacrificed, either wholly or in part, in order to produce F.   This is the difference so constantly referred to between a causal *nexus* of events and the merely formal connexion of conditions and consequences.   Our ontological discussions proved to us that, in the simplest case of causation, at least two factors, a and b, must enter into a relation c, and that the result which takes place consists in this, that a becomes changed into a, b into β, c into γ.   Every effect, therefore, is the effect of two elements acting upon each other, neither of which can inflict upon the other a change in its condition, without paying a definite price for it by a corresponding change in its own.   If a wishes by acting at a distance, whether in the way of attraction or of repulsion, to change the place of b, it can only do so by displacing itself in the opposite direction to a corresponding distance.   There is no reason for excepting any single operation of nature from this general law ; it holds good even in those cases of communicated motion when the process cannot be observed in all its details.   It is not possible for a motion of one element, after imparting a certain velocity to a second, to persist unchanged in the first, ready to produce the same result again, and so increase its effect to infinity ; its influence is exhausted in proportion to the degree in which it has been exerted.

212. Certain corollaries, of different degrees of certitude, arise out of this general conclusion.   If we assume that the course of nature includes occurrences differing in kind from each other, and not admitting of being represented as mere quantitative or formal modifications of a single homogeneous process; we shall not be justified in asserting that every occurrence, A, calls into existence every other, C,

or admits of immediate application to its production. It would be quite conceivable that there was no way from A to C except through the medium of a third, B, A and C remaining unsympathetic to each other. If therefore it cannot be said that there is necessarily any reciprocal action between every A and every C, it is equally clear on the other hand that if such a relation does take place, a specific amount of A must be sacrificed in order to produce a specific amount of C. Nor is it logically necessary, or self-evident, that every connected succession of two occurrences A and C must be convertible. No doubt, whatever is lost to A in the process of producing C, testifies to an effect of C upon A; but this effect is merely to impede A, and it is not a matter of course that every C which is able to do away with an A should therefore be able to call into existence an A which does not exist. That none the less it seems natural to us that this should be so, is due to the assumption which unconsciously we make to ourselves, that the economy of nature has no other object than self-conservation. In a process which implied progress, the order of events might easily be so determined as that A should lead to C, but that there should be no way back from C to A.

It cannot therefore be asserted *a priori*, and as a self-evident truth, that all the processes in Nature must be mutually convertible backwards and forwards ; how far this convertibility extends can only be learned by experience. But, even in those cases in which it holds good, it is still by no means certain that the same amount c of C, which was produced by the amount a of A, and which therefore caused a to disappear, would now reproduce exactly the same amount, a, as was spent in its own production. That could not be unless it had been previously proved that there is in Nature no tendency towards progress; if there is progress, there can be nothing to make it impossible that each stage in a series of occurrences, $a\,c$, $c\,a_1$, $a_1\,c_1$, $c_1\,a_2$, should contain the condition of an advance in the next stage. This assertion is at variance with ordinary ideas ; as, however, I do not intend to apply it to explain the actual details of the course of Nature, I shall merely repeat by way of justification what has previously been suggested, viz. that the nature of being and process is not limited by any premundane system of mechanics, but that it is the very import of this process which determines all the quantities in which the elements make themselves felt, and the consequences which their relations entail.

Finally, if, proceeding on the assumption of the unlimited convertibility of mutually productive activities, we suppose that a and b enter into a varying relation c, the sum of the effects which one is able to exercise on the other will be, within certain determinate values of c, a constant quantity. As each intermediate amount of c is reached, the capacity for action continues as regards that part of the possible total amount which it has not yet produced; on the other hand, it has lost so much of its force as was required to produce the result thus far achieved; this loss can only be made good by restoring the elements to their original state, that is, by doing away with the results already obtained. If we call this capacity for future action potential energy[1], in contrast with kinetic energy[2] which is active at the moment, the sum of these two forces, when the two elements are related as above, forms a constant quantity.

In the same way a sum of money M, so long as it remains unspent in our possession, has a purchasing power, and loses this power in proportion to the purchased goods which it acquires. Its original purchasing power can only be restored and applied to other objects by re-selling the goods. This example throws light upon the difficulties raised above. It would be impossible for us to know *a priori* that the potential force which the possession of the money would imply, could put us in possession of other objects by being itself got rid of; this exchangeability depends, in fact, on highly complex relations of human society. Nor should we be any more justified in taking for granted that the goods, G, by being similarly got rid of, would put us again in possession of the money; and as a matter of fact, this convertibility, which in like manner presupposes the connexion of human wants, has its limits; for it is well known that by buying goods and selling them again, we are equally likely to gain and to lose. It is not the case, then, that in the conversion of trade every quantity reproduces the same quantity as that by which it was produced. It is of course obvious, and need not be urged as an objection, that this result is due to conflicting circumstances, and to the influence which the nature of human business has in determining the relation of M to G; it is to these dealings of men with each other, not to any essential peculiarity of M and G, that the fact of their standing in any relation is due. But this is the very point which I would urge against the over-confident procedure of natural science. It does not appear to me self-evident that a perfectly adequate ground can be found for the mutual relations of the elements of Nature either, merely by considering their fixed characters M and G; here, too, their exchangeable value may depend partly on some larger commerce of the world. At the same time, I have no doubt

---

[1] [Spannkraft]   [2] [Lebendige Kraft.]

as to the practical truth of the principle of the Conservation of Force, within the limits of our experience. Merely in the interests of Metaphysic I felt compelled to speak of these difficulties, and I wish now to make mention also of some accessory notions which have formed round this general principle.

213. We are often told with enthusiasm how it has at last been shown that all the various processes of the natural world are produced by a single indestructible force never varying in its intensity, and that nothing changes except the form in which the ceaseless transformations of this force are presented. It is especially the important corres-pondence between mechanical work and heat, which, by a somewhat hasty generalisation, has given rise to this idea of a transition of forces into one another, and of a universal primitive force to which they are all subordinate. The satisfaction thus given to that feeling which compels us to comprehend the infinite multiplicity of things and events under some single principle, seems to me to be illusory. Lichtenberg once contrasted the early ages of the world, when man-kind was equally ready to believe in God and in ghosts, with the present age, which denies both; he feared compensation in a future when all that would be believed in would be the ghosts. Something like this seems to have happened in the case before us. For after all we are only doing honour to a ghost, when we dream of an ab-solutely nameless primitive force, which, formless in itself, and consisting of nothing but an unnamed number of constant amount, assumes, as a trifling addition that needs no explanation, the changing names under which it is manifested. If, however, we reflect upon and realise the fact that this original force never exists in this naked and nameless shape, but is continually passing from one to another of the forms which it assumes, we are again admitting that what really gives to each phenomenon its character is the concrete nature of that which embodies the quantum of force, either wholly or partly, for the time being. The same reflection would show that what makes the succession of changing phenomena possible, is a unity of meaning which pervades and connects all those concrete forms of being with one another. Finally, it would appear that the persistence of quantity through all this play of forces is only a mode in which the already existent reality manifests itself, and cannot be the source from which that reality with all its various forms originally springs.

The latter view, which would reverse the true order and mistake the shadow for the substance, scarcely needs any further refutation; more serious are the objections which may be raised against the general assertion which we admitted above, that the conservation of the same sum of force is as a matter of fact the rule of experience. In as far as we can reduce two physical processes A and C to comparable primary occurrences consisting in comparable velocities v of comparable masses m, so far it may be shown that C which is produced by A, contains precisely the same amount of energy which A, by producing it, has lost. Where, however, the two elements do not admit of this exact comparison, and we have before us merely the fact that the specific amount a of A produces the specific amount c of C, and, it may be, *vice versa*, it is an essentially arbitrary course to conclude that c and a contain the same amount of energy, merely distributed in each case in a different form. All that can be said is, that a and c are *equivalent*, not that they are *equal*. It is possibly a just expectation that all the various processes of external nature will admit of being ultimately referred to variously combined motions of infinitesimal elements, and as regards these particular processes, the arbitrary interpretation referred to might be defended on this ground; but the general conception which underlies the principle of the conservation of Force must without doubt apply to one case in which no such expectation can be entertained; I allude to the interconnexion of physical and psychical processes.

Whatever effect is produced on the organs of sense by an outward irritation I, whether it is simply received, or transmitted, or diffused, or changed, there must always be left over from the physical process a residuum i, to which the psychical process of the sensation s will succeed immediately; nor can we doubt that the strength of the sensation will change with the changes in the strength of i. Again, no matter what constitutes an act of will W, or how it may act upon other states of consciousness, or be limited by them, there must be ultimately a part of it w, from which the first motion, f, of the body and all its consequences take their rise, and in this case we do not doubt, any more than in the other, that the extent of the physical effect is determined by the varying intensity of w. Now according to all ordinary views of what happens in such cases, by itself the mere fact that there is an i or w, considered as an opportunity or occasion, is not enough to entail the existence of s or f. In order that the reaction may vary with the varying amount of the stimulus, the stimulus must be perceptible by that which it affects, in other words, must produce in it a change of state of definite amount. In the two cases before us, as in all others, it will be found that no effect can take place, i. e. neither that of the last physical movement upon the sensitive subject, nor that of the last mental excitation upon the first nerve-element which it acts upon, without a corresponding loss ; here, too, the productive energy is consumed, in whole or in part, in bringing about the result. But never will it be possible to refer *i* and *s*, or *w* and *f*, physical and psychical processes, to a common standard ; the members of each of the two groups may be compared with each other, but the unit of measurement in the one has nothing in common with that in the other. Granting, then, that here is compensation for physical energy by psychical or for psychical by physical, still in such a case as this there ceases to be any meaning in saying that one and the same *quantity* of action or work is maintained throughout; all that is open to us is, to speak of an equivalence of two activities, such that a specific amount $s\mu$ of the one, measured by the unit μ, corresponds to a specific amount im of the other, measured by the unit m. No one, however, can say whether these two activities are equal in quantity, nor which of them is the greater.

214. These considerations suggest certain others. In the first place, we may attempt to generalise from what has been discovered as regards these processes ; in all cases, we may say, the simplest fact, the fact which first meets us in experience, is this relation of equivalence between two processes or forces. We do not first discover that two forces are equal and like[1], and therefore produce equal and like[1] effects; but what we do first is to observe that they balance each other, or, that under the like circumstances they produce the like motions. From this equivalence which has been found to obtain between them in certain special cases we infer their quantitative equality; at the same time we assume for the elements to which the forces in question belong, the qualitative identity which enables us to apply to them the same standard of measurement. I have no motive for entering here into all the indirect reasons and proofs which show in what a number of physical processes this assumption holds good ; I would refer especially to the idea of homogeneous mass and of its conservation understood as it has been above. Confining myself to the metaphysical aspect of the question, I wish merely to point out that the principle of the conservation of Force, or, as I prefer now to say, the equivalence of different effects, does not impose on us any obligation to reduce all processes in Nature to the single class of material motion. So far as the principle applies to this latter

[1] ['Gleich,' cp. note on § 19.]

class it is only a special instance of that more general correspondence, existing between heterogeneous things as well, which we express by this wider term of 'Equivalence.' Far, therefore, from being a monotonous transmission of the *same* unchanging process, it might be that the course of nature is for ever producing unlike by unlike; though the equivalence which the sovereign purpose of the world has established between these several disparate activities, would make the 'incidental view' practicable and fruitful, according to which we reduce the concrete varieties of phenomena to mere quantitative values of a single, abstract, uniform principle, just as we determine the value of the most different things by the same artificial standard of money.

I know well how stubbornly this view will be contested. The very analogy we have used will appear defective ; the prices of things, it will be said, only admit of comparison because the things all serve more or less to satisfy human wants which themselves admit of comparison ; and this implies that the effects of the things on us, and ultimately therefore that which is the source of those effects, must be homogeneous. I on my side am not less stubborn in the defence of my own view. I do not deny that in so far as different things have like effects upon us, we are able by means of an artifice to ignore their specific differences for the time being, and to regard them as differing only quantitatively ; but the things themselves are not therefore like because they admit of this justifiable fiction. Even if all qualitative differences are pronounced to be mere appearances, yet the difference of this appearance still remains, and belongs no less to the sum-total of reality ; the utmost, therefore, that we can do will be to exhibit the external world as a mechanism of homogeneous parts which produces in us these appearances ; but by no process of Mathematics or Mechanics would it be possible to deduce analytically concrete magnitudes from abstract ones, or magnitudes of different denomination from magnitudes of the same denomination. The process of the world is no mere combination of identical elements, but a synthesis of elements differing in quality and only connected by unity of plan.

215. But are we really correct in what we have laid down with regard to physical and psychical processes ? Is it true that in this case also, the activity which occasions the result must necessarily be sacrificed in the process ? Long before the principle of the Conservation of Force had excited its present interest, I had pointed to this conclusion ; but it is not self-evident except upon the assumption which we adopted above, viz. that isolated elements can only be

influenced by one another if they are capable of acting upon one another, and that no one element will adapt itself to another without requiring compensation from its amenability. But, it may be said, if all the elements, a and b, must be regarded as moments of the one M with no independence of their own, why should not the change of a into a suffice to give the signal, which is simply followed by the change of b into β, according to the theory of Occasionalism? Why should any special effort be required in order to bring about an affinity between the elements which already exists? Still it is clear that if what this theory demands is conceded it cannot apply exclusively to the interaction of physical and psychical processes as an exceptional case. The same consideration would apply also to all that takes place between the elements of the external world. Even the atoms would find in M a constant bond of union, and what was experienced by one atom would be the simultaneous signal for changes in another, which would follow like premisses from their conclusion, without involving any self-sacrifice on the part of the first. If, however, we find that this sacrifice does as a matter of fact take place, as it certainly does in the external world, though it can scarcely be proved by experience in the case of physical and psychical processes, all that remains to us is to suppose that this fact too, is a constituent element in the purpose which finds or ought to find expression in the real world; at the same time, we must not represent it as a condition imposed by some inscrutable necessity, without which the world as it is would not be possible. My only object in making this remark was to repeat, that if all conditions continued to exist simultaneously with their consequences (which is what would follow from the principles of Occasionalism), the world would appear again as a merely systematic whole, from which all change was absent. If, however, Becoming, the alternation between Being and not Being, is the very characteristic of the real world, it appears to me that the absorption of the cause in the effect is quite as necessary to that world as persistence is necessary to the conception of motion. For those signals which we spoke of could themselves have no signals for their occurrence except in the succession of effects; they would be produced by one set of effects, they must disappear again in producing another.

216. Amongst the general habits described as characteristic of the course of Nature, it is common to hear Principles of *Parsimony* mentioned. The conception is a very vague one, and even in the principle of least action the way in which it has been formulated is

not without ambiguity. What it signifies is only clear in cases where there is some end in view which admits of being equally realised by different means, each however involving a different amount of expenditure. But the standard by which this amount is estimated is still dependent on circumstances, which make in one case the saving of Time, in another that of distance, in another that of material, the more important, or cause us to prefer an habitual method to the trouble of learning a new one. In order, therefore, to settle with any certainty the question as to the procedure which involves the least expenditure of means, a statement of the direction in which economy is most valuable must be included in the original definition of the end.

This alone is enough to show what ambiguities are likely to be involved when this conception is transferred to the operations of Nature. Assuming that Nature follows certain ends, we do not know what these are, nor can we determine what direction her parsimony must take. The one thing which we should perhaps assert would be this, that nature is not sparing in matter or in force, in Time, in distance, or in velocity, all of which cost her nothing, but that she is sparing in principles. It is this kind of parsimony which we do in fact believe to exist in Nature, especially in the organic world; by variations of a few original types, by countless modifications of a single organ the variety of organic beings, we believe, is produced, and their different wants supplied. Here Nature seems to us, if it may be permitted to our short-sighted wisdom to say so, to be wasteful of material and Time, and to reach many of her ends by long circuitous routes which it would have been possible, by departing from her habitual and typical course, to have shortened. These ideas do not hold good of mechanics, since mechanical laws apply, not to any particular type of effect, but to any and every type. We know that, within certain limits, the various elements in a mechanical effect are convertible; thus increase of velocity may make up for decrease of mass, and increase of Time for decrease of force. There cannot therefore be an economy in all elements at once for the attainment of a given end e; we must look for the least expenditure in that combination of all the different elements which amounts to less than any other combination equally possible under the circumstances.

But this gives rise to a fresh ambiguity. If we look at the matter fairly, it appears that e, which we just now described as the end or aim, is nothing more than the particular occurrence e, and it need not be said that the modes of activity which led to this result must have been exactly adequate to produce it.   But, under the special circumstances in the given case, the modes of activity were at the same time the only possible ones which could give rise to *e*.   For in order to follow a given path, it is not enough that it presents no obstacles, there must also be a positive impulse to follow it.   It is therefore quite idle to excogitate different methods by which, theoretically, the end *e* might have been arrived at; that would require that we should analyse precisely the starting-point *A* from which the effect is supposed to proceed, and then, after considering all the several possibilities contained in *A*, that we should be able to determine that in this particular case the other methods were still equally possible.   But this we shall never succeed in doing, for it involves a contradiction; it is true that the other methods may be, even in this particular case, all equally free from impediment, but there could not be positive inducements to follow them all equally; otherwise what would eventually take place would be, not *e*, but *E*, the resultant of all these different inducements.   If therefore we find on comparison that the method *m* by which the result *e* is actually reached, is the shortest of many conceivable methods, what makes the possibility actual reality is not that this method has been chosen out of many others equally possible; rather we should say that *m* was in this case the only possible method, because any other direct method *M*, which might have led to the same result, lacked the conditions for carrying them into effect; in a different case, where these conditions were present, the result *E* would be different and the shortest way to it would be *M*.   We must not therefore speak of parsimony in the sense of an act of choice, the exercise of which is merely a peculiar habit, not a causal necessity, of nature.   The utmost that we could venture to assert is, that the Laws of Nature are so devised that the shortest way to any given result is in every case a necessary result of the laws themselves.

Yet even this statement would be no better than ambiguous. For the new truth which it seems to contain, and which makes it appear more self-evident than the preceding one, is similarly dependent on our arbitrary determination to regard as an end what is really only a result.   It is true that according to the known law of reflexion a ray of light transmitted from the point *a* and reflected by the surface *S*, takes the shortest way to a point *b* which lies in the line of its reflexion, or again that according to the known law of refraction, if refracted by an intervening body, it takes the shortest way to a point *b* in the line of its exit from the refracting medium.

But by whose command did the ray proceed from *a* precisely towards this point *b* and no other? That it arrives at this point is not to be wondered at, since it lies in the line of direction which the laws above mentioned prescribe to light; but for this very reason the ray is not transmitted to any of the other innumerable points *c*, which lie outside that direction, and which might yet deserve to be illuminated no less than *b*. If we conceive the attainment of the point *b* as a sort of end which in some way or other reacts upon the means to its attainment, the shortest way would have been for the ray at once to change its direction at *a* and traverse the straight line *a b*; this, however, was forbidden by the general laws to which it is subject, and the ray was compelled to follow a course not absolutely the shortest, but only the shortest conditionally upon the necessity of its reflexion. If, again, by an equally arbitrary assumption, we suppose a point *c* as that which has to be illuminated, those same laws of reflexion now appear in the light of hindrances which do not allow of the attainment of the end except by a longer way, not perhaps until the ray has been several times reflected upon many different surfaces. Hence, the only thing quite certain is this. In passing from any fully determined point *A* to the consequence *E* which flows from it, Nature makes no circuits to which she is not compelled but always takes the way which under the given conditions is the only possible but therefore also the necessary one. The parsimony of Nature consists in the fact that groundless prodigality is a mechanical impossibility.

Something more, however, remains. We can conceive laws of reflexion, e. g. which would require that each of the points on which a ray of light is to touch, though lying in the line of its projection, should yet be reached by a longer way than that by which they are reached as a matter of fact. That reflexion, once assuming its necessity, takes place according to the known law of nature in the shortest possible geometrical line, this and other like considerations may confirm the opinion above expressed, that the concrete laws of Nature are so constituted that it is a necessary characteristic of their operation to effect their results at the smallest cost. It will not, however, be doubted that the law of reflexion in question is itself a mechanically necessary consequence of the motion of light, not a codicil subsequently imposed upon that motion by Nature from free choice and preference for parsimony. All that we come to finally, therefore, is the quite general conclusion, which is also perfectly obvious, that the order of Nature does not rest on a disconnected heap of isolated ordinances. There is contained in the fundamental

properties of reality, taken together with the necessary truths of
Mathematics, a wonderful rationality which at countless different
points gives the impression of an elaborately concerted plan and fixed
aims.   That even the most axiomatic principles serve a purpose, is
due not to any property implanted in them, as in some strange soil,
after they have come into being, but rather in these axiomatic principles themselves there is a deep and peculiar adaptation to purpose,
which might well furnish an attractive subject for further enquiry.

CHAPTER VIII.

The forms of the Course of Nature.

217. I GAVE to this second book the name 'Cosmology,' intending to show that it would be devoted to the consideration only of those general forms and modes of behaviour, which enable us to represent to ourselves how manifold phenomena are connected together so as to form an ordered universe; it remained for the facts themselves to determine with which amongst the various possible formations the outlines thus sketched should be filled in, and these facts which are what constitute reality in the full sense, it was proposed, therefore, to leave to Natural Philosophy. Yet after all, how easy is it to invent well-founded titles for sciences of the future. If only it were as easy to discover the facts which would fill up their framework! But indeed we have not been able to establish much, even as regards those general tendencies of Nature, in spite of their seeming to be so near to the region of necessary truth. We found that they too were really dependent on the plan which is working itself out in the world. Still less shall we be able to show as long as we are in ignorance of that plan, that concrete processes and products, which can depend on nothing but it, are elements and stages in a systematic development. Such a hope was once entertained by Idealism; light and weight, magnetism and electricity, chemical processes and organic life were all made to appear as necessary phases in the evolution of the Absolute, the innermost motive of whose working was supposed to be known; not only so, but bold attempts were made to represent the varieties of plants and animals as following each other in a regular succession, and where a link was missing, to deduce it from the pre-supposed order of development, explaining the previous oversight of it as an accident. I see no reason for repeating the criticism that history has passed upon these attempts. It was a delusion to suppose that the forms of reality, while still inaccessible to observation, could be

deduced from a single fundamental principle : all that could be done with such a principle was to *reduce to* it the material already given by experience, with its attendant residuum of peculiarity which cannot be explained but must be simply accepted as a fact. It did not of course follow that the interpretation of given facts which these theories had to offer were wrong throughout, and they gave rise to many fruitful suggestions which subsequent science has thankfully followed, though they had to be put in a new light before it could utilise them. At the same time, there is one direction in which even the scientific views now prevalent require to be on their guard against the continuance of a similar illusion.

The later exponents of those Idealist doctrines lived, like ourselves, under the influence of the cosmographical views which recent scientific enquiry had developed; far from participating in the fanatic notions of antiquity, according to which the earth was the centre of the Universe, and all things besides were merely subsidiary to it, they admitted the Copernican discoveries, and realised that they and all the exercise of their observation were fixed at an eccentric point in the small planetary system. Yet in spite of this they persuaded themselves that the spiritual development of their absolute was confined to the shores of the Mediterranean, and that its plastic force in the physical world was exhausted in producing the forms of plants and animals, neither of which, as they knew, could exist except upon the earth's surface. Now it is certainly an idle and profitless task to attempt really to imagine what the forms of existence and life might have been, had the circumstances been wholly different; all such attempts result in mere clumsy reduplications of the forms of existence which experience presents to us. The just general conviction that Spiritual Life, the ultimate end of Nature, does not stand or fall with the earthly means which it uses for its realisation, cannot call to its aid any creative imagination capable of actually picturing another life of which we have had no experience. But, however mistaken may be the attempts which are made in this direction, the general conviction which inspires them will always remain valuable ; supposing physical science to be justified in assuming that certain physical processes prevail without variation over the whole universe, it would still be premature to assert a universal uniformity, which excluded any idea of forces peculiar in character and unexampled on the earth. So much the less ground is there for placing the concrete forms of reality, which no man can number, on the same footing with conceptions which, under the head of cosmology, we endeavoured to form of the universal rules of action to which Nature conforms. The former, therefore, I leave to be dealt with by natural philosophy, and renounce the prevailing fashion of relieving the dryness of Metaphysical discussion by picturesque illustrations selected from the experimental sciences.

218. It might, however, be truly objected, that though it may be impossible to deduce the concrete forms of nature, the reduction of them to the universal laws mentioned above is just one of the duties of metaphysic. I admit this duty, and only regret that it is one which no one can fulfil, not at least to the extent which the objection would require. The two points in which we seemed to run most counter to the ordinary view are, firstly, that of the phenomenal character of Space, secondly, that of the inner activity of Things, to which, instead of to external changes of relation between fixed elements, we ascribed the origin of events. Now, I have not neglected to insist in general terms on the necessity of starting from these inner states in order to explain even the possibility of that causative force which external circumstances appear to exercise. A more minute investigation of them, however, seemed to be forbidden, by the admitted impossibility of knowing them; and this would be the same even if more use were made than has yet been done of the hypothesis that their nature is spiritual. But this practical inapplicability does not impair the value of an idea which we found to be necessary, and to which no objection can be found either in itself or in the facts of experience. With respect to the Phenomenality of Space, I have argued at equal length and with a minuteness which has probably seemed tedious, that the appearance both of Space itself and of the changes which take place in it, is to be referred to real events which do not take place in Space, and I reserve for the Psychology what remains to be said by way of supplement to this; but, in this case also, it seems to me quite unfair to require my view to be worked out in detail. Such a requisition, if it applied to the particular perceptions of everyday life, would be as extravagant as the demand not merely to see what takes place before us, but at the same time to know the physical causes which make all that we see present itself to sight just as it does; only that here what we should ask to see through would be not physical causes but the supersensuous relations which the elements assume in the universal plan, and to which their appearance in Space is due.

Perhaps, however, no more is required than that in the case of the various main groups of natural processes, the hypotheses which had been constructed to explain them on the supposition of the reality of

Space, should now give place to others equally capable of explaining the facts, on the understanding that true being does not
exist in Space. If this is what is meant, I think the demand will
in the future certainly be complied with, but at present this is impossible, or, if approximately possible, is not to be regarded as
a slight addition to what has been already done.   In order to make
such a translation of physics into metaphysics possible we should
require first of all to have the whole text which is to be translated,
incontrovertibly fixed and settled.   Nothing can be further than
this is from being the case at present.   As things stand now, every
hypothesis which is used in explanation of the several branches of
natural phenomena, is compelled, in order not to ignore any peculiarities of the object in question, to assume a plurality of original
facts, which, though they may not be mutually inconsistent, exist only
side by side, and are not derivable the one from the other.   Still
more untrustworthy is our knowledge of the border-lands in which
these various spheres of natural phenomena meet.   What use then
would it be to show—what would be a difficult task in itself—that
these hypotheses can be replaced in all points, with equally fruitful
results, by a view which substituted for the supposed objects and
motions in Space, determinate supersensuous relations and excitations
in the inner elements of true being?   We should still have no other
way of determining these internal states than that by which physics
discovered the corresponding external ones : we should have to assume them as primary facts, which the phenomena in question required for their explanation.   But Metaphysic, if once she set herself
to this task, would have to do more than this ; she must be in a
position to show that all these necessarily assumed individual facts are
at the same time the logical consequences of those inner states, and
that the nature and character of true being justifies the attribution of
those states to it.   As long as this, which is again in fact a kind of
deduction of reality, is impossible for us, there can be little good and
small hope of reward in the attempt to reduce sensible facts to super-
sensible ones.   Leaving therefore any such attempt for another occasion, I will merely add a few general observations on the relation of
speculation to the ordinary methods of experimental science.

219. Man must make the best of what he has, and not decline
valuable knowledge merely because it does not at once offer him the
whole truth which he wishes to know.   In every science there will
always be a considerable gap between the most general points of
view from which we should wish to regard the given objects, and the

actual knowledge which we can possibly acquire about them; and this gap proves nothing either against the rightness of those ultimate points of view or against the value of the methods by which we succeed in investigating particular facts. We must beware of that doctrinairism, which will allow no conclusion to be valid, unless it is reached by the method of a logical parade-ground, reminding us of Molière's physician, who only demanded of his patient, 'qu'il mourût dans les formes.' In respect to applied Logic it must be granted that there is some truth in the cynical remark of the Emperor Vespasian. Every method is praiseworthy which leads to a sure result; even the most monstrous hypothesis, if it really enables us to connect the facts together and to explain their mutual dependence, is better than the neatest and trimmest theory, from which nothing follows. Holding these views, I can have no sympathy with the often repeated attempts of philosophers to show that the fundamental ideas of Physical Science are inadequate, disconnected, and frequently inconsistent. Without attempting to determine how much there is of justice or injustice in this indictment, I readily admit that it is in the main true; but I am not so much struck by these defects, as filled with sincere and unmixed admiration at the manifold variety of consistent and reliable results, which, with such imperfect means at her disposal, science has established by unwearied observation and by brilliancy of invention.

I hope and believe, also, that if science continues to work with the same conscientiousness, many truths, which now appear only in necessary juxtaposition, and many others which are seemingly opposed, will enter into a nearer and better relation, as different results of one and the same original process; in fact that, as at the end of a long and complicated reckoning, a simple total will be left over, which the philosophy of the future will be able to apply to the satisfaction of its own special wants. This much to be desired result, however, can only be obtained in the first instance by means of clearly outlined hypotheses, framed so as to meet the observed facts, and modified and transformed so as to keep pace with each fresh discovery: it matters not that the expression which our suppositions assume in this intermediate stage of discovery is imperfect in form; the wished for simplicity and clearness of statement can belong only to the finished result. No other method can be substituted for this; not that of Positivism, which bids us be content with general formulæ for the observed connexion of facts without introducing ideas about the inner connexion of things, advice which at first sight commends itself, but which is entirely fruitless in practice: not a lofty philosophic intuition which only a great poetic genius could delude men into regarding as an actual means to the discovery of truth ; not any speculative deduction, which hears only part of the evidence before rushing to its conclusion.   These leave us where we were: Moses may stand on the mountain of speculation and pray that the laws of thought may be faithfully observed ; but facts can only be brought into subjection by what Joshua is doing in the valley.   After this confession, my present object can only be to analyse those conceptions by the help of which philosophy distinguishes the wealth of natural processes into groups, seeing in each group either the operation of a specific principle, or a particular application of general principles, and regarding them at the same time as contributing in different ways to the realisation of the all-embracing plan of Nature.

220. The word *mechanism*, which has so many meanings, is used by modern schools of thought to describe sometimes a particular mode of action, sometimes a class of effects produced by this action : in either case, the mechanical aspect of Nature is spoken of in terms of marked disparagement, as compared with another and different aspect, to which it is deemed inferior.   What the word means is more easily learned from the customary use of language than from the conflicting definitions of the schools.   All modern nations speak of the mechanism of government, of taxation, of business of any kind. Evidently, what is signified by it is, the organization of means either with a view to realising a particular end, or to being prepared for carrying out different but kindred objects.   We do not, however, speak of a mechanism of politics; we expect political ends to be effected by an art of statesmanship, and this we should blame, if we saw it working by mechanical rules.   This distinction in the use of the term clearly expresses the limitation that the mechanical organization of means is only calculated for general conditions, common to a number of kindred problems, and meets the requirements in question by working according to general laws.

Now, it is impossible to conform to a law in a merely general way ; every application of the law must give rise to a determinate result depending on a determinate condition, whereas the law in its general expression makes the dependence only general.   It seems, therefore, up to a certain point to be part of the very essence and conception of a mechanism to take account of the differences in the particular instances to which it applies.   In the first place, the laws themselves which it obeys require that its effects shall be proportionate to the given circumstances ; next, the circumstances themselves, their peculiar nature, resistance, and reaction, modify the action and combination of the forces which it sets in motion—also according to fixed laws—and so enable it to produce the designed effect even under unforeseen conditions. The technical industry of the present day furnishes many examples of this self-regulation of machinery ; but whatever advances it may make in many sidedness and delicacy, it never escapes the limitations which popular language, as we saw, imposes upon the capabilities of mechanism. It is the ingenuity of the inventor to which alone the handiness of the machine is due ; it is his calculation, his comparison of the end with the means and the hindrances to its realisation, which has enabled him so to combine the forces of Nature, that they must now lead of themselves to the desired result according to universal laws of their own which are independent of him. His penetration may have enabled him to see disturbing causes in advance and to meet them by a combination of the means at his disposal so that the disturbances themselves liberate the reacting forces which are to compensate them ; even disturbing causes which he has not foreseen may by good luck be neutralised by the internal adaptation and power of self-adjustment of a machine. But all these favourable results have their limits. If they occur, they are the necessary consequences according to universal law of the joint action of the machine and its circumstances ; if they fail to occur, the machine is destroyed ; the power of resisting the conditions has not been given it from without, by the genius of its inventor or by a lucky chance, and it is incapable of generating such a power of itself.

Here lies the difference of statesmanship and every other practical art from what is mechanical. Every art, following as it does ends which cannot be realised of themselves, is confined to the use of means which it cannot make but can only find ; it cannot compel any one of these means to produce effects which are impossible or extraneous to its nature ; it can only combine together the means at its disposal in such a way that they will be compelled by the universal laws of their action to produce necessarily and inevitably the desired result. Every higher form of activity, consequently, which we are inclined to assume in Nature, even the most perfectly unrestrained freedom must, if it would be operative in the world, take just that mechanical form which is supposed at first to be inconsistent with its nature. The only privilege that distinguishes it, is the power of varying according to its aims the combination of the several mechanical elements, and of taking first one and then another part of the mechanism for its base of operations, thus making each part yield its own results. But its capabilities come to an end as soon as its object is one which cannot be produced by any combination of mechanical operations, or as soon as it can no longer bring about that particular combination which would have the result in question.

221. As regards the special meaning attached to the term 'mechanism' in their explanation of natural phenomena, philosophers undoubtedly understood by it primarily a peculiar mode of activity, the range of which was still undetermined. But it was distinctly believed, at the same time, that there was a certain special class of natural products, which was subject to the single and undisputed sway of the mechanical principle. I cannot subscribe to either of these two theories, except with essential reservations. Mechanism could only be defined in the sense in which it is employed in current language. Always determined by the given circumstances and general laws which lie behind it, never by the nature of an end which lies before it, it was contrasted (I shall return to this contrast later) as a concatenation of blind and irrevocable forces with those organic activities which seemed to follow ends with a certain freedom though they were also liable to fail in their attainment. But even within the limits of what was called the inorganic world, mechanism was opposed and deemed inferior to chemism. While in the chemical sphere, owing to the elective affinities of the elements, the specific qualities of bodies were continually destroying old forms and properties and creating new ones, thus co-operating decisively in determining the course of events, mechanical action was depreciated as a mere external process, which never gives a hearing to the distinctive nature of things, deals with them all as mere commensurable mass-values, and therefore produces no other effects but various combinations, separations, movements, and arrangements of inwardly invariable matter.

But Philosophy ought never to have believed in the reality of a mode of activity which it regarded in this light. A man or an official might be reproached for executing general laws and regulations without regard to exceptional cases, which deserve special consideration and forbearance. Such action, which we blame as mechanical, only succeeds because the combined force of human society deprives the ill-treated exceptions of the power of resisting. But things are not hindered from defending themselves by any such considerations, nor can there be anything in nature to prevent them from asserting their special peculiarities in the production of each effect, to the precise extent to which, if we may speak of them as human beings, they have an interest in so doing. It will be objected, however, that it is not meant to conceive of this mechanical agency, after the analogy of the inflexible official, as an authority of nature imposing itself autocratically upon things from without; what is meant is merely a process which is indeed developed from the interaction of things themselves, but which derives its character from the very fact that the things have no interests of their own, that they have not reached the point of letting their individuality be seen and heard, but are content to behave as samples of homogeneous mass; so far as this indifference of things extends, so far does mechanism extend. But even when stated in this improved form, the doctrine is not tenable unless either a physical process can be pointed out which takes place without being in any way influenced by the distinctive idiosyncracies of things, or it can be shown that results in the final form of which such influences though really operative seem to have vanished, are to be considered as preconceived elements in the plan of Nature. All attempts to establish the first case are from our point of view based on a wrong foundation. After having maintained that a change of outer relations is only possible as a consequence of mutual solicitations in the inner nature of things, we can only regard a mechanism which combines things in mutual action without taking account of this inner nature and its co-operation, as an abstraction of Science, not as a reality. Science, no doubt, has need of this abstraction. Whatever distinctive differences there may be between things, at any rate the contributions which they make to the production of a single event must admit of being expressed in values of comparable action. In order to be able to estimate their effects, we must refer the laws which govern them to certain ideally simple instances, zero values or *maxima*, of their effective differences, and then, after calculating our result upon this basis, subjoin such modifications as the concomitant conditions of the given case require.

It is in this way that we arrive at the indispensable conceptions of mechanics; the conception of a rigid immutable atom, from which every qualitative change is excluded; the conception of an absolutely fixed body, from which we have eliminated any alteration of form and all other effects of composition; at the principle, lastly, which may serve to express in the shortest form what we mean by mechanism, the principle that, if several forces act together upon the same object, no one of them has any effect on the tendency to action of the rest, but each continues to operate as if the rest were not present, and it is only these several and singly calculable effects which combine to form a resultant. Now none of these conceptions expresses anything which we can regard as occurring in actual fact, not even the principle last named. But supposing that this principle were not valid—and indeed the limits within which it holds good cannot be fixed *a priori*—supposing that the tendency to act of a force were altered by its relation to other forces working simultaneously, we should still require to make use of the principle, for we could not estimate the nature of the alteration, unless we first knew what the action would be unaltered; for even though it does not occur in its unaltered form, it would still help to condition the variation which does occur. So far, however, as the principle does hold good, it merely allows us to measure results when they take place, it does not tell us how they take place: it is not the case that the forces have been indifferent and taken no account of one another: the truth rather is that they, or the inner movement of things which correspond to them, *have* taken this account of each other, only it happened that the resolution at which they arrived in this particular case was to the effect that each should maintain its former tendency to act, just as in another case it might have been that this tendency should be changed. From this it appears that these very processes which, as far as the form of their result goes, exhibit all the characteristics of mechanism, are not produced mechanically in this sense at all, and the whole conception of *mechanism* as a distinct type of action, based on the mutual indifference of things, must be banished entirely from the philosophical view of Nature.

Nor does it receive more than a semblance of support from observation. Even in cases of impact, to which most of the so-called mechanical processes are reducible, there are produced along with the imparted translatory motion permanent or elastically neutralised changes in the form of the body impinged upon, besides inner vibrations which make themselves known as Sound or Heat. The number of these secondary effects, and the completeness with which the translatory motion is imparted, depends in every case on the inner interactions which hold together the ultimate elements of the bodies, depends, i.e. on forces which have their origin in the heart of things, and which differ from each other according as things themselves differ in quality. These inner effects we are accustomed, and for purposes of science obliged, to regard as secondary, and as disturbances of the theoretically perfect instance; but in taking account of them as corrections to be added to the result which strict rules would give us, we are really correcting our own abstract conception of a pure mechanism, which, as such, has no real existence in Nature.

222. As this is the case with regard to the first of the two alternatives[1] proposed, it remains that the Philosophy of Nature can only
undertake the second. Looking only at the ultimate form in which
processes result, it would be possible to arrange the facts of Nature
in groups according as the qualitative nature of things, which is a
constant factor in each process, was more or less apparent in the
results. And this is naturally the course which Idealism would have
followed, had it been consistent with itself. Its object being to point
out phenomena in which as a series the ends of Nature were successively realised, it might have entirely disregarded the question how
all these phenomena are produced, and have considered them solely
from the point of view of their significance, when once in existence.
All the misunderstandings which have arisen between Idealism and
the Physical Sciences, have been occasioned by this error of confound-
ing interpretations of the ideal significance of phenomena with explanations of the causes which have led to their existence. Imposing on
ourselves, then, this restriction, we might seek, in the first place, for a
department of processes where there seemed to be no trace at any
point of the constant silent influence of the qualitative differences of
things; or where, in case the elements producing the result were
homogeneous, there was no sign of the perpetual return of the process into, and its reproduction out of, the inner nature of things. It
would be in such a group of activities that we should have to look for
the *semblance* of a perfect mechanism.

In the small events which every day pass before us in changing
succession, in the motions which partly at our instance, partly owing
to causes which remain unobserved, bodies communicate to each
other—we do not find this mechanical action exemplified. In these
cases, though varying in distinctness, those secondary effects are
never wholly absent, in which the diversity of the co-operating
elements manifests itself. We find what we are looking for only in
the process of gravitation, or, more properly, in the revolutions in
closed curves, which result from the attraction of the heavenly bodies
and an original tangential motion. Attraction itself cannot be considered as an external appendage to the constituent elements of the
planets: as these elements are different, the degree of attraction
would have to vary to suit the nature of each part. But the different
distribution in different planets of elements varying in the degrees of
their reciprocal action, determines what we call the mass of the planets;

[1] [The treatment of 'Mechanism' a as a mode of Action, β as a kind of Effect,
v. Sect. 220 init.]

and so after having included in the conception of this unchanging
mass everything which related to the qualitative nature of the elements,
we find ourselves able to calculate the subsequent motions of the
heavenly bodies without assuming anything beyond their original
velocities and directions, and the general law of the variation of force
with distance ; and without being obliged to recur again to the inner
nature of the elements, though it is from this that the whole result
springs. It was this great spectacle of the universe maintaining
itself perpetually the same, that claimed the attention of Philosophy,
which saw in it the first stage of the self-development of thought in
Nature, the exhibition of the universal order, which remains undisturbed by any inner movements of the particular.

Next and in contrast to Gravitation or Matter, which was strangely
identified with it, was placed Light, or rather (since a name was
wanted which would include not only Light but also Sound) those
undulatory processes, by means of which impulses diffuse themselves
on all sides, without any considerable translatory motion. It was
not altogether without reason that in these phenomena of Nature an
analogy was found to Mind; for it is through them, no doubt, that
things convey to each other their fluctuating inner experiences, each
as it were reflecting itself in the other; so that a communication
between them is established, similar to that which exists between the
knowing subject and its object. It was owing to a misconception that
speculative Philosophy refused to allow these processes to be classed
under mechanism and treated mechanically. The equal diffusion of
light compels us, no doubt, to explain the force with which each
particle of ether communicates motion to the adjoining particle, as
due to inner experiences arising from their constant and sympathetic
relationship : but as it also leads us to assume that the ether consists
of none but homogeneous elements, the further progress of this
occurrence of transmission admits of being treated in precisely the
same way as the motions of the heavenly bodies. It is only when
these undulations come into contact with material bodies, i.e. when
they are reflected, refracted, dispersed, that the quality of particular
bodies makes itself felt in effects, which necessitate a number of new
truths derived from experience, and serving as starting-points for
analytical deductions. I have no intention of discussing in this
place the validity of those fruitful hypotheses, on the basis of which
optics has raised her imposing edifice ; nor do I wish to replace them
by others. I wished merely to justify to some extent the older speculation in its view, that these phenomena exhibit a new, characteristic,

and important form of Nature's activity, a form in which the influences of the specific qualities of things are not indeed quite neutralised, but do not appear to dominate the whole process : the general form, in fact, of a still inoperative affinity between diverse and changing elements.

223. A different impression was produced by the phenomena of *electricity* and *chemistry*. Philosophy here encountered the doctrine of the two electric fluids, which had already been fully developed by Physics, and was thus confirmed in regarding this as the first case in which the qualitative opposition of things appears as really determining the course of events. The further development of this branch of Physics will certainly not be able to dispense with the special presuppositions, which have been framed in consequence of this view. There seems at any rate no prospect at present of explaining that peculiar notion of absorption or neutralisation, in which forces, once in full activity, evanesce without leaving any trace of themselves, as due to a mere opposition of motions, similar to the absorption of Light by inter-ference. Such an explanation would still leave the question, what is the principle on which these conflicting motions are distributed amongst the bodies, from which the electric appearances are elicited ? And this question could hardly be answered without reinstating—though perhaps in a different form and connexion—the conception of a polar opposition of a qualitative kind. But this also does not con-cern us here : it is sufficient that electric phenomena, whatever may be their origin, in the form of their manifestation express precisely this idea of an opposition inherent in the nature of things.

This influence of the specific nature of agents was believed to be much more distinctly apparent in the case of *chemical* phenomena, which had likewise been already connected by Physics with electricity. The idea that in chemical as opposed to so-called mechanical action, the individual nature of things for the first time awoke, co-operated, and underwent inner transformations, was not, strictly speaking, sup-ported by observation. Striking changes were frequently seen to take place in the sensible qualities of things, in consequence of mere changes in their composition. Hence it was possible to suppose that other changes, the origin of which was not similarly open to experimental proof, might also be due to differences not directly perceptible in the arrangement of the ultimate particles and their resulting interactions. But the chemical process, according to that view of it which was favoured by Philosophy, was that, out of *a* and *b*, a third new and simple product *c* resulted, in which both *a* and *b* are merged, though by reversing the process, they may again be produced

out of it. This view, which obviously implies a constant and complete
interpenetration of the active chemical elements, expressed the idea
of which the phenomena of chemism furnished sensible illustration.
As a Physical theory it remained barren, because it failed to explain
how similar combinations of elements can give rise to permanently
different products, as also because it left out of account the manifold
analogies between combinations of essentially distinct elements.

To this view, there succeeded an exclusively atomic conception of
chemistry. The elements *a* and *b* were supposed to subsist unchanged
in the result *c*, and the properties above-mentioned were accounted
for by the different positions which the various samples of *a* and *b* may
assume in the product *c* of their combination. I do not understand
why the pictures which we often see of the structure of such chemical
combinations should be accompanied by the warning that they are not
to be understood literally. If they are only symbols, they at once
lead to a metaphysical view, according to which we should speak, not
of positions in Space, but of intelligible relations of varying intensity
between the actions of the absolute, which present themselves to us
singly as chemical elements. If we shrink from making use of these
certainly impracticable notions, of which I have spoken previously,
and make up our minds to follow the ordinary view of the reality of
Space, it seems to follow that either these graphic representations
must be understood quite literally, or that they have no intelligible
meaning at all. It is not, however, my purpose to describe the
consequences which the atomic view of chemistry has had in general,
and especially of late the hypothesis of Avogadro, in itself an entirely
improbable one. I would only call attention to the fact that after all
that can be said, our knowledge is limited on the one hand to the
elements which enter into composition, on the other to the actual and
probable typical forms which the composition finally assumes; the
process by which the combination takes place, i.e. the true chemical
process, still escapes us. Our conceptions of it cannot be made to
fit with the rest of our mechanical notions, unless we admit as new
data both the original difference between the elements, not reducible
to physical modifications of a common matter, and the special elective
affinities of these elements, which determine their general capacities
of combination and the proportions in which they will permanently
combine.

Even then one phenomenon still remains dark, that which gave to
chemistry its old name 'Scheidekunst' (art of division), the analysis
of the combinations. Let us suppose that between all the elements,

a b c . . . s, the only affinity that exists is that of attraction in varying degrees of intensity. In that case, if there is no new condition introduced, any reciprocal action between the two pairs *a b* and *c d* can only lead to their amalgamation *a b c d*, never to their fresh distribution into *a c* and *b d*. And even if the affinities between *a* and *c*, *b* and *d*, be ever so much closer than those between *a* and *b*, *c* and *d*, there cannot be any separation of the elements: the most that can happen is this, that an external force, *if* it were brought to bear upon the whole combination *a b c d*—which would be the necessary result of mere forces of attraction—would detach *a* from *b* or *c* from *d* more easily than *a* from *c*, or *b* from *a*. Any repulsion, therefore, must come from elsewhere than the results of attraction ; and as there is no evidence of direct repulsion between the single elements it can only be looked for in the circumstances which accompany the chemical process, or, as is probable, actually constitute it. These may consist in motions which disconnect the elements, or in the affinity of the elements to the different electricities, the polar antagonism of which may require them to move in these particular ways.

But however that may be, my only purpose was to show that Philosophy was right in ascribing to the qualitative differences of things a decisive influence in the sphere of chemistry, wrong in denying any such influence in that of mechanics: and that therefore though the opposition between these processes of nature is not without some reason in it, it is practically impossible to draw a sharp line of distinction between them, such as would separate their spheres, and assign to them two different principles of action.

224. But all this has now scarcely more than an historical interest; the relation of forces to *organic* activities is still the subject of conflicting opinions. In an essay on ' Life and Vital Energy,' which forms the introduction to Rudolph Wagner's Hand-Dictionary of Physiology, I defended, six-and-thirty years ago, the claim of the mechanical view to a place in the science of Physiology, a claim which was at that time still much disputed. Scientific taste has now to some extent changed ; at present, not merely all the practical investigations of Physiology, but to a great extent also the formulation of its theories are dominated by the mechanical spirit ; those who are opposed to it, repeat the old objections, for the most part in the old form. If, though weary of going back to these matters, I proceed now to recapitulate shortly the conclusions which were developed in the above-mentioned essay, and subsequently in the ' General Physiology of Corporeal Life' (Leipsic, 1851), it is chiefly for the sake of a remark which has been often overlooked, at the end of the essay, and which is to the effect that it necessarily contained only the *one* half of the principles which a complete biological theory implied. The other half would have touched on the question, how the mechanical treatment of vital phenomena, necessitated by the facts, harmonises with those requirements of an opposite kind, which the primary instincts of philosophy will never cease to make, as in times past. For this dispute is, in fact, an old . one. I should have been able to go back to Aristotle, whose 'substantial forms' extended the dominion of the activity of Thought far beyond living things, to which in the modern controversy it is confined, while already in antiquity the Aristotelian view was elaborately opposed by the Epicurean physics, which denied the activity of thought no less unrestrictedly. The question did not, however, become one of pressing importance, until, with the development of modern science, a definite formulation had been given to the group of ideas, the application of which to explain life meets with so much opposition. Putting aside the more ethical, æsthetic, and religious grounds for this aversion, which it is not necessary here to examine, the theoretical motive which has prompted it has always been the same. The scanty knowledge which we possess of the formative influences active throughout the rest of nature, did not seem sufficient to explain the complex and yet fixed forms of organic life; their germs at any rate, it was thought, must have an independent origin, even if in their subsequent development they were subject to the Universal Laws of Nature. But further, the peculiar phenomena of growth, nutrition, and propagation, the general fact of the interdependence of continuously active functions, and that of self-preservation in presence of repeated disturbances, all this seemed to demand the continued presence and operation of that higher principle, to which had been attributed at first only the initial formation of the germ. Finally, the undefined but overpowering general impression of pervading adaptation, witnessed to the presence of an end which guided organic nature, rather than to a past which blindly compelled it. The conception of a *vital force* was the first form in which these ideas were united.

225. As long, however, as this expression was merely thrown out in a general way, it could not serve to solve the difficulty, but only to indicate its existence. It was not allowable to follow the example of Treviranus, and explain everything from the byssus to the palm, from the infusorium to the monster of the sea, as living by Vital

Force : the difference between the palm and the byssus had also to be taken account of; every species of living things required its own special vital force, and every individual of the species needed its own share or its particular sample of the force. The general name Vital Force indicated, therefore, merely a formal characteristic, which could attach to many different real principles yet to be discovered. It was besides an improper use to make of the term *force*, which had been applied by Physics in quite a different sense; the appropriate word was *impulse* (Trieb). For when the general characteristic in question had to be described, the contrast was obvious. Every physical force always produces under the same conditions the same effects, under different conditions, different effects ; it is always conditioned by a general law, irrespectively of the ensuing result; everything that under given circumstances the force can effect, it must necessarily effect, nor can any part of the effect be kept back, nor any addition be made to it which would not have been inevitable under the existing circumstances. To Impulse, on the other hand, we ascribe the power of changing its manner of operation, not indeed without regard to existing circumstances, but with regard at the same time to a result which does not yet exist; a power of leaving undone much that it might do, and of beginning something new instead which it is not bound by the given conditions to do at all. It had to be admitted, however, that the vital impulse never produces anything in a vacuum, but only works with the materials supplied to it by nature ; and thus arose the ordinary view of vital force as a power, which, though dependent in a general sense upon material conditions, is superior to the physical and chemical laws of matter, and gives rise to phenomena which those laws will not explain.

226. I must take permission to refer to the above-mentioned essay for many details, which here I can only lightly touch on, but could not altogether omit without leaving constantly recurring fallacies only half-refuted. We are continually being told that no application of the improved means which we now have at our command will enable us to manufacture artificially a product which even remotely resembles a living organism. The fact must simply be granted. Neither cellulose nor albumen, nor any other of the tissue-forming substances of organic bodies can be produced by chemical art, although the distinction between the ternary and quaternary combinations of organic life and the binary combinations of inorganic nature, which was once so much insisted on, has long since lost its meaning: nor are we any longer under the delusion that these combinations last only so long as

the vital force lasts, a delusion which any thoughtful student might have been disabused of from the first, if he had only thought of the wood of the table, at which he was writing, or of the pens and paper. Still, it is true that in none of our artificial productions is there any such connected series of chemical transformations, form-modifications and functions as could be compared with the growth, nourishment, and propagation of an organic Being: even the recently observed formation of cells out of inorganic substances, though worthy of all consideration, is not likely to prove the starting-point for new discoveries in this direction. But all that this proves is that in the present course of Nature, Life is a system of processes self-maintaining and self-propagating, and that outside its sphere there is no combination of materials, such as would make the development of such phenomena possible. Nothing is thus decided as to the conditions under which this play of forces is sustained *after* it has begun, and yet these must first be known before it can be determined what requirements a theory as to the first origin of Life has to meet. But neither the question concerning the origin of the whole organic world, nor, the consideration whether in the future it may not be possible to add to it by artificial means, must be allowed to confuse the discussion here. The only point to be considered is, whether the vital force which organic beings as a matter of fact exercise in developing themselves and resisting external injury, requires us to assume a principle of action, which is strange to the inorganic world; and whether that other vital force, which such a principle of action is assumed to be, is conceivable in itself, and adequate to explain the given facts?

227. We shall require, in the first place, for the sake of clearness, to be definitely informed as to the nature of the subject, to which the activities included under the name of vital force are supposed to belong. There has been no lack of theories which endeavoured to meet this question fairly. Some have spoken of a universal substance of Life, which they found either in a ponderable matter, or in electricity, or some other unknown member of the more refined family of ether. Others regarded the soul as the master-builder and controller of the body, assuming at the same time that plants had souls, which was, to say the least, not a fact of observation. I will only mention briefly the common defect in all these theories. It is impossible to deduce difference from a single homogeneous principle, unless we have a group of minor premisses to show why the one principle should necessarily develop *a* at one point, *b* or *c* at another. As has already been said, we should always have to assume as many

different material bases of life as there are different kinds of living things; or else it would have to be shown to what subsequently arising causes it was due that such different forms as an oak tree and a whale could be produced out of the one substance. In the latter case the development of Life would be at once brought again under the general conception of a mechanism. For mechanism in the widest sense of the term may be said to include every case in which effects are produced by the reciprocal action of different elements, of whatever kind, working in accordance with universal Laws; and such conformity to law would have to be assumed by all these theories; they could never leave it open to doubt that, under the influence of an accessory condition *a*, the single principle of life would take shape in the product *a* rather than in *b*.

But metaphysic has no interest in maintaining the claims of the mechanical principle, except in this very general sense; nor, on the other hand, will physics be so narrow-minded as to insist that it is precisely from these materials and forces which we now know, and according to the exact analogy of inorganic processes, that we are to conceive of the phenomena of organic Life. All that physics claims, is, that whatever kinds of matter, force, or energy remain yet to be discovered, must all fall within the compass of her investigations, must all be connected together according to Universal Laws. Further, however, experience did not at all show that the choice between these accessory conditions was so unrestricted. It is not the case that every organic kind requires as the basis of its existence peculiar kinds of matter which it places at the disposal of the one vital force. The most different products of Nature are all constructed from the same storehouse of material elements, which are found on the surface of the earth. Hence, however peculiar the principle of Life may be in itself, it can never have been free from interaction with that same matter which we know to be also controlled by physical laws of its own. The principle might issue what commands it pleased, but could only carry them out (supposing the materials in question not to obey them spontaneously) by exerting those forces to which the matter is naturally amenable. We know that in all cases the contribution which is made by the several co-operating factors, to a result in the final form, may be of the most different amounts. Thus it *may* be that the form which Life is to assume in any given case is already traced by anticipation in some specific kind of substance; but the actual existence of this life is always the result of mechanical causes, in which the original substance would be only *prima inter pares*, contributing just so much to the

result as can arise according to general laws from its coming into
contact with the other factors. But that that is actually the case, at any
rate in the sense that there are certain kinds of matter specially privileged in this respect, could not in any way be proved; the natural conclusion which the facts suggest is, that the phenomena of Life arise
out of a special *combination* of material elements, no one of which has
any claim to be called exclusively, or, in the degree suggested above,
preeminently, the principle of life. The very fact which has been taken
to imply a special vital principle, the fact that Life is only maintained
by successive self-propagation, ought rather to lead to the conclusion
that the germ of its development can only be found in a certain peculiar
combination of material elements, which maintains and reproduces
itself in unbroken continuity. It is, therefore, quite a matter of
indifference, whether we shall ever succeed in giving a name to the
general form, or in exhibiting in detail the development, of such a
material combination in which life is implicit; the point is, that the
supposition of a single Real principle of Life is both impossible in
itself and quite barren of results, whilst on the other hand, the only thing
which the mechanical view leaves unexplained is the ultimate origin
of Life. I will reserve what I have to say on the Soul till later; as
it neither creates the body out of nothing, nor out of itself, it can have
no special dignity as regards the construction of the body (whatever
other dignities it may have) except that of being *prima inter pares ;*
it must work jointly with the material elements which are supplied to
it. The conception of mechanical action, however, is wide enough to
include that of a co-operation, according to universal laws, between
spiritual activities and conditions of matter.

228. It is the way of mankind to meet a theory not by direct refutation, but by expressing general dislike and pointing out the defects
in the working out of it, and to magnify striking though unessential
differences until they seem to be impassable gulfs. I should certainly
never of my own motion speak of the living body as a machine, thus
nullifying the distinction between the poverty of even our most ingenious inventions and the mighty works of Nature; but those who
are so morbidly anxious to leave out of account in their consideration
of life all those operations which they can stigmatise as mechanical,
need to be reminded that the living body and not inorganic Nature
furnishes the models of the simple machines, which our art has imitated;
the pattern of pincers is to be found only in the jaws of animals; that of
the lever in their limbs which are capable of movement. Nowhere else
are there instances of motions produced in articular surfaces by cords

such as the muscles are, and of their guidance by ligaments in definite directions: it is the living body alone which utilises the production of a vacuum and the consequent inhalation of atmospheric fluids, the pressure of containing walls[1] upon their contents, and the valves which prescribe the direction of the resulting motion. How little does all this resemble that mysterious power of immediate agency which is most eagerly claimed for the vital force !

The exaggerated pictures of the superiority of living machines to artificial ones do not rest on any better foundation. The comparison of an organism to a self-winding clock altogether ignores the drooping plant which can find no substitute for water, if water will not come to it, and the hungry animal which is indeed able to seek its own food, but yet dies of want if none is found. Irritability, or the power of responding to impressions, is said to be a distinguishing characteristic of organisms; when a given stimulus is applied to them, they are supposed to react in ways which are not explicable from the nature of the stimulus; at the same time, it has been assumed that in mechanical action the cause and effect are precisely equal and similar, though not even in the simple communication of motion is this really the case, while organic life has been contrasted with it on the ground of a supposed peculiarity which is in fact the universal form of all causative activity. For it is never the case that an impression is received by an element ready made, merely to be passed on in the same form; each element always modifies by its own nature the effect of the impulse experienced. In a connected system of elements, the effects which will follow a stimulus will be more various and striking in proportion as the intermediate mechanism is more complex, which conducts the impression from point to point and changes it in the process. The same must be said of the power of recovery from injury which is supposed to belong peculiarly to organisms, and to prove clearly a continuous adaptivity superior to anything mechanical. But if it were really the case that this force of resistance raised organic Beings out of the sphere of physical and chemical necessity, why was it ever limited? If once it had become independent of mechanical influences there was no task which it need fail in accomplishing. But the numberless cases of incurable disease indicate plainly enough its limits. No doubt, when once its combinations of elements and forces have been fully matured the body is so well furnished for its purpose that even considerable changes in its environment produce reactions in it which avert or remove the disturbing influences which

---

[1] [Of the heart and blood vessels.]

threaten or have begun to act upon it. But as in every mechanical product, there are limits to this power of self-preservation. There is no such power, where the body has not been blest at starting with these particular provisions, nor do we ever see the want supplied by the subsequent creation of fresh means; we much more often see the means already at its disposal forced into a reaction, which under the special conditions of the moment can only lead to further dissolution.

229. I shall not continue this polemic further, having devoted sufficient attention to it before. I simply adhere now to the decision which I then expressed. In order to explain the connexion of vital phenomena, a mechanical method of treatment is absolutely necessary; Life must be derived, not from some peculiar principle of action, but from a peculiar mode of utilising the principles which govern the whole Physical world. From this point of view, an organic body will appear as a systematic combination of elements, which, precisely because they are arranged together in this form, will be able by conforming to fixed laws in their reciprocal action, and by the help of external nature, to pass through successive stages of development, and within certain limits to preserve the regularity of its course against chance disturbances. This makes me the more sorry that Physiologists should regard this view, which embodies the necessary regulative principle of all their investigations, as being also the last word upon the subject, and should exclude every idea which is not required for their immediate purposes, from all share in the formation of their ultimate conclusions. But they will never remove from the mind of any unprejudiced person the overwhelming impression that the forms of organic life serve an end; nor will men ever be persuaded that this marvellous fact does not call for explanation by a special cause. I know full well that as a thesis it may be maintained that every result which presupposes mechanical agency presupposes nothing more than this. Nor is this new; long ago Lucretius declared that animals were not provided with knees in order to walk, but that it was because the blind course of things had formed knees, that they were able to walk. It is easy to say this, and it may be that it sounds particularly well when expressed in Latin verse; but it is impossible to believe it; there is no more tedious product of narrow caprice than such philosophy of the schools. Yet it is unfortunately true that the conviction of a higher power working for an end, and shaping life with a view to it, has too often intruded itself rashly and confusingly into the treatment of special questions; and this explains the unwillingness of conscientious enquirers to recognise what to them must seem a barren hypothesis. It cannot, however, be ignored that many of our contemporaries are animated by a profound hatred of everything that goes by the name of Spirit, and that, if a principle were submitted to them which seemed to bear traces of this, even though it was not opposed to any postulate of science, they would, none the less, turn away from it in indignation to enjoy their feast of ashes, and delighted to feel that they were products of a thoroughly blind and irrational necessity. Such self-confidence it is impossible to reason with; we can only consider the difficulties which stand in the way of the acceptance of the opposite view.

230. We must not stop short at those general accounts of the matter, which merely represent a higher power in any indefinable relation of superiority to mechanical laws without making the obedience of those laws intelligible; in speaking of this, as of all other forms of rational activity directed to an end, the first thing to do is to give a name to the subject from which the action is supposed to proceed. Now we certainly cannot speak of ' ends' with any clearness, except as existing in a living and willing mind, in the form of ideas of something to be realised in the future. Hence it was natural to look for this highest wisdom in God; and not less natural was the desire to bring again into an intelligible relation the unlimited freedom of action involved in the conception of the divine essence, and the fixed course of Nature which seems to bear no traces of that freedom. Thus arises the theory upon which sooner or later Philosophy ventures, the theory that the world was created by God and then left to itself, and that it now pursues its course simply according to the unchangeable laws originally impressed upon it. I will not urge the objection that this view provides only a limited satisfaction to our feelings; in its scientific aspect it is unintelligible to me. I do not understand what is meant by the picture of God withdrawing from the world that He has created, and leaving it to follow its own course. That is intelligible in a human artificer, who leaves his work when it is finished and trusts for its maintenance to Universal Laws of Nature, laws which he did not make himself, and which not he, but another for him, maintains in operation. But in the case of God I cannot conceive what this cunningly-contrived creation of a self-sustaining order of Nature could be; nor do I see what distinction there can be between this view and the view that God at each moment wills the same order, and preserves it by this very identity of will. The immanence of God in the course of Nature could not, therefore, be escaped from by this theory; if Nature follows mechanical laws, it is the Divine action itself, which, as we are accustomed to say, *obeys* those laws, but which really at each moment creates them. For they could not have existed prior to God as a code to which He accommodated Himself; they can only be the expression to us of the mode in which He works.

This unavoidable conclusion will not be at once nor willingly admitted : however much the world may be primarily dependent on God, the desire will be felt, that it should contain secondary centres of intelligent activity as well, not entirely determined in their effects by the mechanical system of things, but themselves supplying to that system new motives for developed activity. It was this wish which was expressed by Stahl's theory of the soul, when he spoke of it as moulding the body to its own ends. This theory was in so far correct that it conceived of the soul as a living and real Being, capable of acting and being acted upon with effect : but it missed its mark, because the formation of the body, in its most essential and irreversible features, is concluded at a time when the soul may perhaps have some dream of its future aim, but certainly cannot as yet have knowledge enough of the external world to be able to adapt the body to the conditions which life in that world imposes. Thus the advantages which the soul might seem to derive from its consciousness and power of taking thought for the proper development of the organism, are all lost ; and the only power of adaptation which it remains to ascribe to it is an unconscious one. Though this conception is very frequently misapplied, it does not seem impossible to attach to it a definite meaning. All along, we have considered things as distinguished from each other by manifold differences : and although we cannot fully realise to ourselves what constitutes the essential character of any single thing, there is nothing to prevent us from assuming a certain difference of rank between them, such that when two things were subjected to the same external conditions, the one would manifest its nature in simple and uniform reactions, the other in complex and multiform ones ; and these latter reactions might be such that each gave rise to some entirely new capacity in the thing, or that they all united to form a single development directed to a definite end. In that case, we shall possess in the soul a real principle at once active in the pursuit of ends and yet unconscious, such as would not be at variance with mechanical laws ; for none of the possibilities that lie latent in the soul would be realised, except through stimuli acting upon it according to fixed laws, and eliciting its development step by step.

Clearly, however, in this case, the soul will no longer imply anything peculiar or characteristic; once get rid of consciousness, and it becomes a mere element of reality like other elements; and that superiority of nature, which made it so pregnant a centre of manifold forms of life, might equally well be ascribed to any other element (making allowance for differences of degree) even though it possessed none of the characteristic properties of the soul. The question as to the true origin of the soul, leads to the same conclusion. If it is conceived as eternally pre-existent and prior to the Body, it must still be confined within the limits of the course of Nature ; what then is it, and where? For to suppose that it suddenly becomes a part of Nature without having previously been so, is virtually to assign it an origin. If then it is always a part of Nature, we cannot help regarding it as one among other natural elements; and as there is no reason for supposing the other elements inferior, we must ascribe to them too, and in a word to all elements whatever, the same inner capacity for organic development. And here it seems as if we were once again brought back to the unfruitful idea of a common material basis of life. For the manifold forms which these elements assume, would depend on the different modes in which they were combined by the course of Nature ; hence, the form which is actually realised at any given moment, must be either the result of mere mechanical agencies —though these may be of a higher type than any with which we are familiar in Physico-Chemical processes—or else, supposing that traces of an independent activity still remain, the soul, which concentrates the different active elements upon this particular development, must come into existence afresh at the moment that they unite ; and the question then arises, Whence does it come ?

231. This difficulty of finding a real subject, capable of formative activity for an end, has led to attempts to dispense with a subject altogether; it was thought that the generic *Idea* or *Type* would be sufficient to account for such activity. Aristotle set the example with the unfortunate but often repeated remark, that in living things the whole precedes the parts, elsewhere, the parts precede the whole. This saying, no doubt, gives utterance to the mysterious impression which organic life produces; unluckily, it has been regarded as a solution of the mystery. And yet what truth can be more simple than this, that Ideas are never anything else but Thoughts, in which the thinker gathers up the peculiar nature of an already existing phenomenon ; or of one which he knows will necessarily exist in the future as soon as the data exist which are required to produce it? It may be allowed that Reality is so constituted, that from our point of

view it is always exhibited in subordination to certain Ideas, general notions, or Types; and we may accordingly go on to say that these Ideas hold good in reality and dominate it; but their dominion is only like that of all legislative authorities, whose commands would remain unobserved if there were no executive organs to carry them out. Never, therefore, in Organic Life is the whole before the parts, in the sense that it is before *all* parts; it only has existence in so far as an already formed combination of parts guarantees that existence in the future as a necessary result of the germ here present, and not of the germ only, but also of favourable external circumstances acting upon it. Anyone who is not satisfied with this development of the whole from the parts, and desires to reverse the relation, will be required to show who the representative of the generic Idea is, who stands outside the parts and gives to the Idea, which in itself is merely potential, a real power in the real world. It must be shown where these Ideas reside, before they initiate a development, and how they find their way thence to the place where they are attracted to an exercise of their power.

Quite recently, an attempt of a different kind has been made by K. E. von Baer. We could have wished that this deservedly popular investigator had succeeded in making out his point to satisfaction; I cannot, however, persuade myself that his proposal to conceive of Nature as striving towards an end, really carries us any farther. If all that it means is, that the different forces, which are active in the construction of organisms, converge in different directions towards a common result, this fact has never been doubted; nor, considered merely as fact, is it the subject of the present controversy. The question at issue is rather this; is the cause which determines this combined action to be found merely in the course of things after they have once been set in motion? i.e. does the convergence occur when there is this motion to produce it, and not occur when there is no such motion? or is there anywhere a power not subject to this constraint of antecedent conditions, which, on its way to the attainment of an end, brings together things which but for it would exist apart? Naturally, it is this latter view which is preferred here. Yet it is not clear, how this supposed tendency to an end would differ from that which might be ascribed, e.g. to falling stones, which, while converging from all quarters of the globe towards its centre, move merely in obedience to a universal law. It is the presence of purpose alone which could constitute that difference, converting the mere end of a process into an aim, and motion to that end into an impulse. Such a purpose Baer's theory accepts, and yet by banishing consciousness, which is presupposed by it, at the same time rejects. Finally, to whom is this tendency in the direction of an end to be ascribed ? It would not suit the character of the individual elements, which, varying as they do in capability, tending now to one end, now to another, need some power outside themselves to inform them upon what point they have to converge in any given case ; and it is, in fact, from *Nature* that such a tendency is supposed to proceed. But, where is this Nature ? It is allowable in ordinary discourse, no doubt, to use this term in such a merely general sense ; but in the particular cases in which the designation of Nature as an efficient cause is intended to decide in its favour the choice between it and other agents, there should be some more accurate determination of the conception of it, as well as of the metaphysical relation in which, as a whole, it stands to its subordinate parts. We propose now to supplement the theory in this point, and thus to bring our investigations to a close.

232. The grounds which have led me to my final conclusion have been expounded at such length throughout my entire work, that what I shall now add with regard to this much debated question will be only a short corollary. Men have created for themselves a false gulf, which it has then seemed impossible to bridge over. It is not with any special reference to the opposition which has to be reconciled between living Beings and inanimate Matter, but on much farther-reaching and more general grounds that I have all along maintained the inconceivableness of a world, in which a multitude of independent elements are supposed to have been brought together subsequently to their origin, and forced into common action by Universal Laws. The very fact that laws could hold good in the same way of different elements, showed that the elements could not be what they pleased. Though not directly homogeneous, they must be members in a system, within which measurable advances in different directions lead from one member to another ; on this condition only could they and their states be subsumed under the general Laws, as instances of their application. But the validity of general Laws, so established, was not enough to explain the possibility of their application in particular cases ; in order that they should necessitate one event at one time and place, another at another, the changing state of the world as a whole had to be reflected at each moment in those elements, which are working together for a common result. It would be idle, however, to suppose that the elements, being originally separate, required the mediation of some 'transeunt' agency which should convey to them

the general condition of the world and stimulate them to further activity: rather, what is experienced by one element must become *immediately* a new state of another. Hence we saw that every action that takes place necessarily presupposes a permanent and universal relation of sympathy between things, which binds them together in constant union, and which itself is only conceivable on the supposition, that what seems to us at first a number of independent centres of energy, is, in essence, one throughout. It is not, therefore, to bring about any specially privileged and exalted result, that the assistance of the infinite Being M, which we have represented as the ground of all existence, is required; every effect produced by one element on another, even the most insignificant, is due to the indwelling vitality of this One Being, and equally requires its constant co-operation. If there is a class of processes in Nature, which, under the name of mechanical, we contrast as blind and purposeless with others in which the formative activity of the One Being seems to stand out clearly, the contrast is certainly not based on the fact that effects of the former kind are left to be governed by a peculiar principle of their own, whilst only in the latter does the one universal cause attempt after some incomprehensible fashion to subdue this alien force. In both cases alike the effects proceed solely from the eternal One itself; and the difference lies in *what* it enjoined in each case, in the one case, the invariable connexion of actions according to universal laws which constitute the basis of all particular conditions, in the other, their development into the variety of those particulars. But, instead of repeating this line of thought in its generality, I shall endeavour to show how it applies to the special question now before us.

233. The germ of an organic growth is not developed in empty Space, in other words, not in a world of its own which has no connexion with the whole of Things. Wherever the plastic materials are present, there the absolute One is likewise present; not as an idea that may be conceived, not as an inoperative class-type, not as a command passing between the elements of a group, or a wish without them, or an ideal above them; but as a real and potent essence present in the innermost life of each element. Nor is it, like divisible Matter, distributed among them in different proportions. It manifests itself in each one in its totality, as the unity that embraces and determines them all, and in virtue of the consistent coherence of its entire plan, assigns to each of these dependent elements those activities which ensure the convergence of their operation to a definite end. But the Absolute is no magician; it does not produce Things

in appropriate places out of a sheer vacuum, merely because they correspond to the import of its plan. All particular cases of 'its operation are based on a system of management according to law, adapted to its operation as a whole. But I must repeat: it is not here as it is with man, who cannot do otherwise; rather this conformity with general principles is itself a part of what is designed to exist. Hence it is, that each stage in the development of organic Life seems to arise step by step out of the reactions which are made necessary for the combined elements by their persistent nature; nor, is there anywhere an exception to the dependence of Life on mechanical causes.

At the same time, we are never justified in speaking of a merely mechanical development of Life, as if there were nothing behind it. There is something behind, viz. the combining movement (of the absolute, the true activity that assumes this phenomenal form. We may even admit that it apparently breaks through the limits ordinarily assigned to mechanical action. I have before mentioned, and I now repeat, that the principle of mutual indifference, which Mechanics has laid down in respect to forces working concurrently, is, if strictly taken, by no means justified as a universal law. It should rather be laid down as true universally that an element a when it is acted upon by the determining circumstance p, has, by this very fact, become something different, an a which $= a^p$, and that a new force q will not exercise the same kind of effect on this modified element, which it would have exerted on it if unmodified; that the final result, therefore, will not be a^{pq}, or $a^{(p+q)}$ but, a^r. But this r could never be obtained analytically out of any mere logical or mathematical combination of p and q; it would be a synthetic accession to those two conditions, and thus not deducible except from the import of the entire course of things. This is expressed, according to ordinary views, thus—the combination of several elements in a simultaneous action may be followed by effects, which are not mere consequences of the single effects produced by the reactions between every pair of them. That which we now wrongly regard as the universal and obvious rule, viz. that effects should be summed up in a collective result without reciprocally influencing each other, would be only one special case of the general characteristic just mentioned. I shall not now enquire whether and in what direction Biological science will find itself compelled to recognise the possibility of this modification of effects; we must, however, leave a place for it in our own theory. Its admission would not in any way invalidate our conception of the mechanical order, but only extend it further. For it would be our first position even with regard to these new grounds of determination, which intrude upon the course of events, that neither did they arise without a reason, but according to rules, though rules which are more difficult for our apprehension to grasp. But at the same time we should escape from regarding Life as a mere after-effect of a Power, which having formed the mechanism, had left it to run its course. The Power would rather continue to manifest its living presence and constant activity, as operative in the phenomena of Life.

What direction our thoughts might have to take beyond this point, I am not now called on to decide. There is nothing more to be added, which could be urged with absolute certainty of conviction against those who regard the whole sum of the effects produced by this ultimate agency, not less than the inner activities whence they proceed, as still but mere facts of Nature, a tendency which the course of things has followed from all eternity ; but which includes no element resembling what we understand by intention, choice, or consciousness of a purpose. Our view, it must be admitted, is no such very great advance upon the mechanical explanation of Nature, from which a refuge was sought. The development of the world would on it be no less a necessary concatenation of cause and effect ; excluding all free initiation of new occurrences. Only the most extreme externalism would be avoided. The mechanism would not consist, at starting, of an unalterably fixed complement of forces, which would only suffice to effect changes of the position of existing elements. The mechanism would itself produce at certain definite points those new agencies which would be the proximate principles governing organized groups of connected phenomena. For my own part, I cherish no antipathy to the opposite view, which insists that this whole world of forces, silently arrayed against each other, is animated by the inner life of all its elements and by a consciousness which is that of an all-embracing spirit. I shall not even shrink from attempting, in the proper place, to show that there is a real Freedom which can give rise to truly new departures, such as even this latter belief does not necessarily involve. But such a demonstration would transcend the limits of Metaphysic. It would lead us to consider a mysterious problem, which our discussions down to this point have bordered upon. I have already expressed the opinion that we must not merely credit things with a persistent impulse to self-preservation ; but are justified in assuming (as an hypothesis, and in order to explain the phenomena) an impulse to the improvement of their state. Now, if this hypothesis is conceivable in regard to the individual elements, it becomes almost necessary when we no longer speak of them as individuals, but conceive of them, both in their nature and in their actions, as manifestations of a single and all-embracing supreme cause whose mandates they execute. I should at the same time most unquestionably admit that this assumed tendency towards improvement, though it may be the ultimate *ratio legis* from which all special laws of action of things are derived, could never furnish us (since we cannot define this 'improvement') with anything more than the final light and colour of our view of the world; it could never serve as a principle from which those laws could be deduced. But here, the same question which we asked concerning the vital energy, suggests itself once more—If this endeavour after improvement is a fact, why does it not everywhere achieve its end? Whence come all the hurts and hindrances by which the course of Nature, as it is, so often prevents from being fully satisfied the impulses which it nevertheless excites? The conflict of forces in Nature, like the existence of evil in the moral world, is an enigma, the solution of which would require perfect knowledge of the ultimate plan of the world. Metaphysic does not pretend to know what this plan is; nor does she even assert that it is a *plan* that rules the course of events; for this would be inseparable from the idea of the purpose of a conscious being. But, if it limits itself to the belief, that existence has its cause in a single real principle, whatever its concrete nature may be, no considerations concerning these ultimate enigmas can affect the certainty of such conclusions. For, I wish here most distinctly to assert, that though I am old-fashioned enough not to be indifferent to the religious interests which are involved in these problems, the views for which I have been contending rest on a purely scientific basis, quite without reference to Religion. No course of things, whether harmonious or discordant, seems to me conceivable, except on the supposition of this unity, which alone makes possible the reciprocal action of individual existences. The disturbing effects which things exercise upon each other witness to this unity, not less clearly than the joint action of forces with a view to a common end.

234. Similarly, the limits within which metaphysical enquiry is confined compel us to exclude from its sphere the much debated question as to whether the conception of a *kind* has really that objective validity in the organic world which we ordinarily ascribe to it. It will not be supposed that we are going to fall back into thinking that the type of a kind is a real self-subsistent principle, which makes its influence felt

in the world by its own inherent force. The only question is—does the disposition of things as a whole require that the forms of combination which the forces active throughout the world assume in the production of Beings capable of existence and growth, should be limited to a certain fixed number? or, on the other hand, may there not be innumerable forms intermediate between these types, and partaking in different degrees of their permanence and power of self-preservation, while the types only represent points of maximum stability? We must leave this question to be decided by the sober evidence of Natural History. Philosophy will do well to regard every attempt at an *a priori* solution of it as a baseless assumption. The bias of our minds in this case is determined by our own preconceived unverifiable opinions regarding the course of the world as a whole. Suppose, however, we assume that not merely self-conservation, but also Progress is a characteristic of the world as a whole, yet, even then, it would be conceivable that in the age of the world's history in which we now live, and of which we cannot see the limits, the forms of Life established by Nature might be incapable of addition, just as the quantities of those permanent elements which Nature uses in order to construct her products, are incapable of addition. According to this view, any forms in which things combined, owing to the influence of circumstances other than the forms determined by Nature, would have only a passing reality, and would be subsequently dissolved owing to the influence of the same circumstances which had produced them. On the other hand, nothing hinders us from introducing the alleged development within the limits of the epoch which we can observe, and regarding it as possible that new forms may come into Being and old forms pass away, and that what went before may gradually be transformed into what follows. The present aspect of the discussion on this subject forms part of a larger and more general question, the question, as to whether the world is finite or infinite.

235. It is needless to discuss at length the question as to whether the succession of events in time is finite or infinite. We cannot represent to ourselves in thought, either the origin of reality out of nothing, or its disappearance into nothing, and no one has ever attempted to take up this position without assuming, as existent in the Nothing, an originating principle or agency, and ascribing to it previous to its creative act a fixed existence of its own which has had no beginning in Time. Hence, whatever difficulties may be involved in the attempt to conceive of the course of events in Time as without beginning or end, the idea itself is inevitable. Nor need we occupy ourselves at any greater length with the question as to the limits of the world in Space. If Space is to pass for a real existence, the only difficulty is in the infinity of Space itself, which in that case is the infinity of something real. I leave this assumption, therefore, to be dealt with by those who are interested in maintaining it. On the other hand, it does not at all follow, even if Space is infinite, that the world need occupy the whole of it, as long as the content of that world admits of the predicate ' finite.' It would be quite sufficient to say with Herbart that Space sets no limits or conditions to the world, but that it occupies just so much room in Space as it requires for its movements, and that thus its boundaries are perpetually shifting. My own view of the matter is almost to the same effect. Every change in the true reactions of real elements must find room within the infinity of our Space-perception for its phenomenal manifestation as shape, position, and motion. But there is nothing to compel the real existences to fill up at every moment all the empty places which our Space-perception holds in readiness for impressions that may require them.

The question therefore resolves itself into this, whether the sum of real existence in the world is limited or unlimited, a question in reference to which we follow alternately two opposite impulses. On the one hand, the idea of infinity gratifies us just because we cannot exhaust it in thought, by enabling us to marvel at the immensity of the universe, of which we then readily acknowledge that we are but a part ; though, at the same time, by making it impossible for us to comprehend the world as a unity or whole, this infinity perplexes us. On the other hand, by conceiving of the world as finite, we are indeed enabled easily to grasp it as a whole; but it vexes us to think that a hindrance to its being greater than it is should have been imposed from without. This last supposition, at any rate, is plainly absurd. The world of reality is the sole source from which, in the minds that form a part of it, the notion of these countless unrealised possibilities springs. Hence arises the false idea that the Real world is limited and conditioned by what it does not produce, though it is the Real world alone that does produce this empty imagination in our minds. And this misconception has then absolutely no limits. What would be the use of assuming an infinity of real elements, if each one of them was finite ? Surely it would be still better that each element should be infinite. Yet even then we should still have only an infinite number of infinite elements. Why not, in order to get rid of all limitation, assume the existence of an infinite number of worlds, both of infinite magnitude themselves, and composed of elements whose magnitude was infinite? There is therefore *prima facie* no objection to the finite character of real existence—whereas, the character of infinity is opposed by Physics, not merely as inexhaustible by thought, but also as involving certain special mechanical difficulties. The unlimited distribution of matter would make impossible a common centre of gravity. No one point would have any better claim to be regarded as such than the rest. But what is our motive in looking for a centre of gravity? and what exactly do we mean by it? The supposition could not be entertained, unless it were regarded as self-evident that the same laws of reciprocal action which obtain between the particles of matter in our planet, and which we call Laws of Gravitation, obtain also throughout the whole range of existence. I well know how little precedent there is for doubting this fact. It is, indeed, ordinarily taken for granted without the slightest misgiving. And yet, in the absence of positive proof derived from observation, it can only be a bold argument from analogy. It seems to me by no means a self-evident fact, that all the real elements which are contained in the infinity of space, including even those which are stationed at the furthest points, are held together according to a single law by the uniting force of gravity, just as if they were mere samples of the mass to which it applies, and without individuality of their own. The Law of Gravitation is only known to apply to the bodies of our own planetary system. Besides this, there is only the conjecture, which may be a true one, that certain of the binary stars are kept in their courses by a similar mutual attraction, the law of which we do not as yet know. But that the same influence by which one system of material elements is made to cohere, extends as a matter of course to every other coherent system in the Universe of Space; this is by no means such an established and irrefutable truth as is, e. g. the uniform diffusion of the undulations of light through all Space.

For a reason which has already been several times touched upon, I am forced to proceed at this point by a different path from that which is ordinarily followed in the physical Sciences. If I really thought that the number of the real elements, or of the systems which are formed from such elements, was infinite, then, though I should certainly not regard them as having no connexion with each other, I should just as certainly not imagine that the relation subsisting between them was so monotonously uniform that they should be treated as mere samples of homogeneous mass endowed everywhere with the same force, so as to raise the question of their common centre of gravity. In each of these several systems the inner relation of the parts might be essentially peculiar, depending on the plan which governed its structure. Similarly, the several systems might be united by different kinds of relations into the one universal plan. Not that, in insisting on this point, I have any wish to maintain that Real existence *is* infinite, any more than I wish to maintain that it is finite. I have no sympathy with the point of view from which this question thus conceived seems to be one of real importance. I have more than once expressed my conviction that everything is subject to mechanical Laws ; but I have at the same time asserted the essentially subordinate character of these Laws, when considered with reference to the Universe as a whole. I do not know if my expressions have been understood in the sense in which they were intended. Certainly they were not meant to imply that previous to the creation of the world there existed a fixed sum of real elements, along with a code of absolute mechanical Laws, and that an organizing power then entered on the scene, and had to make the best of these resources. I have throughout taken as my starting-point the living nature of the real existence, that unity whose essence can only be expressed, if we are to attempt to realise it to our intelligence, as the import of a thought. Out of this import there arose (what was not prior to it) the funda-mental system of most general laws, as a condition which Reality imposes on itself and its whole action. But just because dependent on this import, the system possessed a wealth of meaning and power of accommodation, adequate to provide not merely for the uniformity of processes which never vary, but also for the manifold variety of activities which are required by the animating idea of the Whole. I should be the last to deny the necessity and value of the other point of view which, as represented by modern mechanics, conducts calcu-lations based on the abstract conceptions of Mass and its constancy : Force, and the conservation of Force, the inertia and invariability of the elements. Not only do we owe to this method the greater part of our present knowledge of Nature, but we may also safely assume it as a guide throughout the whole range of our possible observation. At the same time, I should be the last to ascribe to these notions, being as they are abstractions out of the fraction of the world's course which is accessible to us, that metaphysical certainty which would fit them to serve as a key to the solution of questions which are such as to transcend all experience of this kind.

What I have now to say in regard to the question of infinity has been already indicated in several passages of my work.   If the reality of the world is to be found in a thought which fulfils itself in every moment, the question as to the finite or infinite character of this thought is as meaningless as the question as to whether a motion is sweet or sour.   As regards, however, the different and ever-changing related points, by means of which the thought realises itself, we would remark, in the first place, that their number is not absolutely either finite or infinite.   It is not, indeed, a fixed quantity at all.   It is, at each moment, precisely what the realisation of the thought demands and its living activity produces.   This heterodox assertion I have already ventured on, thereby placing myself in opposition to the dogma of the constancy of Mass.   Supposing we fancied that we had a standard in terms of which the sum of real existence at any given moment of its history $= m$, it might very well at the next moment be found to $= \mu$.   In the same way as the world might take up just so much space as it should require at any given moment, so the Idea which animates it would create for itself just so many elements as are needed in order to accomplish its development.   Not as if there had been some material substance present from all eternity, which was afterwards merely differently distributed according as the Idea might require, nor yet as if the Idea created new elements out of nothing.   These new creations would spring from the Idea itself, which is the cause of all things.   Enough, however, has been said on this point.   It would be hopeless to attempt to bring these thoughts home to anyone who was convinced that a fixed quantity of matter had been ordained to exist from all eternity.   Whether, at any given moment, the number of the real and active elements is unlimited, or whether there are certain fixed limits within which the numbers vary, I confess myself unable to say.   The question itself involves confusion, until we have fixed on the unit the number of whose recurrences is sought.   It could have absolutely no meaning for those who have admitted the infinite divisibility of matter. It would be intelligible only, if it were held to apply to individual atoms or to separate and distinct groups of elements, as, e. g. the number of the stars.   Here I will only say quite shortly that I am content to assume that the number of material existences is limited, provided it is understood that this number must suffice to enable them to carry out the behests of the Idea, and that if this same condition is fulfilled, I am equally content to conceive of their number as infinite.   In this latter case, the impossibility of reckoning their number would be due

merely to a defect in us. It would not be a fault on their side, or inconsistent with their reality.

238. The progress of observation has led us to the conviction that
the formation of the earth's crust took place gradually, and that
organic life could not have existed throughout the stages of this
process in its present state. This imposes on us the necessity of
attempting to show how the forms of life at present existing were
developed out of earlier and simpler ones. In the heat of the controversy on this subject, care should have been taken not to confuse
two questions which ought to be separated. Only one of them
belongs to Metaphysic, that, viz. as to the determining principles
which have been active throughout the course of this development.
I feel all the less inducement to make any addition to the rapidly
increasing literature which the discussion of this question has called
forth, inasmuch as, before this controversy had begun to rage, I endeavoured to bring together whatever seemed to admit of being said,
with any claim to respect, in favour of explaining all cases of
adaptation as due to a fortuitous concourse of accidents, a view which
has a recognised place in the History of Philosophy. In the second
chapter of the fourth book of the Microcosmus, I treated expressly of
this derivation of the Cosmos from Chaos, and I cannot convince
myself that the more recent arguments from the same point of view
add anything of importance to those well-known ones of former
times which are there mentioned. I content myself with referring to
what I then said in regard to the details of the question. My
conviction on the matter as a whole needs not again to be stated
here. The controversy will become milder with time ; at least this
will be so, in so far as it is conducted in the interests of Science and
not from a feeling of invincible repugnance to every Idea which is
suspected of favouring the cause of Religion. An improvement in
this respect is already to some extent visible. Those who pray too
much are destined, says the proverb, to pray themselves through
heaven and to keep geese on the other side. A better fate has
befallen those who, out of a conscientious regard for the interests of
Science, have felt themselves compelled to derive Organic Life from
blind chance and purposeless matter. They have invested their
original principles with so much reason and power of internal
development, that nothing but the caprice of their terminology which
keeps to the names of Matter, Mechanism, and Accident, for what other
people call Spirit, Life, and Providence, seems to prevent them from
relapsing into notions which they have before strenuously opposed.

237. On the other hand, as regards the second question to be distinguished, that, viz. as to the actual development of Organic Life, this is purely a matter of Natural History. Philosophy is not concerned to dispute or to deny any results of observation on this subject, which are based on sufficient evidence. Not even Religion should presume to prescribe to God the course which the world's development must have followed subsequently to its creation. However strange the path may have been, we might be sure that its strangeness could not remove it from His control. Considering that the human body requires to be kept alive each day by absorbing into itself nourishment derived from common natural substances, there can be no reason in claiming for it a manner of origin so exceedingly distinguished. And with regard to the whole matter we would say that man esteems himself according to *what* he is, and not according to that *whence* he arose. It is enough for us to feel that we are now not apes. It is of no consequence to us that our remote and unremembered ancestors should have belonged to this inferior grade of life. The only painful conclusion would be that we were destined to turn into apes again, and it was likely to happen soon. It seems to me, therefore, that from the point of view of Philosophy these scientific movements may be regarded with the most perfect indifference. Each result, so soon as it had ceased to be a favourite conjecture and had been established by convincing proof, would be welcomed as a real addition to knowledge. The very remarkable facts of Natural History accumulated by the unwearied research of Darwin, might be provisionally welcomed by Philosophy with the warmest satisfaction, whilst, on the other hand, the pretentious and mistaken theories based on those facts might be not less completely disregarded. All that Philosophy herself can contribute towards the solution of these questions is, to warn us against making unfounded assumptions, which, whilst they are themselves to some extent of philosophical origin, rob Science of its fairness. Whatever may have been the state of the earth's surface, which first occasioned the production of organic life, it cannot but be improbable that the required conditions should only have been present at a single point; equally improbable, considering the diversity of the terrestrial elements which were subjected on the whole to uniform influences, that organic germs of the same kind only should have been generated at all points; and finally, it is extremely improbable that this productive period should have lasted only long enough for the occurrence of an instantaneous creative act, instead of being so protracted that the

conditions, slowly altering while it still lasted, might superadd fresh
creations to the earlier ones instead of merely developing their further
phases. Nor is there any difficulty in imagining that these various
organic beings, though produced at different times and on different
spots of the earth, would still present numberless analogies of structure.
The equation which contained the conditions of the union of elements
so as to be capable of life would restrict all possible solutions within
determinate limits. Hence, according to what is at any rate the
most probable supposition, Organic Life is derived from an original
multiplicity of simple types having a capacity for development.

Here we break off. We cannot pursue further the attempts which
are now being made to arrive at an explanation of the first beginnings
and the final destiny of things. Our knowledge of the present state
of the globe and of the forces that act upon it, does enable us to
form an idea, imperfect indeed, but not contemptible, with regard
to its fate in the future; and it is of importance for Science to consider to what end the processes which we now see in operation would
lead, supposing them to continue unchecked and to follow the same
laws. From this point of view, we are able to appreciate those
ingenious calculations which draw conclusions as to the final state of
the world from our experimental knowledge of the economy of heat.
They are, however, nothing more than the indispensable computations
which draw out this portion of our physical knowledge into its results.
For this purpose we are obliged to assume the continuance of the
conditions which are operative at present. Whether this hypothesis
will be verified, or, whether the end towards which things now seem
to point, will not sooner or later be shown by fresh discoveries in a
new light, no one can decide. At the same time, however, the fate
which most attempts to forecast the future by means of statistics have
hitherto met, has been of the latter kind. Hence, we must be on our
guard against crediting as a prophetic announcement with regard to
the future, conclusions which follow, no doubt, necessarily on the
arbitrary assumption that the given conditions are the only ones to be
taken into account. Still less do we intend to busy ourselves with
the fancies of those who relate to us, just as if they had been themselves present, how things were first produced; how, e. g. the
inorganic elements of the earth's crust found themselves united in the
form of crystals capable of imbibition, and in systems endowed with
life and growth; or, again, how the atmosphere of the primitive
world settled upon the earth in the shape of protoplasm, and there

struck roots of the most various kinds. This insatiable desire to get beyond the general principles which still admit of being applied to the investigation of these problems, and actually to conjecture those special circumstances which are simply inaccessible to our knowledge, may, by way of palliation, be considered to be characteristic of that historical sense by which the present age is distinguished, thus contrasting favourably with former ages, when, owing to their speculative bias, men sought for truth not in matters of fact but in ideas that had no reality in Space and Time. Yet I do not know in what the worth of history would consist, if facts were in truth only described as having occurred in this or that place, without any attempt being made to pass beyond the facts and their succession, and to lay bare the nerves which govern the connected order of things always and everywhere. But for this purpose history must above all things be *true*. Every fact of the Past which can be demonstrated by certain proof we shall esteem as a real and valuable addition to our knowledge. On the other hand, those rash anticipations of knowledge, entertaining at first, but wearisome in their recurrence, have nothing to do with this laudable 'historical sense,' but spring from the dangerous inclination to anecdote simply for its own sake. It is thus that our own generation, maintaining its opposition to Philosophy, endeavours to console itself for its want of clearness in respect to general principles by a vivid exercise of the sensuous imagination. If we come upon piledwellings in some forgotten swamp, we piously gather together the insignificant remains of a dreary Past, supposing that by contemplating them we shall grow wiser and learn that which a glance into the affairs of everyday life would teach us with less trouble. Compared with such objects as these how small a chance of notice have the Philosophical ideas, which represent the efforts of long ages to obtain a clearer insight into eternal truth. If only these ideas could be stuffed ! Then it might be possible that beside a fine specimen of the Platonic idea and a well-preserved Aristotelian entelechy even the more modest fancies which in these pages I have devoted to speedy oblivion, might attract the attention of a holiday sight-seer.

BOOK III.

ON MENTAL EXISTENCE (PSYCHOLOGY).

— ◆ —

CHAPTER I.

The Metaphysical Conception of the Soul.

THE old Metaphysic of the Schools reckoned among its problems the construction of a Rational Psychology. This name was not meant to imply that the science in question could dispense with such a knowledge of its object as should agree with experience; the design was merely to bring the general modes of procedure which were observed in that object into connexion with metaphysical convictions as to the possibility of all being and happening. I will not ask here how much or how little the science accomplished; but I accept the end it set before itself as a limit for my own discussions. There is at present a strong inclination towards the empirical investigation of psychical phenomena, in all their manifold complexity, and I am not opposing this inclination when I confess some want of confidence in the trustworthiness of its results. Speaking generally no great doubt can be felt as to the nature of those associations of impressions, by means of which the whole of our sensuous view of things as well as the riches of our mental culture are in the last resort acquired; but the ingenious attempts which have been made to demonstrate the way in which particular portions of this total sum actually came into our possession, have not the same certainty. Often, instead of being founded on empirical evidence, they are merely descriptions of the modes in which we can without any great difficulty imagine the material in question to have originated; sometimes they are accounts of processes of the possibility of which we persuade ourselves only because we use as self-evident means of explanation mental habits which it is really our first business to explain. It is not my purpose, however, to lessen the deserved sympathy which these valuable efforts have won ; but this book must come to an end somewhere, and therefore they are excluded from it ; and my wish here is simply to overcome, for a moment at least, the disfavour which any metaphysical treatment of these subjects is apt to encounter.

When we say that we adopt an empirical stand-point we must mean more than that we wish to stand still at this point; we really intend it to be no more than the starting-place from which we may appropriate the field of experience around us. Now, considered as such a point of departure, the knowledge of those facts which are furnished by experience is indispensable to every psychology alike ; and even those attempts which have been especially stigmatised as transcendent, are in the end simply interpretations of the material supplied by observation. The divergence of opinion does not really begin till we ask by what method we are to appropriate in the form of theory that which, from the empirical position, we all see with the same eyes. In speaking of the physical investigation of nature I pointed out how slight and how arduous its progress would be if it confined itself to bare observation and refused to connect the given facts by framing hypotheses respecting that nature of things which cannot be observed. And I may now appeal for confirmation to the excellent attempts which have been made in psychology to reach, at least at one point, the beginnings of an exact science—the point I refer to is the question how the strength of a sensation is related to that of its external stimulus. For these attempts have at once become involved in a mass of theoretical and speculative problems, to the settlement of which a future experience may perhaps contribute much but which it will certainly never completely solve. If then we are compelled to use as a basis *some* hypothesis respecting the connexion of physical and psychical phenomena, why are we to take the first hypothesis that comes to hand? Why not go back to the most general ideas that we necessarily form respecting all being and action, and so attempt to define the limits within which we can frame suppositions, sometimes trustworthy and at other times at all events probable? But, further, even supposing it were possible, in the investigation of this special subject, to find a point of departure which should be productive of results and yet should imply no fixed pre-judgment as to the nature of the subject, a difficulty would still remain : for though this freedom from pre-suppositions would be possible for this particular investigation, it would still be unacceptable to us as men. We may be warned to abstain from discussing

questions which do not seem to be soluble by the special methods of a
particular science ; but the warning will never deter the human race
from returning to these riddles ; for a consistent opinion about them
is not less important and indispensable to it than are those explanations of observed facts which in this field can never be more than
fragmentary.   I shall therefore attempt to extend these metaphysical
considerations to the sphere of Psychology, and so to bring them to a
conclusion.   For the elaboration of many particular points I may
refer to the corresponding sections of the *Mikrokosmus ;* here I wish to
bring together the essential points treated in the *Medicinische Psychologie,* (Leipzig, 1852), which I shall not reissue, and to show the
metaphysical connexion which in those two works could not be
sufficiently brought out.

238. Let us leave out of sight, to begin with, anything which the
earlier part of this enquiry might offer by way of foundation for what
is to follow.   If we do this, we shall have to confess that mental life
is given us, as a fact of observation, only in constant connexion with
bodily life.   Accordingly the supposition at once suggests itself that
this mental life is nothing but a product of the physical organization,
the growth of which it is observed to accompany.   Yet such a view
has never been more than a doctrine of scientific schools.   We meet
with the word 'soul' in the languages of all civilised peoples ; and
this proves that the imagination of man must have had reasons of
weight for its supposition that there is an existence of some special
nature underlying the phenomena of the inner life as their subject or
cause.   It is, I think, possible to reduce these reasons to three, of
very different value.   The first I will refer to, the appeal to the
freedom which is said to characterise mental life, and is distinguished
from the necessity of nature, has no weight.   It is a conviction with
which we begin our enquiry and to which we hold, that all events in
external nature form an uninterrupted series of causal connexion
according to universal laws; but this necessity is not a fact of observation.   There remain always vast tracts of nature, the inner connexion of which is simply unknown to us and which can therefore
furnish no empirical verification of that presupposition.   But, when
we come to mental life, not even those for whom freedom is in itself a
possible conception can regard it as the *universal* characteristic of that
life.   They can demand it only at one definite point, viz. the resolutions
of the will.   Everything else, the whole course of ideas, emotions, and
efforts, is not only, in the souls of animals and men alike, manifestly
subject to a connexion according to universal laws, but the denial of

that connexion would at once destroy the possibility of any psychological enquiry; since it, like every other enquiry, can be directed to
nothing but the discovery of conditions universally valid.

239. The second reason which led to the conception of the soul
was the entire *incomparability* of all inner processes—sensations,
ideas, emotions, and desires—with spatial motion, figure, position, and
energy; that is, with those states which we believe we observe in
matter, or which we can suppose it to experience if we see in it only
what the physical view of nature gives it out to be. It is a very long
time since philosophy recognised this incomparability, and it needed
no new discovery or confirmation. It has escaped no one except
those who, out of their prejudice in favour of a desired conclusion,
have not been afraid of the logical error by which two different things
are held to be of the same kind simply because as a matter of fact
they are connected with one another. We may imagine a quantity of
movements of material elements, and we may attribute to them whatever degree of complexity we choose; but we shall never reach a
given moment at which we can say, Now it is obvious that this sum
of movements can remain movements no longer but must pass into
sweetness, brightness, or sound. The only obvious change we could
ever anticipate from them would be into a fresh set of movements.
We shall never succeed in analytically deducing the feeling from the
nature of its physical excitant; we can only connect the two synthetically; and the physical event does not become a condition of the rise
of the feeling until the sum of motions in which it consists meets with
a subject which in its own nature has the peculiar capacity of producing feeling from itself. In this fact a limit is at once placed to all
physiological and psychological enquiry. It is utterly fruitless to
attempt to show how a physical nervous process gradually transforms
itself (as we are told) into sensation or any other mental occurrence.
There remains only the different but extremely important task of discovering *what* psychical event *a* and *what* physical stimulus *a* are as
a matter of fact universally connected in the order of nature, and of
finding the law by which *a* undergoes a definite change and becomes
β, when *a* by a change equally definite (but definable only by a
physical standard and not a psychical one) becomes *b*. This is a point
at which the professedly empirical method and the metaphysical change
their *rôles*. The former, in pursuing the dream of an identity of
physical and psychical processes, leaves the field of experience far
behind it and does battle with our most immediate certainty that they
are not identical: the latter, when it refrains from describing an event

which cannot occur at all, is not denying the connexion between the two series of events; but it limits itself to a more useful enquiry, it investigates the laws according to which the results of that connexion change, and it forbears to ask questions, which to begin with at any rate cannot be answered, regarding the mode in which that connexion is in all cases brought about.

240. On the other hand we must beware of drawing conclusions too definite from this incomparability of physical and psychical processes. All that follows unavoidably from it is that we should reserve for each of these two groups its own special ground of explanation ; but it would be going too far to assert that the two principles, which we must thus separate, necessarily belong to two different sorts of substances. There is nothing to be said at starting against the other supposition, according to which every element of reality unites in itself the two primitive qualities, from one of which mental life may arise, while the other contains the condition of a phenomenal appearance as matter. On this view, instead of having, on the one side, souls destitute of all physical activity and, on the other, absolutely self-less elements of matter, we might suppose that the latter, like the former, possess in various grades an inner life, though a life which we cannot observe nor even guess at, so long as it has no forms of expression intelligible to us. And with regard to the cause which would unite these two attributes in what exists, this theory would be as much within its right in refusing to discuss it as ours was in simply appealing to the fact of a connexion between two series of incomparable processes. It seems to me that every mode of thought, which calls itself Materialism, ultimately rests on this supposition, or on a little reflexion must be led to it ; the matter from which such modes of thought would deduce mental phenomena, is privately conceived by them as something much better than it looks from outside. So it comes about that it can be held a fair problem, to deduce the mental life of an organism from the reactions of the psychical movements of the corporeal elements in the same sense in which its bodily life arises as a resultant from the confluence of the physical forces of those elements. And if we were confined to the external observation of a psychical life not our own, I do not know of anything perfectly decisive that could be alleged against this supposition. But, according to it, every psychical manifestation would be merely the final outcome of a number of components destitute of any centre : whereas our inner experience offers us the fact of a *unity of consciousness.* Here then is the third and the unassailable ground, on

which the conviction of the independence of the soul can securely
rest. The nature of this position I proceed to explain.

241. It has been required of any theory which starts without presuppositions and from the basis of experience, that in the beginning
it should speak only of sensations or ideas, without mentioning the
soul to which, it is said, we hasten without justification to ascribe them.
I should maintain, on the contrary, that such a mode of setting out
involves a wilful departure from that which is actually given in experience. A mere sensation without a subject is nowhere to be met with
as a fact. It is impossible to speak of a bare movement without
thinking of the mass whose movement it is; and it is just as impossible to conceive a sensation existing without the accompanying
idea of that which has it,—or, rather, of that which feels it; for this
also is included in the given fact of experience, that the relation of the
feeling subject to its feeling, whatever its other characteristics may be,
is in any case something different from the relation of the moved
element to its movement. It is thus, and thus only, that the sensation
is a given fact; and we have no right to abstract from its relation to
its subject because this relation is puzzling, and because we wish to
obtain a starting-point which looks more convenient but is utterly un-
warranted by experience. In saying this I do not intend to repeat
the frequent but exaggerated assertion, that in every single act of
feeling or thinking there is an express consciousness which regards
the sensation or idea simply as states of a self; on the contrary,
everyone is familiar with that absorption in the content of a sensuous
perception, which often makes us entirely forget our personality in
view of it. But then the very fact that we can become aware that
this *was* the case, presupposes that we afterwards retrieve what we
omitted at first, viz. the recognition that the perception was in us, as
our state. But, further, there are other facts which place in a clearer
light what in the case of single sensations might remain doubtful.
Any comparison of two ideas, which ends by our finding their contents like or unlike, presupposes the absolutely indivisible unity of
that which compares them : it must be one and the same thing which
first forms the idea of *a*, then that of *b*, and which at the same time is
conscious of the nature and extent of the difference between them.
Then again the various acts of comparing ideas and referring them to
one another are themselves in turn reciprocally related; and this
relation brings a new activity of comparison to consciousness. And
so our whole inner world of thoughts is built up; not as a mere
collection of manifold ideas existing with or after one another, but as

a world in which these individual members are held together and arranged by the relating activity of this single pervading principle. This then is what we mean by the unity of consciousness; and it is this that we regard as the sufficient ground for assuming an indivisible soul. As compared with the thousand activities of this unity involved in every act by which two ideas are referred to each other, it is a matter of indifference whether at every moment that particular act of relation is explicitly performed by which these inner states are apprehended in their true character, as states of this active unity. Although this reflexion is possible, we can think of many conditions which frequently prevent it taking place. But that it can take place at all proves to us the unity of the active subject which performs it.

242. Further discussion is, however, needed, in order to show the necessity of our conclusion and to explain its meaning. First, as to its necessity: even if we admit the unity of consciousness, why are we bound to trace it back to a particular indivisible subject? why should it not resemble a motion which results from the co-operation of many components; seeing that this resultant, like the unity of consciousness, appears perfectly simple and gives no indication of the multiplicity of elements from which it arose? I answer: such an idea seems possible only because we state the mechanical law, to which we appeal, in slovenly short-hand. We must not say, 'From two motions there comes a third simple motion:' the full formula is, When two different impulses act simultaneously on one and the same material point, they coalesce at this point into a third simple motion *of this point;* they would not do so if they met with different elements, nor would the resultant have any significance if it were not a motion of that very same element in which they met. If we wish then to make an analogous construction of consciousness, it is indispensable that we should mention the subject whose states we have to combine. Thus if $a, b, c, \ldots s$ are the elements of a living organism, each of them may have at once a physical and a psychical nature and each of them may be capable of acting in accordance with its two natures; but the fact still remains that these actions cannot stream out into the void and be states of nobody, but must always consist of states which one element produces in other elements. Supposing this reciprocal action took place equally among them all, then the impressions received and imparted would be equalised, and the end of the process would be that each one of the elements would reach the same final state Z, the resultant of all the single impulses. If then Z were a consciousness, this consciousness would be present as many times as there were

homogeneous elements : but it would never happen that outside, side by side with, or between these elements a new subject could be formed, privileged to be the personified common spirit of the society of inter-acting units. Doubtless, however, the homogeneity we have assumed will not be found to exist; the constituents of the organism will differ from one another; they will be conjoined in accordance with their nature, and will have different positions, more or less favourable to the spread of their interactions; and at whatever moment we sup-pose the course of these interactions to be finished, the result will probably be that the different elements will have reached different final states $A, B, \ldots Z$, depending on the degree of liveliness with which each element has received the influences of the others, and on the measure in which it has succeeded or failed in concentrating those influences in itself. In this case it becomes still more impossible than before to say which of all this array of resultant consciousnesses is the object of our search, the soul of the organism : but in this case as in the former, it is certain that there cannot arise, outside of and beyond all these elements, a new subject which in its own conscious-ness should bring together and compare their states, as we who are investigating can compare them in the unity of *our* consciousness.

Our only remaining resource would be to fall back on the idea of Leibnitz and to say that although the countless monads which com-pose the living creature are essentially homogeneous, there is never-theless among them a *prima inter pares*, a central monad, which in virtue partly of its superiority in quality and partly of its favourable position between the rest, is capable of the intensest mental life and able to over-master all the others. This central monad would be what we call our soul, the subject of our one consciousness; the others, though they too have psychical movements of their own, would be for our direct inward experience as inaccessible as the inner life of one person in a human society is for that of any other. Thus the end at which this attempted construction would arrive would not be that it set out to reach. It too would have to recognise the absolutely indivisible unity of that which is to support our inward life : and, instead of the hope of showing this unity to be the resultant of many co-operating elements, there would remain the more moderate as-sumption that these many elements stand to the one being in manifold relations of interaction. Such a view has no longer any special pecu-liarity, beyond, first, the idea that all elements of the body have a soul-life, although this soul-life has not much significance for ours ; and secondly, (though this applies only to the hypothesis I am describing, and not to Leibnitz) the doubtful advantage of being able to attribute to the one element which is the soul not only psychical predicates but the predicates of an element which is operative after the fashion of matter.

243. I said that the *meaning* of the unity of consciousness, as well as the necessity of assuming it, needed some further explanation. My remarks on this meaning ought to be saved by their connexion with the rest of a metaphysical work like the present from the misunderstanding with which my previous accounts of the subject have met. The conclusion we have now reached is usually expressed by saying that the soul is an indivisible and simple substance; and I have used this formula in all innocence, as an intelligible name. How it can be misunderstood I have learned from the way in which my esteemed friend Fechner in his *Atomenlehre* characterises my view in opposition to his own. It was natural to him as an investigator of Nature, and probably his intimacy with the most eminent representatives of the Herbartian philosophy made it still more natural, to understand by substance a physical atom or one of the simple real ' existences ' of that school. But I had given no special occasion for this misunderstanding : on the contrary I had put forward the proposition which was censured and therefore could not have escaped notice ; ' It is not through a substance that things have being, but they have being when they are able to produce the appearance of a substance present in them.' I have discussed this point at sufficient length in the Ontology, and have now only to show its consequences for our present question. When from the given fact of the unity of consciousness I passed on to call the subject of this knowledge existence or substance, I could not possibly intend by doing so to draw a conclusion which should deduce from its premises something not contained in them but really new. For my only definition of the idea of substance was this,—that it signifies everything which possesses the power of producing and experiencing effects, in so far as it possesses that power. Accordingly this expression was simply a title given to a thing in virtue of its having performed something; it was not and could not be meant to signify the ground, the means or the cause which would render that performance intelligible. Was substance to be one or many ? It would have been too absurd to suppose this power of producing and experiencing effects in general to have its ground in *one* universal substance, and then to expect that a grain of this substance, buried in each individual thing, would quicken this general capacity into the particular ways of producing and experiencing effects which distinguish that thing from all other things. On the other hand the supposition that each thing, instead of being carved out of the matter of the universal substance, is a substance on its own account would have at once led us back to our starting-point, and we should have recognised the name substance to be, what it really is, simply the general formal designation of every way of producing and experiencing effects, but not the real condition on which in each particular case the possibility of doing so and the particular way of doing so depends. I was therefore very far from sharing the view of those who place the soul in the mid-current of events as one hard and indissoluble atom by the side of others or as an indestructible real existence, and who fancied that its substantiality, so understood, offered a foundation from which the rest of its phenomena could be deduced. The fact of the unity of consciousness is *eo ipso* at once the fact of the existence of a substance: we do not need by a process of reasoning to conclude from the former to the latter as the condition of its existence,—a fallacious process of reasoning which seeks in an extraneous and superior substance supposed to be known beforehand, the source from which the soul and each particular thing would acquire the capacity of figuring as the unity and centre of manifold actions and affections.

The reason why, in spite of this, I thought it worth while to designate the soul as substance or real existence, I shall mention hereafter when I come to oppose the pluralistic view suggested by Fechner: my point was not so much the substantiality as the unity of the soul, and I wished to emphasize the idea that it is only an indivisible unity which can produce or experience effects at all, and that these words cannot be applied in strictness to any multiplicity, —an idea which I attempted to bring out more clearly in the *Mikrokosmus*, (i. p. 178). But, relying on the fact that the imagination is accustomed to connect this idea of unity with the name 'substance' or 'real existence,' I considered that these two expressions, even in that meaning of them which I have described and repudiated, might still, when once the true account of the matter had been given, be used as serviceable abbreviations of it.

244. It is natural at this point to think of Kant's treatment of that Paralogism of the pure Reason which seeks to establish the substantiality of the soul. We may sum up his criticism thus: It is a fact that we appear in our thoughts as the constant subject of our states, but it does not follow from this fact that the soul is a constant substance; for even the former unity is in the end only our subjective

way of looking at things, and there are many things which in themselves may be quite different from what they must needs seem to us to
be. This last idea is certainly incontrovertible, but it does not affect
the point which constitutes the nerve of our argument. I repeat once
more, we do not believe in the unity of the soul because it appears as
unity, but simply because it is able to appear or manifest itself in
some way, whatever that may be. The mere fact that, conceiving
itself as a subject, it connects itself with *any* predicate, proves to us
the unity of that which asserts this connexion; and, supposing the
soul appeared to itself as a multiplicity, we should on the same
grounds conclude that it was certainly mistaken if it took itself really
to be what it appeared. Every judgment, whatever it may assert,
testifies by the mere fact that it is pronounced at all, to the indivisible
unity of the subject which utters it.

But, I am well aware, I shall still be reproached with having
neglected the fine and subtle distinction which Kant draws between
the subject of our inward experience and the unity of the Soul considered as a thing in itself; he admits the unity of the former, but
prohibits any conclusion to that of the latter. It is a difficult task,
and one in which I have no interest, to dissect Kant's final ideas in
this section of the Critique of Reason; I shall content myself with
explaining clearly the difference between my view and that which I
conjecture to be his. Kant is without doubt right when he is opposing
that traditional argument for the substantiality of the soul, the object
of which was to make that quality, when it had been inferred, a *medius
terminus* for fresh consequences, as, for instance, that of immortality;
but he was mistaken when he looked on this inference as a further
goal which it is our misfortune that we are unable to attain. In
the very prohibition he utters against a conclusion from the unity of
the subject to that of the substance, he admits that this conclusion
would have an important bearing, if only it could be drawn; and all
that seems to him to be wanting is the links of argument which might
justify us in bringing the soul under this fruitful conception of
substance and all the consequences it legitimately involves. That
Kant cannot free himself from this idea, is shown by a foot-note
which in the first edition of the Critique is appended to the doctrine
of the Paralogisms. It runs as follows: 'An elastic sphere which
collides with another in a direct line, communicates to it its whole
motion and, therefore, (if we regard nothing but their positions in
space) its whole state. Now if, on the analogy of such bodies, we
suppose substances, one of which imparted to the other ideas together

with the consciousness of them, we can imagine a whole series of these substances, of which the first would impart its state, together with the consciousness of that state, to the second, the second would impart its own state, together with that of the preceding substance, to the third, and this again would communicate to another, not only its own state with the consciousness of it, but also the states of all its predecessors and the consciousness of them. Thus the states of all the substances which had undergone changes, together with the consciousness of these states, would be transferred to the last substance: and in consequence this last substance would be conscious of all these states as its own, and yet, in spite of this, it would not have been the same person in all these states.' In this way, according to Kant, the identity of the consciousness of ourselves in different times would be possible even without the numerical unity of the soul.

The various assumptions, which are made at starting in this note, are so strange that a criticism of their admissibility would be unbearably prolix: one can only say of them, Certainly, if it were so, it would be so. But if the communication of a completed state together with the consciousness of it is possible, why should we not go further and make an approach to the actual state of affairs by assuming that, over and above this, the fact of this communication will be an object of consciousness for the soul receiving it? In that case the process would resemble the propagation of culture by tradition and instruction. It is in this way, at least, that the busy soul collects by industry the thoughts of its predecessors; but then it is at the same time conscious that the thoughts it receives are not its own, but what it has received. And fortunately there is another point at which the comparison fails; for the original possessor does not lose his thoughts by communicating them. All this, however, matters nothing: but what is the meaning of the conclusion, ' and yet there has not been the same person in all these states'? The fact is the very reverse; it was not the same sphere that served an abode for the personality; but the person is one, in the same sense in which it is possible for any substance capable of development to be one, although at the beginning of its history it is naturally poorer in recollected experiences than it afterwards becomes: and what Kant maintains is nothing but a strange transmigration of the soul, in which the personality, while it grows in content, passes from one substratum to another. I will not dwell longer on the oddities of this unfortunate comparison; but it shows—and this is its only serious interest—that there seemed to Kant to be some meaning in the idea, that beneath the concrete nature or content of anything there lies in the intelligible world a thing in itself, destitute of content, but serving as a means of consolidating the reality of the concrete thing, or useful to it in some other way, I know not what; and that it makes some difference to the unity of consciousness, whether its substratum consists of the first, or second, or third of these things in themselves, whether it is always the same one, or whether it is many of them in succession; and this although there were even less difference between them than there is between those elastic spheres, the positions of which in space at least gives a reason for supposing that there is more than one of them. Nor was this at all the object which the Paralogism criticised sought to reach. No one who wished the doctrine of immortality to be assured, could concern himself with anything but that continuity of his consciousness which he desired not to lose; he would be heartily indifferent to the question whether the thing in itself which was to be the substratum of that continuance occupied in the series the position n or $(n+1)$.

I come back then to the point, that the identity of the subject of inward experience is all that we require. So far as, and so long as, the soul knows itself as this identical subject, it is, and is named, simply for that reason, substance. But the attempt to find its capacity of thus knowing itself in the numerical unity of another underlying substance is not a process of reasoning which merely fails to reach an admissible aim; it has no aim at all. That which is not only conceived by others as unity in multiplicity, but knows and makes itself good as such, is, simply on that account, the truest and most indivisible unity there can be. But in Kant's mind, so at least it seems to me, the prejudice is constantly recurring, that a thing may in a certain peculiar sense *be* unity, and that this is metaphysically a much prouder achievement than merely to make itself good as unity, since this last capacity may perhaps also belong to that which is not really or numerically one.

245. A further question now becomes inevitable. On what does this living unity of self-consciousness rest? Or, to put the problem in its customary and shorter form, what is the soul, and how are we to decide respecting its destiny, if our decision can no longer be drawn from the claims which might be advanced in favour of every substance as such, according to its traditional conception? Here again I need only answer by recalling the preliminary convictions to which our ontology has led us. We know that when we ask 'what' anything is, we commonly mean by this word two different things; firstly, that which distinguishes it from other things, and, secondly, that which makes it a thing, like other things. The error which it was our object to avoid lay in the belief that, corresponding to these two constituents of our conception, there exist in reality two elements capable of entering into an actual relation to each other. But we found our most serious obstacle in the habit of adding to these two constituents of our idea a third, which though foreign to them is supposed to guarantee their connexion: this third constituent is that empty 'matter' of existence on which the content of things is supposed to depend. To anyone who is disposed to agree with me in these ontological conclusions, it must seem utterly inconceivable that we should ask for the ' what' of a thing, and yet look for the answer in anything except that which this thing is and does; or that we should enquire as to its ' being,' and yet seek this anywhere except in its activity. And in the same way here it must seem equally unintelligible that we should suppose we do not know the soul, because, although we know all its acts, we are unluckily ignorant of the elastic sphere to which, according to Kant's comparison, the nature manifested in these acts is attached ; or that instead of seeking the living reality of the soul in its production of ideas, emotions, and efforts, we should look for it in a nameless ' being,' from which these concrete forms of action could not flow, but in which, after some manner never to be explained, they are supposed to participate. But I have already disposed of these generalities, and will not return to them. Every soul is what it shows itself to be, unity whose life is in definite ideas, feelings, and efforts. This is its real nature : and if it were alone in the world, it would be idle to ask how this reality is possible, since we have long ago decided that the question how things are made is not admissible. It is only the fact that the soul is involved in a larger world, and meets with various fortunes there, that makes it necessary to seek within this whole the conditions on which its existence, and the origin or preservation of that existence, depends. Within this sphere the soul shows itself to be to a certain extent an independent centre of actions and re-actions ; and in so far as it does so, and so long as it does so, it has a claim to the title of substance : but we can never draw from the empty idea of substance a necessary conclusion to the position which the soul occupies in the world, as though its modes of action had their ground and justification in that idea.

It will be obvious against what view this remark is directed. A pluralism which considers the order of the world derivable from a number of elements, perfectly independent of one another, and subject only to a supplementary connexion through laws, naturally includes in its idea of the original nature of these elements indestructibility and immutability. Unless then the soul is to be connected with the juxtapositions of these stable atoms as a perishable side-effect, the only resource of this view is to include it among the number of such eternal existences. Thus the soul can rely upon its rights as a premundane substance, and rest assured that in no changes of the world, whatever they may be, can either an origin or an end be ascribed to it.

The fact that this reasoning leads to a double result is, on the face of it, inconvenient. We might be glad to accept its guarantee for immortality, although no great satisfaction is given to our desires by a mere continuity the nature of which remains undecided ; but the other conclusion which is forced on us at the same time, the infinite pre-existence of the soul before the earthly life we know, remains, like the immortality of the souls of all animals, strange and improbable. Our monistic view has long since renounced all these ideas. The order of the world, the existence of all things and their capacity for action, it has placed wholly and without reserve in the hands of the one infinite existence, on which alone the possibility of all interactions was found to rest ; and it has nowhere recognised a prior world of ideal necessity, from which things might derive a claim to any other lot than that which the meaning of the whole has given them in order that they may serve it. Our first and foremost result is therefore this : the question of the immortality of the soul does not belong to Metaphysic. We have no other principle for deciding it beyond this general idealistic conviction ; that every created thing will continue, if and so long as its continuance belongs to the meaning of the world ; that everything will pass away which had its authorised place only in a transitory phase of the world's course. That this principle admits of no further application in human hands hardly needs to be mentioned. We certainly do not know the merits which may give to one existence a claim to eternity, nor the defects which deny it to others.

246. We cannot pass quite so quickly over the question of the *origin* or genesis of the soul. How it can be brought about, or how the creative power of the absolute begins to bring it about, that an existence is produced which not only in accordance with universal laws produces and experiences effects and alterations in its connexion with others, but also in its ideas, emotions, and efforts, separates itself from the common foundation of all things, and

becomes to a certain extent an independent centre,—this question
we shall no more attempt to answer than we have others like it.
Our business is not to make the world, but to understand the inner
connexion of the world that is realised already; and it was this
problem that forced us to lay down our limiting idea of the absolute
and its inner creation of countless finite beings. This idea we found
it necessary to regard as the conception of an ultimate fact; and we
cannot explain the possibility of the fact by using the images of processes which themselves spring from it in a way we cannot explain.
But when the life of the soul does arise, it arises before our eyes in
constant conjunction with the physical development of the organism:
and thus questions are suggested as to the reciprocal relations of two
series of events which, as we have already remarked, cannot be compared, and which therefore might seem inaccessible to one another.
Where, we may be asked, does the soul arise, and in what way does
it come into this body which is just beginning to be, and which was
destined for it; since we are forbidden to regard it as a collateral
effect of the physical forces, and as having its natural birthplace in
this very body? The question may seem natural, and yet it is only
an imagination accustomed to strange images which can ask it. We
are not to picture the absolute placed in some remote region of extended space, and separated from the world of its creations, so that
its influence has to retraverse a distance and make a journey in order
to reach things; for its indivisible unity, omnipresent at every point,
would fill this space as well as others. Still less ought we, who hold
this space to be a mere phenomenon, to imagine a cleft between finite
beings and the common foundation of all things, a cleft which would
need to be bridged by miraculous wanderings. Wherever in apparent
space an organic germ has been formed, at that very spot, and not
removed from it, the absolute is also present. Nor, I must once more
repeat, is it simply this class of facts which compels us to assume such
an action of the absolute. We may regard the process by which
things that possess a life and soul are formed as something unusual
and superior; but the presence of the absolute which makes this
process possible is no less the basis necessarily implied in the most
insignificant interaction of any two atoms. Nor again do we think of
its presence as a mere uniform breath which penetrates all places
and this particular spot among them, like that subtle, formless, and
homogeneous ether from which many strange theories expect the
vivification of matter into the most various forms: but the absolute is
indivisibly present with the whole inner wealth of its nature in this

particular spot, and, in obedience to those laws of its action which it has itself laid down, necessarily makes additions to the simple conjunctions of those elements which are themselves only its own continuous actions, simple additions where the conjunctions are simple, additions of greater magnitude and value where they are more complicated. Everywhere it draws only the consequences, which at every point of the whole belong to the premisses it has previously realised at that point. It is thus that it gives to every organism its fitting soul; and it is therefore needless to devise a way or make provision for the correct choice which should ensure to every animal germ the soul wh'ch answers to its kind. Again, so long as the soul was regarded as indivisible substance, it could only be supposed to enter the body at a single instant and in its entirety: whereas, if we renounce these ideas of an external conjunction, we need no longer wish to fix the moment at which the soul enters into a development which at first is supposed to produce only physical actions.

We have all along regarded the interaction of the absolute with all the elements of the world as eternal and incessant. It is present just as continuously in the first development of the germ; and in the same way there is nothing to prevent us from looking at the formation of the soul as an extended process in time, a process in which the absolute gradually gives a further form to its creation. Doubtless we shall never be able to picture this process to ourselves; but at any rate there is no force in the possible objection that such a gradual development contradicts the unity of the soul. For we are speaking, not of a composition of pieces already present in separation, but of the successive transformations of something established at the beginning of the process. And if this again should seem to contradict the idea of one unchangeable substance, I recur to my previous assertion; it is not because the soul is substance and unity that it asserts itself as such, but it is substance and unity, as soon as, and in so far as, it asserts itself as such; and if it does this gradually in a greater degree, and with a growing significance, I should not hesitate to distinguish in its substantiality, and in the intensity of its unity, countless different grades which it traverses by degrees when first it is being formed, and the last and highest of which it may perhaps be incapable of reaching during the whole of its terrestrial and superterrestrial existence.

And now, after our picture has been thus altered, collecting its various traits, I may return to an earlier statement: if anyone were in a position to observe the first development of the soul, just as with the microscope we can observe the physical development of the germ, the result would infallibly be that everything would look to him exactly as materialism believes it actually to take place. As the structure progressively differentiated itself, he would see appearing, not all at once, but by degrees, the faint and gradually multiplying traces of psychical activity; but nowhere would he meet with the sudden irruption of a power, which seemed foreign to the play of the elements active before his eyes : he would see the whole condition of things which has been thought to justify the view that all psychical life is a side-effect of the physical process of formation. This condition of things we admit; and the view based on it we reject. All the single manifestations which could thus be observed might no doubt be regarded as products of the interaction of the physical elements; but the unity of consciousness, to which at a later time our inward experience testifies, cannot, in the absence of a subject, be the mere result of the activities of a number of elements, and just as little can this subject be created by those activities. Nor again is it out of nothing that the soul is made or created by the absolute ; but to satisfy the imagination we may say it is from itself, from its own real nature that the absolute projects the soul, and so adds to its one activity, the course of nature, that other which, in the ruling plan of the absolute, is its natural completion.

247. I know well that our metaphysical enquiries are constantly and jealously watched by certain side-thoughts of our own ; and here they raise the question whether we are not in the interests of the intellect laying down positions which will afterwards prove fatal to the requirements of the emotional side of our nature. In subjecting the origin of psychical life to the dominion of law, are we not once more reducing the whole course of the world to that necessary evolution of a mere nature in which no place remains for any free beginning and, therefore, none for any guiding providence? I admit that there is ground for such doubts, but not that it is my duty to meet them here. If the need that is expressed in them is a justifiable one, still it is only where its justification is successful that we can attempt to satisfy it without cancelling what we have previously found to be necessary for the theoretic intelligibility of the world. So long then as psychical life is realised in countless instances after the same universal patterns, and so long as the same processes are repeated countless times in every single soul, we cannot refuse to admit a connexion which follows universal laws and which here, as elsewhere, shows like results following on like conditions, and the same changes in the former following on the same changes in the latter. We may put aside the question whether this connexion is all that the reality of things conceals or includes : whatever may be necessary to complete it, it cannot itself be denied.

There are two directions, therefore, in which a mechanical point of view may extend its claim over these subjects. It has been attempted long since in the case of the inward life of the soul, and the conception of a psychical mechanism is no longer unfamiliar to us : I have met with less sympathy for that other idea of a physico-psychical mechanism, the object of which was to base the commerce between soul and body on a series of thoughts similar to those which we apply to the interaction of physical elements. Accepting with gratitude the pleasanter name 'psycho-physical mechanism,' which by Fechner's ingenious attempts has been introduced into science, I will once more attempt to defend those outlines of my theory which I sketched in the *Medicinische Psychologie* (1852). According to some views my proposal is impossible ; and according to others it is superfluous. The essence of it lay in the attempt to regard the soul as an existence possessing unity, and the body as a number of other inter-connected existences, and to regard the two as the two sides, neither identical nor disconnected, from the interaction of which mental life proceeds, that life being *in posse* based on the proper nature of the soul, but stirred to actual existence by the influences of the external world.

248. I need not be prolix in opposing those who adduce the incomparability of things psychical and material as an objection against the possibility of any interaction between them. Admitting this incomparability, it would still be an unfounded prejudice to suppose that only like can act on like, and a mistake to imagine that the case of an interaction of soul and body is an exceptional one, and that we are here to find inexplicable what in any action of matter upon matter we understand. It is only the false idea that an action or effect [1] is a complete state, transferable from one substrate to the other, which misleads us into demanding that any two things which are to influence one another should be homogeneous : for, if that idea were correct, it would of course follow that b, to which the effect passes, and a, from which it issues, must be sufficiently similar to give it admittance in the same way. But, as a matter of fact, the form of any effect proceeds from the nature of that on which the external cause acts, and is not determined exclusively by the latter ; and no species of conditions can be adduced, the presence of which

[1] [Cp. § 57, *supra.*]

is indispensable to enable one thing *a* to excite another thing *b* so to manifest its own nature. To our sensuous imagination, it is true, no interaction but that of similar elements (similar at least in their external appearance) presents itself as a connected image ; but it is only our sensuous imagination that seeks to retain for every case of action the homogeneous character which it fancies it understands to be an essential condition in this particular case. And this is just where it deceives itself. I have frequently pointed out how often we suppose ourselves to understand something, when our senses are simply occupied with a variegated and unbroken series of phenomena. So long as we are merely looking at the outside of a machine we do not imagine that we comprehend it : but when it is opened and we see how all its parts fit into one another, and how at last it brings out a result utterly unlike the impulse first imparted, we think that we understand its action perfectly. And it really is clear to us, in so far as the explanation of a complicated process means its reduction to a concatenation of very simple actions which we have made up our minds to consider intelligible ; but the action which takes place between each pair of the simplest links of the chain remains just as incomprehensible as before, and equally incomprehensible whether those links are like one another or not. The working of every machine yet known rests on the fact that certain parts of it are solid and that these parts communicate their motions ; but how the elements manage to bind one another into an unchanging shape, and how they can transmit motions—and this is what is essential in the process of the action of matter on matter—remains invisible, and the similarity of the parts concerned in the action adds nothing to its intelligibility. When then we speak of an action taking place between the soul and material elements, all that we miss is the perception of that external scenery which may make the influence of matter on matter more familiar to us, but cannot explain it. We shall never see the last atom of the nerve impinging on the soul, or the soul upon it ; but equally in the case of two visible spheres the impact is not the intelligible cause of the communication of motion ; it is nothing but the form in which we can perceive something happening which we do not comprehend.

The mistake is to desire to discover indispensable conditions of all action ; and we are only repeating this mistake in another form when we declare the immaterial soul, as devoid of mass, incapable of acting mechanically on a dense material mass, or conceive it as an invulnerable shadow, inaccessible to the attacks of the corporeal world.

We might without hesitation take an opposite point of view, and speak of the soul as a definite mass at every moment when it produces an effect measurable by the movement of a corporeal mass. And in doing so we should be taking none of its immateriality from it; for with bodies also it is not the case that they are first masses and then and therefore produce effects or act ; but according to the degree of their effects they are called masses of a certain magnitude. The soul again is no less capable of *receiving* effects through the stimulus of material elements than they are from one another, although it does not stand face to face with them in an equally perceptible shape ; for as between those elements themselves shape and movement, impact and pressure, determine nothing but the external appearance behind which, and the scene on which, the imperceptible process of action goes on.

And, lastly, in our present metaphysical discussion we need not have entered on these objections at all. We have given up that simple and thorough division of reality, which places matter on one side and the mind on the other, confident and full of faith in regard to the former, timid and doubtful about the latter. Everything we supposed ourselves to know of matter as an obvious and independent existence, has long since been dissolved in the conviction that matter itself, together with the space, by filling which it seemed most convincingly to prove its peculiar nature, is nothing but an appearance for our perception, and that this appearance arises from the reciprocal effects which existences, in themselves super-sensuous, produce on one another and, consequently, also upon the soul. There may, therefore, be some other way in which the soul is separated from these existences ; but it is not parted from them by the gulf of that incomparability which is supposed to be a bar to all inter-action.

249. So long as we believe this gulf to exist, we naturally try to bridge it, and therefore raise the pointless question respecting the bond which holds body and soul together. What is the use of a bond except to hold together things which, being perfectly indifferent to each other and destitute of all inter-action, threaten to fall asunder? And how is a bond to do its work except through the connexion of its own parts, a connexion which one cannot suppose to be in its turn effected by new bonds between these parts, but which must rest in the end on their own inter-actions? And if in this instance it is clear that the binding force of the bond consists simply in the inter-actions which flow from the inner relations of its parts to one another, why should the case be different between the body and soul? Their union consists in the fact that they can and must act on one another, and no external bond which embraced them both could supply the place of this capacity and necessity, unless its inclusion of them were already based on their own natures. Besides, how poverty-stricken is the idea of this single bond, which in our parsimony we fancy will suffice us! Even supposing it to exist, where are we to find the positive ground of the nature and form of those actions or effects which, as a matter of fact, take place? The reason for their existence cannot be found by another appeal to the indifferent bond; it would have to be sought in the peculiar natures of the things connected. Whatever number of different inter-actions body and soul can effect in virtue of the relation of their natures, so many bonds are there which unite them and hold them together : but to look for the one nameless bond which should take the place of all these, is vain, absurd, and wearisome. Even if we understand it to be merely a *conditio sine qua non* for the exercise of capacities based on something else, we still must refuse to admit it; for the body and soul were never separated from one another like two bodies which cannot act on one another chemically until they are brought together. One word, lastly, on the sarcasm which reproaches us with forming the personality of man by adding two ingredients together. It is just this addition that is made by the one external bond ; and what we want is not it but the multiplicity of a complex double and united life. But in spite of this unity we do not look for man's personality in body and soul alike, but in the soul alone. We seek in the body only the echo or appearance of its action ; for the body is and remains for the soul a part of the external world, though that part which it can most directly rule and to whose influence it is most immediately susceptible.

250. There is another question on which I wish to touch, and these remarks at once suggest it. If the inter-action of body and soul is an easy matter, why not go a step further, instead of still maintaining a separation into two interacting sides? At how many points have we come close to an opposite view! We did not regard the soul as something steadfast in itself from eternity, something which enters as an indissoluble substance into the machinery of the body's formation ; we admitted that they arise together. Even the supposition that the soul arises gradually according as the bodily organization approaches its completion, did not seem to us impossible. What is there now to hinder the confession that it is simply a consequence of this physical concatenation of atoms? And if on the other side it is conceded that, so long as we abide by the customary physical ideas, we cannot deduce the origin of a psychical process from the co-operation of material atoms, why need we hold to those ideas ? Why not adopt that wider view, which holds that if a number of elements meet together, then, according as the number of the connected parts and the multiplicity of their relations increases, perfectly new effects or actions may be connected with those meeting elements, effects which do not follow on the inter-action of two atoms alone, and which therefore we never can discover, so long as we try to find the conclusions of such complicated premisses by merely adding together the inter-actions of each pair of them?

In answering this question I must first go back to an earlier statement. Even supposing we could unreservedly approve of these ideas, still the only purpose we could put them to would be to deduce from them what is given us in experience ; that must not be put aside as a matter for doubt, on the ground that our presuppositions are not found to lead to it. Now what is connected with these associations of many elements is not merely psychical states, phenomena, events, or whatever we like to call them. For each of these results inexorably demands a subject, whose state or stimulation it is ; and psychical life, so far as it is a given object of inward experience, includes for us the fact of a *unity* of this subject, to which the events we have spoken of are or can be referred as something that befalls it. I will not repeat my demonstration that the analogy of the formation of physical resultants can never lead us to this unity, unless we take beforehand as a fixed point the unity of the subject in which a variety of elements is to combine : I will only add that the ideas I have been mentioning offer no new expedient which could lead us beyond that deduction of resultants. Since so much that is new has to arise from the combination of the atoms, it seems to me that we should have to make up our minds to the final step, and maintain that from a certain definite form of this combination there also arises, as a new existence, that one subject, that very soul which collects in itself the states previously scattered among the subjects of the individual atoms. But the mere admission that psychical unity springs from physical multiplicity is no merit in the theory; it simply states the supposed fact, and so gives expression to a very familiar problem, but it offers us no further explanation of it. On the other hand, the expression employed is scarcely peculiar to the view in question ; for the psychical unity of which it speaks is simply what we mean by the word substance. It is under this title then, as substance, that the soul would become the foundation on which our account of the rest of its life would be based ; for by nothing short of this should we have complied with the postulates which experience imposes on our attempts at explanation. And at this point I should take leave to pursue the same point of view still further. According to it, it is possible that a certain state of things may be the real ground of a consequence which we cannot analytically deduce from it but can only conjoin to it as something new ; but if this is so, it is possible that the soul, once arisen, may go its own way and unfold activities which have their sufficient ground in it alone (when once it has come into being), and not in the least in those other facts which led to its creation. There would remain therefore not a shadow of necessity for the proposal to connect with every activity of the soul as its producing condition a corresponding activity of the body, and we should simply come back to that psycho-physical mechanism which allows each side a sphere of inter-action, but at the same time accords to each a field for an activity of its own in which the other has no constant share.

251. I have still something to add to our hypothesis. ' When the elements p, q, r .., are combined in the form F an effect or action Z is conjoined with them, which does not follow from the single effects of the elements when taken in pairs': this is a pleasing expression, and one that satisfies the imagination. But who has conjoined the effect with them ? Or, not to insist unfairly on the words, how are we to conceive the fact that a law holds good for the various elements p, q, r .., which determines the effect Z for their form of combination F? How are we to conceive this other fact, that those elements take notice that at a given moment this F is present, i. e. that a case has arisen for the application of the law which was not present the moment before ? Or lastly, if we recollect that that form of combination signifies nothing but an affection of those elements already present in them, in consequence of which they are no longer p, q, r .., but π, κ, ρ .., still the question would remain, how did this change in the state of each become noticeable by every other, so that they could all conspire to produce the further action Z? I have already raised these questions more than once, and the necessary answer to them has seemed to be that the course of the world is not comprehensible by a pluralism which starts with an original multiplicity of elements reciprocally indifferent, and hopes afterwards through the mere behest of laws to force them to take notice of each other. Apart from the unity of the encompassing Reality which is all things at once and which determines their being and nature, it is impossible to conceive the arising of any action at a given place and time, whether that action be one of those the content of which we believe to be deducible from the given circumstances, or one of those which can only be regarded as a new addition to them. I repeat this here in order to defend the hypothesis of the preceding paragraph. For I should certainly never set anyone the task, out of ten elements to make an eleventh arise equally real with them. It is not from them that, on this hypothesis, the substance of the soul would spring ; nor would it arise above them, between them, or by the side of them, out of nothing. It would be a new creation, produced by the one encompassing being from its own nature as the supplement of its physical activity there and then operating.

252. To a certain extent no doubt I should be merely disputing about words, if I insisted on these statements still further in opposition to Fechner, considering that his works testify so fully to his enthusiasm for a unity of all things which should be at once ideal and effective. Yet it would not be altogether a verbal quarrel ; I am anxious to take this opportunity of declaring against a point of view which may be at any rate surmised from the expressions he has chosen. After what I have said I need not repeat that, in my eyes, nothing is gained in the way of clearness by the invention of the name ' psycho-physical occurrence,' or ' psycho-physical process.' I admit that the expression may have a meaning when applied to a single element, in which, as I said before, we conceive physical and psychical stimulations to exist together. But when it is used to explain that life of the soul, which is supposed to develope itself from the co-operation of a system of elements, it seems to me to be attractive only because of its indistinctness. Where we find it difficult to define the connexion between two members of a relation which must be kept apart and distinct, we all feel some weakness for ideas which represent the two as an original unity and thereby dismiss the object of our enquiry from the world. In the present case I can find no clear account of the definite single subject to which each single instance of this process is ascribed, and no statement of the manner in which these actions or effects work into each other and form a composite whole. What is more important, however, to me is the difference between the lights in which we view what is perhaps the same set of ideas. I allude to the general remarks at the end of the second volume of Fechner's *Elemente der Psychophysik* (p. 515). In this passage I find that he observes upon and supplies though in a peculiar form, what I looked for in vain in other statements of the pluralistic hypothesis. I do not doubt at all that, for those who are accustomed to the terminology, the waves and principal waves of the psycho-physical activity, like its sinking or its rising over certain thresholds, are something more than short and pictorial designations of actual facts in the life of the soul ; that they are signs which, through their capacity of taking a mathematical form, may lead to more definite formulations of reciprocal relations of those facts. But I cannot help feeling that in these descriptions of what happens the real condition of its happening is also looked for ; or, if this is a misunderstanding, that at any rate there is much provocation for it. For if no idea of this kind had had a hand in the matter, many of the explanations that are given would be in reality nothing but elegant transcriptions of familiar thoughts into this sign-language, transcriptions which do not directly advance the enquiry : and the reader will not suppose that he has gained anything by them unless he is allowed to take these images for the discovery of something hitherto unknown, of the instrumentation, so to speak, on which the realisation of the psychical processes rests.

One of the last sentences of this celebrated book (p. 546) may explain what it is I object to. The substrate of what is psychical, we are told, is something diffused through the whole world and connected into a system by universal forces ; the quantity of consciousness depends simply on the quantity, and not on the quality of the psycho-physical motion ; and this quality should rather be connected only with the quality of the phenomena of consciousness. Thus every motion, whatever its form and whatever its substrate, would, on reaching a certain specified value, contribute something to consciousness, whether that consciousness be our own or that of another person or a general consciousness ; and every particular form of motion—i. e. every particular collocation and series of velocity-components—would carry with it its appropriate psychical phenomenon of the appropriate form, so soon as the components entering into that form all exceed a certain quantitative value.

'In this way we dispense with the magical charm, the *qualitas occulta*, which is supposed to qualify for psychical effects only this or that exceptional form of motion.' 'What is unconscious and what is conscious in the world will represent merely two cases of the same formula, which is the standard at once of their relation and of their transition into one another.'

I maintain nothing respecting the meaning intended in these

words : I maintain only that they may easily be understood, or misunderstood, to recommend a view, the admissibility of which I
certainly contest. However much we may bring the phenomena of
two different series of events under one and the same formula—and
I do not deny that it is possible to do so—still all that the formula
in any case does is to describe the phenomena after they are actually
there ; it is not the reason why they are actually there. If all the
hopes here expressed of the psycho-physical calculus were fulfilled,
we should nevertheless still be unable to dispense with that *qualitas
occulta*, which brings, not to an exceptional kind of motion, but to
every motion the capacity for an activity which does not lie in the
motion itself. I may be told that what I miss is already included in
the character of the motion as *psycho-physical* ; and indeed it is not
so much the meaning of these sentences that I wish to object to as
the manner in which it is expressed. Still there appears everywhere
as something first and foremost a universal mechanism, which of itself
is supposed to produce this result, that, in relation to certain forms
of motion, there arises, as their natural and necessary consequence
and as the consequence of nothing beside them, a mental activity ;
for even the general formula which is to include conscious and unconscious as two cases, must obviously, as the common element *of*
which they are cases, mean not the mere abstract formula, but always
in the last resort that which is itself unconscious, namely, motion.
The beautiful thoughts in which Fechner contradicts this interpretation will be put aside by most of his readers as excusable day-
dreams ; but there are many who will make use of his expressions
in order to shelter under a great name their favourite doctrine of
the *generatio aequivoca* of everything rational from that which is devoid
of reason.

CHAPTER II.

Sensations and the Course of Ideas.

253. OUR mental life is aroused anew at every moment by sensations which the external world excites. But the things without us become the cause of our sensation not through their mere existence, but only through effects which they produce in us; through motions, in which either they themselves approach the surface of our body until they touch it, or which they from their own fixed position communicate to some medium, and which this medium in turn propagates from atom to atom up to that surface. And therefore, though language describes things as objects which we see and hear, we must not allow these transitive expressions to suggest the idea that our senses, or our soul by means of them, exercise some activity which goes out to seek for the external objects and brings them to perception. Our attitude is at first one of simple waiting; and although when we strain our eyes and ears in listening or watching we may seem to feel in those organs something of such an outgoing activity, what we really feel is not this but a different activity,—one by which we place them in a state of the utmost sensitiveness for the impressions we expect.

Now it is self-evident that sensations, which we have at one time and not at another, can only arise from the alteration of a previous state, and therefore only through some motion which brings about this alteration. The old idea therefore that the mere assumption of a specific substance or caloric was sufficient to account for our feeling of heat was, apart from all other objections, intrinsically false: for this caloric, even if it were present, could not, in the absence of any motion, produce either the sensation of heat or those other effects which would prove that it itself was present. But that is one objection which I fear will be raised against the doctrine that all our sensations and perceptions depend on motions of the things which are to be their objects. From an ontological point of view I regarded a certain sympathetic *rap-*

port as the ultimate ground of every possible inter-action. But, I may be asked, if this idea is sound, why should not things exist for one another apart from any physical intermediation; and why should not we perceive things immediately, without having to wait for the impact of their propagated motion on us? That sympathy, I answer, the name of which was borrowed from a dubious quarter, was not such a community of all things as is destitute of order and degree. On the contrary, we found that the elements of the whole stood to one another in relations varying widely in their closeness or distance; and it was to these elements we ascribed an immediate sympathy which needs no artificial means for its production. The degree of this closeness or distance determines for any two elements the number of intermediates necessary for their interaction; necessary, not because the laws of a pre-mundane system of mechanics would render the interaction impossible in the absence of these intermediates, but because, in their absence, it would be in contradiction with the degree and nature of the relation on which it is founded, and with that meaning of the whole which again is the foundation of whatever mechanical laws hold good in the world. Thus in our view, the motions in question, the physical *stimuli* of the senses, are not the instrumental conditions, which place all things for the first time in relations to one another and to us, but *expressions* of that existing and irremovable network of conditions which the meaning of the world has established between the states of those things. We know that in any chain, along which an action or effect is propagated, there is necessarily presupposed in the last resort a wholly immediate action between each link and that which lies next to it. The fantastic idea which extends this direct reciprocal influence to anything and everything, and would accordingly place the soul in a communion, free from all physical intermediation, with distant objects, cannot therefore be theoretically proved impossible. But inability to controvert a point of view lies a long way from belief in its validity. Considering that the whole of the known and waking life of the soul is based throughout upon that physical intermediation, we can only answer asserted experiences of an interruption of this connexion by the most decided disbelief, and these experiences could call for attention only if occasioning causes could be assigned, adequate to produce such remarkable exceptions in the course of nature.

254. On their arrival in the body the external stimuli meet with the system of nerve-fibres prepared for their reception. The change which they set up in these nerve-fibres becomes the internal sense-stimulus,

which is the more immediate cause of our sensation. We leave it to physiology to ascertain exactly what takes place in this nervous process. The answer to that question could have a value for psychology only if it were so complete as to enable us to deduce from the various modifications of the process the corresponding modifications of the sensation and to express the relation in a universal law: whereas the mere subordination of the nervous process under a specific conception is only of importance for the question whether we have to consider it as a mere physical process or whether it is itself something psychical. The latter view is frequently met with. The sensation is said to be formed already in the nerve, and to be transferred by it to consciousness. If this assumption is to have any clearness it must name the definite subject to which it ascribes the act of sensation; for sensations which nobody has cannot be realities. Now this subject of sensation could not be found in the whole nerve, as such, which is an aggregate of unnumbered parts: it is only each single atom, however many of them we suppose to be strung together in the whole nerve, that could be, by itself, a feeling thing. But to this difficulty must be added a familiar fact. The external sense-stimulus does not become the cause of a sensation in us, unless the nerve remains uninterrupted throughout its whole course, from its peripheral point of stimulus up to the central portions of the nervous system. If its continuity is broken by a cut, the influence of the external stimulus on consciousness is removed. Whether the idea, to which this fact naturally gives rise, is correct or not,—the idea that the soul has its seat in a particular spot to which the incoming impression must be directed,—or in what other way we are to explain the truth that this integrity of the nerve-fibre is an indispensable condition of our sensation, we need not here discuss. In any case there is a propagation of the stimulation in the nerve itself, and all its parts cannot be at once in the state of sensation presupposed. But it is impossible that one and the same sample of sensation can be handed on from one atom of the nerve to another like a packet; all that can happen is that each single element of the nerve becomes, in virtue of its own state, a stimulus to the next to produce the same state in itself. Now that this excitation is not produced by a direct sympathy, is proved by that interruption to its propagation which results from any mechanical breach of continuity. Such a sympathy would pass undisturbed across the point of section, and would feel no effects from changes in physical relations of which it would be from its very nature independent.

We are therefore compelled to introduce a physical connecting link for the effect we have presupposed. Through the external sense-stimulus there is produced in the first nerve-element the physical state *r* and, in consequence, in the same element the state of sensation *s*. By this change the first element is compelled to awake in the second, its neighbour, the same state *r* and, in consequence, the sensation *s*. Thus, through the physical impact of one element on another there would be propagated at the same time the creation of the corresponding sensation. But where would this end? Wherever and however this chain of atoms with their internal excitations may at last connect itself with the soul, the sensation of the soul, *our* sensation, would arise out of the soul itself simply through the influence of the last *r* with which the last nerve-atom stimulates it, in precisely the same way in which this sensation was produced in link after link of the chain. Whatever service then can be rendered by the nerve in aid of the production of our sensation, it can render just as well by transmitting a merely physical change, as if each of its atoms experienced the same psychical state which is to arise in us at the end of the whole process. A piece of news which passes in the form of a letter from hand to hand along a series of messengers, reaches the recipient no more securely and is no better understood by him if each of the intermediates knows and feels it. Doubtless we shall never be able to portray the action of that final *r* on the nature of the soul; but we cannot do so any the more by adding to the physical process *r* the sensation *s*. This *s* in its turn could only occasion the production of our sensation *S* in some perfectly indemonstrable way; it could not itself pass over into us. On the other hand the propagation in the nerve of a physical process *r* up to this mysterious moment, is something which the fact of experience alluded to compels us to assume. It is sufficient, therefore, to regard the nervous process as a propagation of something, taking place in space and time in a definite direction and with a definite velocity; the precise nature of that which is propagated concerns us but little, and, since these are the only forms of its propagation which are of importance, it may be described as merely physical.

255. The conscious sensation itself, the red or blue that we see, the sound that we hear, is the third and last link in this series of occurrences, and it is familiar to us. We know that this content of sensation admits of no comparison either with the external sense-stimulus or with the nervous processes. There is nothing in the redness of red, the blueness of blue, or the sound of the heard tone, which suggests a larger or smaller number of vibrations of a medium; yet science has indirectly discovered such vibrations to be the occa-

sion of these sensations. In the same way they give us no information respecting that which *directly* occasions them, the process which goes on in the optic or auditory nerve at the moment when these sensations are produced in us; they are consequences, not copies, of their stimuli. Thus they are internal phenomena in the soul, and in this sense of the words the doctrine of the subjectivity of all sensations has long been the property of philosophy and required no acquaintance with the functions of the nerves.

There is another sense of the words, according to which the sensations are held to be *merely* internal phenomena, and the external world to be neither resonant nor silent, neither bright nor dark, but to possess only mathematical predicates of number and magnitude, of motions and their complications; and in this sense of the words the doctrine was in antiquity an insufficiently proved inference, and it remains so for the physiology of the present day. None of the proofs which are commonly appealed to in support of it, can close every way of escape to the opposite view. Anyone who wishes to maintain that things themselves remain red or sweet, will affirm, as we do, that it is not through their *being* that they can appear to us as they are, but only through effects which, in accordance with their nature, they produce on us. These effects or actions, which proceed from them and are sense-stimuli to us, are no doubt only motions and themselves neither red nor sweet; but what is there to prevent our supposing that, by acting through our nerves, they make that same redness or sweetness arise, as our sensation, in our souls, which also attaches as a quality to the things themselves? Such a process would be no more wonderful than the performances of the telephone, which receives waves of sound, propagates them in a form of motion quite different, and in the end conducts them to the ear retransformed into waves of sound. Anything which deprives things of the medium through which their excitations could reach us; anything again which has beforehand imparted to the medium motions which prevent the passage of those excitations, would of course either hinder things from appearing to us at all or would make them appear with other qualities, and so would lead us to suppose that none of these qualities belong to things themselves at all.

There are no individual proofs by which these assertions could be controverted; and yet the doctrine of the mere subjectivity of the qualities of sensation is certainly sound. Their own nature makes it impossible for us really so to represent them to ourselves as qualities of things, as we profess to do. There is no meaning in speaking of a brightness seen by nobody at all, of the sound of a tone which no one hears, of a sweetness which no one tastes : they are all as impossible as a toothache which nobody has got. There is only one place in which what is meant by these words can possibly exist, the consciousness of a feeling being : and there is only one way in which it can exist, the way of being felt by that being. Without doubt then, things are red only so far as they appear to us ; *in* itself a thing could only have a particular look if it could look *at* itself.

256. According to a theorem of the doctrine of specific energies, every nerve, by whatever stimulus excited, invariably calls forth sensations of one and the same kind, the special sensations of its own sense ; and it makes no difference whether the stimulus is one appropriate to the nerve or not. If this were a fact, its physical reason would not be hard to imagine. Let us take a composite system of parts. External stimuli, so long as they are not so violent as to destroy the internal connexions of this system, will cause a motion followed by an effort to return to equilibrium ; and these will take place in forms which essentially depend on the structure of the system, which in that case remains unchanged. So with the nerve ; disturbances of a certain magnitude would injure it ; but to less violent stimuli it would always respond with the same reactions, and these reactions would depend on its peculiar structure. But then, if these reactions are to be different in the case of every single nerve, the structure of the various nerves must be different ; and this variety of structure we do not find in the nerves themselves, though we may perhaps look for it in the central portions to which they lead.

But in any case the facts themselves are generalised in this theorem to an extent which actual observation does not justify. We know nothing of waves of sound which produce in the eye a sensation of light, nor of waves of light which produce tones in the ear. The main support of the hypothesis lies in the sensations of light which frequently arise in the eye from impact or pressure, as well as from electrical stimulation. But there are other considerations which compel us to assume in the media of the eye the presence of the same ether which serves for the diffusion of the light outside ; and accordingly, when in consequence of impact the ponderable elements of the tense eyeball fall into oscillation, we can scarcely help supposing that they impart this oscillation at the same time to the ether. Thus the same objective motion of light which commonly, as an adequate stimulus, comes from without, may be excited in the eye by this oscillation of the eyeball, and a similar motion might be excited by electric

currents; such motion not being sufficient to cast any observable
rays outwards, but strong enough to stimulate the nerve to produce a
sensation of light. Again, in the case of the inadequate stimuli which
actually do create a sensation of *sound*, the question is prudently
avoided whether they may not do so by accidentally exciting such
vibrations as form the natural stimulus of the auditory nerves. The
excitation of *taste* by electricity certainly depends on the adequate
stimulus, the chemical processes which are here set up; the notion
that it can also be produced by laceration of the tongue seems to have
been an illusion, and it will be useless for insipid dishes to look for
help in this quarter: and as to the remaining sensations, we do not
know at all what the adequate form of the stimuli is which actually
must reach the nerves in order to produce them.

We may leave it therefore to physiology to decide whether the real
meaning of the present widely-spread doctrine of the division of labour
is not rather this;—that every nerve is excited to its function only by
its own adequate stimulus, and that other stimuli either leave it unaffected or else interfere with it, but that at the same time there are
stimuli of various kinds which, along with their own effects, frequently
produce the adequate stimuli as side-results. The only interest psychology has in the question lies in opposing the fondness for a
mysterious psychical activity which, on the authority of the facts I
have mentioned, is attributed to the nerves and not to the soul, to
which it really belongs. To speak of a substance of the sense of
sight, and to say that this substance converts every possible motion
that reaches it into a sensation of light, is not to describe facts but to
use a piece of physiological metaphysic; of which I am not sure that
it is at all more elegant than the metaphysic of philosophy.

257. However complete the separation may be between sensations
and the stimuli which occasion them, these two series of occurrences
are, as a matter of fact, connected, and we shall not suppose that this
connexion of fact is destitute of any principle. We shall always find
ourselves presupposing that like groups of sensations correspond to
like groups of stimuli, and different groups of the one to different
groups of the other; that the difference of these classes of sensation
is proportional to the difference which exists between the classes of
stimuli; that wherever the stimuli of a given group are arranged in a
progressive series or, in their progress, reach marked points of
eminence, the corresponding sensations are arranged in a similar
series and accordingly reproduce both the progress and the points of
eminence; that, lastly, in the unity of the soul its various kinds of

sensation not only *are* together as a fact, but in their meaning are coherent according to some rule, though that rule may not be expressible in mathematical terms.

But of an 'empirical confirmation of this presupposition we find but faint traces. Not only is it impossible to say why waves of ether must necessarily be felt as light; but, even if this fact were given as a starting-point, no theory, however much it emphasized the unity of the soul, could prove that this same soul must in consistency perceive waves of sound as tones, and other affections as taste or smell. So far as we can see, that unity produces, from a nature of its own which is quite unknown to us, the various classes of sensation, each for itself and apart from the others; and, even after we have come to know them, all that we can connect with their impressions are vague and fantastic ideas respecting the organization of a universal realm of sensations. Again, when we come to the individual groups, the only one which confirms our supposition is the group of sounds. Here the increase in the height of a tone corresponds to an increase in the number of waves within a given unit of time. The ascending scale, which is just as clearly an ascent as is the increase in the number of waves and yet is quite unlike that increase, repeats in its own specific form the progress in the series of stimuli. Wherever this series attains, through the doubling of a previous number of waves, a marked import, there the sensation follows with the marked impression of the octave of the key-note, and thus again in its own particular way represents sensuously the likeness and difference of the two series. ˙ On the other hand the colours, though their prismatic order rests on a similar increase in the number of waves, give no one who is unprejudiced the impression of a similar progress; and the reason of this possibly lies in the peculiar nature of the nervous process which intervenes between the stimulus and sensation, and which we cannot take into consideration because we do not know it. In the cases of the remaining senses we have no exact knowledge of the nature of their stimuli, nor have we succeeded in discerning any fixed relations between their individual sensations. We do not possess even names for the various smells, except such as describe them by their origin or their incidental effects; and among the multitude of tastes the only ones that can be distinguished as well-defined are the four forms of acid, alkaline, sweet, and bitter. Hypothetical theories carry us no further. In the case of sight and hearing alone we know that each sensation rests on the total effect of a very large number of successive impulses, and changes with the alterations of this number within the given unit of time;

whether the single impact of a wave of light or sound would be observable by our senses, and if so in what way, is utterly unknown to us. Still we can generalise this fact with some probability. Perhaps it is true of all our sensations that they rest not on a constant and indiscriminate stream of excitation, but on the number of alternations of excitation and non-excitation included in a certain time; the nature of the process, which thus in the form of oscillation stimulates the soul, might be a matter of less importance, and the same perhaps for all the nerves. But then again this supposition makes it no easier to connect the various kinds of sensation with one another in a progressive series; and we have further to admit the possibility that our human senses do not include the whole range of sensible existence, and that other living beings may have other forms of sensation unknown to us and answering to processes which entirely escape our perception.

258. There is at any rate one point at which the modern psychophysical investigations have resulted in the beginnings of an exact knowledge regarding the relation between sensation and stimulus. The commonest observation of a brightening light or a rising sound shows us that our senses can detect very slight alterations in the strength of an impression. But we never reach a moment at which, judging merely by the direct impression, we could say that one brightness was twice or thrice as strong, or one sound half as strong, as another. In consequence of this inability to reduce to numerical equivalents the more and less which we perceive, it is impossible for us to place a series of values of stimuli side by side with the values of the corresponding sensations, and so to formulate a universal law according to which the intensity of the latter would depend on the strength of the former. There is however one judgment we can pronounce, if not with absolute yet with sufficient certainty, viz. that there is or is not an observable difference between two sensations. To this point accordingly were directed those experiments, the object of which was to discover, first of all, what amount of increase a stimulus requires in order that the sensation which belongs to it as increased may begin to distinguish itself from the sensation of its previous strength; or, again, to discover the limit of slightness down to which the difference between two strengths of the stimulus can be diminished without removing the possibility of the sensations being distinguished. With regard to the moderate stimuli which are strong enough to excite a distinct sensation, and yet do not approach the point at which their intensity disturbs the function of the nerve, Fechner and many others

since, following E. H. Weber's example, have made a very large
number of experiments; and these experiments lead with sufficient
unanimity to the result that that difference between any two stimuli
which makes it possible to distinguish the corresponding sensations
from one another, is not a constant quantity, but, in the case of each
class of sensations, amounts to a definite fraction of the intensity
already possessed by that one of the two stimuli from which we start.
We are not interested in following the various mathematical formulations of Weber's Law, or the corrections which its application has
appeared to render necessary; we may ascribe the latter to the
influence of the particular circumstances which, as in the case of
most natural laws, prevent the phenomena from answering precisely
to a law which in itself is valid.

The experiments themselves give no further result than that
described above; they do not tell us in what way the difference
between the stimuli makes it possible for us to distinguish the resulting sensations—whether it is by producing a difference of *strength*
between these sensations, or whether we are aided by qualitative
changes set up in the content of the sensation and dependent on the
difference of the stimuli. Nothing but our direct impression can
decide this point, and it certainly does not seem to me that this impression speaks quite clearly in favour of the first alternative. A
concentrated solution of an acid does not simply give us the same
taste in a stronger form which a more diluted one gives us in a
weaker form ; it also tastes *different.* Two degrees of heat, though
they rest on differences of intensity in the same stimulus, are felt as
different sensations and not merely as different degrees of strength
in the same sensation. If this is not so clear in the case of slight
differences, the fact is all the clearer that our direct impression makes
us speak of heat and cold as two positive opposites, and does not lead
us to recognise in them mere differences of degree. Lastly, no one
who experiments on degrees of brightness by means of shadows compared with the ground on which they are thrown, feels sure that he is
merely comparing differences of intensity in the same sensation ; the
shadow is not only a less degree of illumination, but it looks *different*
from the brighter ground—black if it is on a white ground.

I do not wish to lay any great stress on these doubts; still they
would have to be removed before we could follow with entire security
the theory which deduces from the experiments I have alluded to a law
respecting the strength belonging to the sensation, and its dependence
on the strength of the stimuli. Supposing them removed, we should

then regard the transition from the point at which two sensations are indistinguishable to that at which their difference is just observable, as an increase, the same in amount in all cases, in the strength of the first of the two,—and so the law in question would take this form : Where the intensity of a sensation increases *by equal differences*, that is, in arithmetical progression, it implies in the strength of the stimulus an increase in geometrical progression. Thus the activity of sensation would be one of those activities which it becomes increasingly difficult to heighten as the degree of liveliness already attained increases.

259. Our present result, according to which the sensation does not follow the growing strength of the stimulus at an equal speed, would not, if taken by itself, present any extraordinary problem. But none of the theories which have been formed on this point explain why the continuous curve of growth in the strength of the stimulus is not *continuously* followed by the slower augmentation in the strength of the sensation,—why, on the contrary, there remains an interval throughout which the stimulus strengthens without showing any result, until at last, on its reaching a final degree of strength, it produces an observable difference in the sensation. This difficulty, I think, is most easily met by the physiological view which attempts to explain it by reference to the mode in which the nerves are excited. It is a problem soluble in mechanics, so to construct a system of material parts that a force which impels continuously is nevertheless prevented by internal hindrances from exerting its influence except intermittently at certain moments. Following this analogy we should have to suppose a structure of the nerve of such a kind that, given a certain attained degree of excitation, a definite concentration and heightening of that excitation is necessary before such a motion of the nerve can be produced as will afford a stimulus to the rise of a new sensation ; thus the sensation would increase in intensity proportionally to these intermittent excitations. On the other hand, we do not in the least know how and where such an arrangement is to be presumed in the nervous system. There is less probability to my mind in the *second* hypothesis, according to which the nervous excitation increases proportionally to the stimulus and continuously. This hypothesis has to look to the nature of sensation itself for the reason both of the slower rate and of the want of continuity in its increase ; there is nothing in the mere idea of sensation which could with any probability be supposed to take the place of the machinery which must, *ex hypothesi*, be absent. Nor is the solution offered by the

third view more convincing. The sensation, it tells us, increases in strength proportionally to the stimulus and the nervous process, but perception brings the actually increased intensity of the sensation to consciousness in a different relation and discontinuously. The separation of these two processes, the sensation and the perception of what is felt, we shall be able to justify later on ; but we certainly shall not be able to find in the nature of a perceiving activity, as such, any reason for its *not* perceiving something. If the idea could be made plausible, that the act of distinguishing two impressions—an act which is always at the same time an act of comparison—is guided not by single differences between them, but by their geometrical relation, still the only deduction.we could draw from this idea would be that, given two pairs of impressions, this act would find an equally great difference between the members of each pair, supposing that in both cases these members stood to one another in the same ratio. But I do not know why that act should fail to distinguish *at all* those which did *not* stand in that ratio.

260. No method has yet been discovered of experimentally determining the consequences which result from simultaneous impressions on different senses ; it is even doubtful what goes on when the same sense is excited in several ways at once. We are accustomed to the notion of a mechanism of ideas ; but the attempt to go further and to oppose to it the notion of a chemistry of ideas, can be met only with the utmost distrust. As long as two external stimuli *a* and *b* are producing effects in the same nerve-element, there must ensue, in this physical sphere, the formation of that resultant *c* which the conjunction of all the mechanical conditions renders possible and therefore necessary. To this resultant *c*, which alone reaches the soul as an exciting motive, corresponds the simple sensation γ; and this γ is not the resultant of the two sensations a and β which the two stimuli, if taken separately, would have produced, but appears *instead of* them, since they are unable to arise. If, on the other hand, we suppose that *a* and *b*, either because they are transmitted in different nerve-elements, or because they do not form one indistinguishable resultant within the nerve, have actually produced the two sensations a and β, the result will be that the contents of the two sensations do not blend in consciousness into a third simple sensation, but remain apart and form the necessary pre-requisite of every higher activity of mind in the way of comparison and judgment.

At the same time I must allow that there are objections to this last view. For though the theoretical assertion that the soul is compelled

by its own one-ness to attempt to fuse all its internal states into an
intensive unity, could decide nothing so long as our inward experience
offered no example of such a result, it is on the other hand indubitable
that the simultaneous assault of a variety of different stimuli on different
senses, or even on the same sense, puts us into a state of confused
general feeling in which we are certainly not conscious of clearly
distinguishing the different impressions. Still it does not follow that
in such a case we have a positive perception of an actual unity of the
contents of our ideas, arising from their mixture; our state of mind
seems to me rather to consist in (1) the consciousness of our inability
to separate what has really remained diverse, and (2) in the general
feeling of the disturbance produced in the economy of our body by
the simultaneous assault of the stimuli. As to the *first* point, I recur
to that distinction of sensation and perception, to which we found the
psycho-physical theory obliged to appeal. The act of distinguishing
two sensations is never a simple sensation; it is an act of referring
and comparing, which may supervene on those sensations, but need
not always do so. Where it is prevented, the result is not that the
sensations melt into one another, but simply that the act of distinguishing them is absent; and this again certainly not so far that the
fact of the difference remains entirely unperceived, but only so far as
to prevent us from determining the amount of the difference, and from
apprehending other relations between the different impressions. Anyone who is annoyed at one and the same time by glowing heat,
dazzling light, deafening noise, and an offensive smell, will certainly
not fuse these disparate sensations into a single one with a single
content which could be sensuously perceived; they remain for him in
separation, and he merely finds it impossible to be conscious of one of
them apart from the others. But, further, he will have a feeling of
discomfort—what I mentioned above as the *second* constituent of his
whole state. For every stimulus which produces in consciousness a
definite content of sensation, is also a definite degree of disturbance
and therefore makes a call upon the forces of the nerves; and the
sum of these little changes, which in their character as disturbances
are not so diverse as the contents of consciousness they give rise to,
produce the general feeling which, added to the inability to distinguish,
deludes us into the belief in an actual absence of diversity in our
sensations. It is only in some such way as this, again, that I can
imagine that state which is sometimes described as the beginning of
our whole education, a state which in itself is supposed to be simple,
and to be afterwards divided into different sensations by an activity of

separation. No activity of separation in the world could establish differences where no real diversity existed ; for it would have nothing to guide it to the places where it was to establish them, or to indicate the width it was to give them. A separation can only proceed from a mixture of impressions which continue to be diverse, and then only if, owing to favourable circumstances, the single constituents of the mixture are, one after the other, raised above the rest by an access of strength, so as to facilitate comparison and the apprehension of the width of the individual differences : if ideas of the single impressions have once been acquired, it may then be possible to dissociate them even in the unfavourable case of such a mixture as that described above. It this way it might perhaps happen that many apparently simple sensations may be dissociated into several sensations of the same kind ; for example, in a colour we might separate the other colours which formed its constituents, or in a tone the partial tones of which we were unconscious at first, or in tastes and smells the elementary sensations which were combined in a variety of different ways and of which at present we have no knowledge. Thus within these narrow limits a real chemistry of sensations, combining different elements into a new quality of sensation, is not inconceivable. But after all our experience up to the present time it remains uncertain whether this intermingling into new resultants has not in all cases already taken place among the physical excitations in the nerve or in the central portions of the nervous system.

From these premises again, a conclusion might be drawn respecting those sensations which attach to others in the way of contrast, and do not need a particular external stimulus. I do not think they can be considered reactions of the soul unoccasioned by anything physical. It might be possible to take that view of the false estimates of magnitude which make a sudden silence ensuing on deafening noise, or a darkness ensuing on dazzling light, appear extra-ordinarily deep ; for these are not sensations, but comparisons. And yet even in these cases the probable cause of the judgment is the distance between the degrees of excitation in the nerve, a distance just as great as that between the sensations. But a colour β cannot attach to another a by way of contrast or complement through a mere reaction of the soul. Even if we imagine in the soul a disturbance which seeks a compensating adjustment, the aim of that search can be no more than an opposite *Non-a*, the whereabouts of which is unknown. That it is β and nothing but β which gives the desired satisfaction we know only from experience ; to seek the reason of the fact in a comparison of the two impressions *a* and *β*, is to seek it in something far from self-evident, it must lie in the way in which the nerve acts, and this activity of the nerve must attach the excitation which leads to *β* to the excitation which produces *a*, in the character of an effort to attain equilibrium.

261. Neither observation nor theory have so far thrown any light upon the interval which intervenes between the occurrence of a sensation and its disappearance from consciousness. If we say that it gradually diminishes in strength until at last it reaches zero or disappears below the threshold of consciousness, we merely describe what we think we can imagine to be going on ; no one can *observe* the process, since the attention necessary for observing it makes it impossible. Whether this hypothetical view has a sufficient theoretical justification, is doubtful. Beside the presupposition that a diminution of the activity of representation, from its strength at a given moment down to its disappearance, must be continuous, the physical law of persistence is called in, in order to make the undiminished continuance of the sensation appear as the natural course of events, and its disappearance from consciousness as the problem to be explained. This last idea is not free from difficulty. A material atom undergoes no internal change during its motion,—at least according to the ordinary view of that motion,—and its state in any new place *q* is exactly what it was in its former place *p*; it follows therefore that it itself contains nothing which would at any point resist a further motion, and that the cause of the change or the checking of this motion must come from outside. The soul, on the other hand, when it feels *a*, falls into an internal state differing from its state when it feels *β* : if we consider it capable of reacting against stimuli at all, we must admit that there may lie in its own nature the permanent motive which stirs it to oppose every one-sided manifestation of its capacity that may be forced on it, and therefore stirs it also to eliminate the state of sensation forced on it by the external stimulus. If indeed it were able *completely* to annul what has occurred, it would be wholly impassive and therefore incapable of interaction ; but might not its opposing effort be strong enough to repress the sensation into a condition of permanent unconsciousness?

If we leave this question, which cannot be decided, we may seek the causes of hindrance or checking partly in the new impressions which arrive from outside, partly in those far less familiar ones which are constantly being brought to the soul by the changing states of the body. The first of these, the struggle of ideas with one another,

served as the foundation of Herbart's theory of the internal mechanism
of the soul-life. I put aside at present the doubts which are suggested
by the metaphysical basis of this theory; the unchangeability of a soul
which yet experiences changing internal states; its effort to fuse them
all into a unity, and the shipwreck of this effort on the differences of
the ideas ; the assumption, lastly, that the soul finds a satisfaction in
at least lessening the strength of the parties whose opposition it has
to tolerate. We accept simply as a hypothesis what Herbart offers us
as the foundation of his theory, the hypothesis that ideas check one
another according to the degree of their strength and of their opposition ; and we utilise his just rejection of figurative modes of speech.
Consciousness, as he tells us, is not a space in which ideas appear
side by side. Even if it were a space, still the ideas are not extended
things which require a definite place to exist in, rigid bodies which are
incapable of condensation, and therefore push one another from this
narrow stage. Nor, lastly, is there any original repulsion of ideas
against ideas ; it is only the unity of the soul in which they attempt to
exist at the same time, that turns their mere difference into a struggle.
The question now is, Does our internal observation confirm these
hypotheses ?

262. We have in thought to separate two things which never
appear apart in the real world ; the content to which the activity
of representation or sensation is directed, and this activity itself
which makes the content something represented or felt : to both of
these we might attempt to apply the conceptions (*a*) of opposition
and (*b*) of variable strength. (*a*) Now I cannot find anything given
in internal observation which testifies to a checking of ideas according
to the degree in which their *contents* are opposed. Doubtless we hold
a simultaneous sensation of opposite contents through the same nerve-
element to be impossible ; but I do not know that the idea of the
positive and of affirmation exercises any special repulsion against the
idea of the negative and of negation ; on the contrary, every possible
comparison of opposites implies that the two members of the comparison do *not* check one another. If, on the other hand, we apply
the opposition to the representing *activity*, it is doubtless self-evident
that two of its acts which are opposed in respect of their action will
cancel one another; but this proposition, if self-evident, is also
fruitless, for we have no right whatever to presuppose that the ideas
of two opposite contents rest on an opposition of the representing
activities in respect of their mode of action. Thus we do not know

where in such action we are to find oppositions which are to have a mechanical value.

(*b*) The conception of a variable strength of ideas suggests similar doubts. In the case of the sensations of an actually operating sense-stimulus, it did not seem worth while to draw the distinction I have just used ; the hearing of a louder noise, or the seeing of a brighter light, is always at the same time a greater activity, excitation, or affection ; and it is not possible to hear loud thunder as loud and yet to hear it weakly, or to feel a brighter light to be brighter and yet to feel it less strongly than a dimmer light. But the case may well be different with our ideas ; by which name I understand, in accordance with usage, the image in memory of an absent impression, as opposed to the sensation of the present impression. The difference between the two is clear enough. The remembered light does not shine as the seen light does ; the remembered tones do not sound as heard tones do, although they reproduce in their succession the most delicate relations of a melody ; the idea of the intensest pain does not hurt, and is nothing compared to the least real injury. I will not enquire whether this difference is due to the fact that an idea, as a remembrance having its origin in the soul only, is not accompanied by any bodily excitation, whereas such an excitation accompanies every sensation and is the cause of its beginning and continuance; or whether that view is correct which, in spite of its not receiving much support from the direct impression of internal experience, assumes that in sensation and idea alike there is always a physical nervous excitation, and that the difference in the two cases is only one of degree.

Now whatever we remember we can certainly represent in idea in all the degrees of which its content is capable ; but it is not so clear that the representing activity directed to this content can itself experience the same changes in magnitude. We cannot represent more or represent less to ourselves one tone of a given height and strength, or one shade of a colour ; the attempt to do so really introduces a change in the *content*, and we are representing a stronger or weaker tone, a brighter or duller colour, instead of merely representing more or representing less the same tone and the same colour. Nor does internal observation give us any more justification for regarding this activity of representation, like the activity of sensation, as *proportional* to the content to which it is directed. The idea of the stronger does not call for or cause any stronger excitation or greater effort than the idea of the weaker. The images of memory resemble shadows, which do not differ in weight like the bodies that cast them. Thus it appears so far as if the conception of a variable strength, when applied to our ideas, may hold good of their content, but not of the psychical activity, to which the mechanical theory at starting certainly intended it also to apply.

263. To this it may be objected that the capacity of being heightened, possessed by the representing activity, cannot be disclosed by a trial made on purpose. Such an experiment, it may be said, naturally brings before us the maximum attainable by that activity in reference to the content chosen, and does not bring to our notice the lower degrees to which it sinks, and through which it passes on its way to extinction. It cannot be denied, we may be told, that the distinction of clearer and dimmer ideas signifies something which really exists in consciousness and which confirms our belief that the activity has various grades although we cannot directly observe them. To this objection I should give the following answer. I cannot convince myself that internal observation testifies without more ado to the reality of dim ideas in this sense of the word. If the image of a composite object in our memory is dim, the reason is not that the image is present, with all its parts in their order, and that consciousness sheds only a weak light over the whole. The reason is that there are gaps in the image; some of its parts are entirely absent; and, above all, the exact way in which those parts that are present are connected, is usually not before the mind, and is replaced by the mere thought that there was some connexion or other between them; and the wideness of the limits within which we find this or that connexion equally probable, without being able to come to a decision, determines the degree of dimness we ascribe to the image. Let us take as an example the taste of a rare fruit. We either have a complete idea of this taste, or we have none at all : and the only reason why we suppose that we really have a dim idea of it is this ;—we know from other sources that fruits have a taste, and the other characters which are present to our memory and which tell us the species of the fruit, move us to think only of that particular class of tastes which belongs to this species ; the number of the tastes which lie between these limits and between which we hesitate, determines again the degree of the obscurity of the idea, which we suppose ourselves to possess though we are really only looking for it.

To take another example ; we try for a long time to remember a name, and then, when one is suggested to us, we at once recognise it to be the right one. But this does not prove that we had an obscure idea of the right one, and now recognise it as the right one by comparing it with the name that is uttered. For on what is this recognition to rest? The name that is uttered might be wrong; so that, before we could proceed, we should have to show that the obscure idea with which that name was found to be identical, is the same idea we are trying to find? Now this idea we are trying to find is distinguished from others for which we are not looking, by its connexions with remembrances of some qualities or other in the object whose name it is or whose content it signifies; for we cannot try to find the name of something, unless this something can be distinguished from other things which we do *not* mean. When, then, the right name is uttered, the sound of it fits these other remembrances of the object without trouble or resistance, and in its turn calls them up anew or extends them; and this is the reason why it seems to us the right one; whereas any wrong one that is uttered would be foreign to the other ideas that come to meet it. And supposing that the word we wanted to remember were one we did not understand, still there must be some memory or other even of it remaining behind, with which the uttered word must agree; whether it be the number of syllables, or the quality of the vowels, or some prominent consonant, or merely the circumstances in which we heard it, or the momentary general feeling with which its sound was once connected. In none of these cases therefore have we an obscure idea; we are merely looking for the idea which we have not got at all, and helping ourselves in the way I have mentioned. But no idea that we really have, whether simple or complex, can be heightened in the strength with which it is represented; and the complex idea only seems to be so, so long as it is imperfect. No one who thinks of all those ideas of parts which together form the idea of the Triangle, and also of the way in which they are really connected, can further strengthen his activity of representing this complete content. If the geometrician seems superior to the beginner in this point, it is not because he represents this content more, but because he represents more than this content, viz. the innumerable relations which are conjoined with this figure in connected knowledge.

264. I am not rejecting what we all regard as a correct interpretation of the facts, the assumption, I mean, that ideas push one another out of consciousness, and change one another into permanent unconscious states of the soul. For these states we retain a name

which is really self-contradictory, unconscious ideas, in order to indicate that they arose from ideas and are capable, under certain
circumstances, of being re-transformed into ideas. But all that this
assumption actually says is that the ideas have exercised a certain
power against one another, and that some of them have come off
victorious over the rest; it does not follow as something self-evident,
though we naturally infer it, that they must have owed their power
to a degree of strength which belongs to them as such. In fact we
had no means of measuring this strength of theirs at all before the
struggle took place; we only attribute it to them by reasoning backwards after we have seen the issue of the struggle. And further, the
victory does not always fall to that side which in itself is the stronger;
favourable circumstances may give it to the weaker. Since then
this assumption of a variable strength is found to apply not to the
activity of representing but only to the content of the ideas represented: and since on the other hand, if we follow experience, we
cannot maintain that the idea of the stronger *content* always overcomes
that of the weaker, but meet with numberless cases of the opposite
event, the result is that we must look for the source of the power
exerted in something that *attaches* to the representing activity and is
in its nature capable of degrees of intensity.

I may say at once that this power rests neither on any strength in
the activity itself nor on that of the content represented, but on the
amount of our *interest* in the latter. If we could observe the first
stirrings of a soul still destitute of experience, we should certainly find
that that sensation[1] which, in its total effect, is the greater agitation
of the soul and therefore the stronger in respect of its content, overcomes the others which, measured by the same standard, are the
weaker. But in the developed life, which alone we can observe, the
strength of the sensation is of far less moment than that which, in the
connexion of our memories, intentions, and expectations, it means,
indicates, or foretells. Many external stimuli, therefore, are unregarded
by us, if the strong sensations which they would naturally produce
have no relation to the momentary course of our thoughts. Very
slight stimuli attract our attention if they are intimately connected
with these thoughts. And this is still more the case with our
mere remembrances which are unsupported by any present bodily
excitation.

This interest of our ideas, which constitutes their power, has a

---

[1] ['Sinnliche Empfindung' translated merely 'sensation,' to avoid the use of
'sensation of sense,' and 'feeling' which has been reserved for 'Gefühl.']

constant element and a variable one. I cannot suppose that any sensuous impression could be originally entirely indifferent to us. Each, it seems to me, as being an alteration of our existing state, must create an element of pleasure or pain; the former, if it occasions an exercise of possible functions within the limits in which this exercise answers to the conditions of the well-being and continuance of the whole; the latter, if it sets up changes which in their form or magnitude contradict those conditions. The general economy of the vital functions may be assumed to be nearly constant; and therefore, when the impression is repeated at later periods, the same element of emotion will always attach to it, just as the same kind of light-waves, repeated thousands of times in succession, always calls forth the same sensation of colour. But this fixed component of the interest is far outweighed by the variable one which an impression acquires in the course of our life through its various connexions with others, connexions which enable it to recall these others in memory. One impression, which in itself is accompanied by an insignificant constant element of emotion, may, if it is connected with a second, the accompanying emotion of which is strong, excite a more lively interest than a third impression, the feeling of pleasure or pain attached to which comes between the two. But this interest of an impression changes not only with the number of those with which it is connected and with the constant emotion attaching to them, but also with our momentary state of feeling at the time when it occurs. And for this state of feeling the total content of the impression has more or less value, according to the closeness or distance of its relationship to that which is moving our feeling at the moment. If in the case of the representing activity as such it was difficult to point out different degrees of strength, it seems not less self-evident that all *emotions*, on the other hand, have various degrees of intensity. The force of ideas therefore seems to me to rest on their concatenation with emotions; and if I spoke of their strength I should use the word merely to express the fact that they are victorious over others, and the understanding that their victory occurs in this way and in no other.

265. Respecting the connexion of ideas, a point to which these remarks have already led us, we have little to recall. We know that, on the renewal of an idea *a*, another idea *b* which we have had before may return to consciousness without requiring any separate external reason for its reappearance. This fact, which alone can be directly observed, we interpret as a *reproduction* of the idea *b* by the idea *a*,

without meaning by our use of the word to give any account of the process through which *a* succeeds in recalling *b*. But then from this fact we infer that, even in the time during which both *a* and *b* had vanished from consciousness, there must have been a closer connexion between them than is given alike to them and to all other ideas by the fact that they belong to one and the same soul. This specific connexion we call the *association* of the ideas *a* and *b*, a name again which denotes a necessary presupposition but gives no explanation of the exact nature of this connexion, i. e. of that which distinguishes it from the more remote connexion obtaining between all the states of one subject. Any attempt to find such an explanation would be fruitless : but there is another question, which ought to be answered, viz. What are the universal rules according to which this inexplicable junction of ideas takes place ? It is customary to distinguish four kinds of association. Two of them I hold to be fictions of the brain, and the other two I reduce to one. The former consist in the assertions that similar or like ideas on the one hand, and on the other hand opposite ideas are preeminently associated ; and to these assertions I find nothing in internal observation to correspond. I do not know, at least, that the idea of one tone usually recalls all other tones to memory, or the idea of one colour all other colours ; or again, that the idea of brightness suggests that of darkness, or the sensation of heat the remembrance of cold. Where anything of this kind seems to occur, it is plainly due to different causes from the simple association of these ideas as such. If we are calculating, and at a given moment are engaged in comparing quantities and referring them to one another, there is a special reason why the idea of the *plus* we affirm should make us think of the *minus* we reject. In the night we who are busied with plans for the future have abundance of reasons for thinking of the day we long for: and so on in many cases not worth counting up. The third and fourth classes are composed of the associations of those impressions which are perceived either at the same time as parts of a simultaneous whole, or one directly after another as parts of a successive whole ; and their existence is testified to in a variety of ways at every moment of our daily life, the connected guidance of which rests wholly on them. But the separation of these processes into two classes seems to me needless. Not because the apprehension even of a spatial whole takes place, as is supposed, through a successive movement of the glance which traverses its outlines : I shall have later on to mention the reason why this movement is necessary in order to make reproduction secure; but it is none the less indubitable that the momentary illumination of an electric spark makes it possible to perceive objects and gives us images of them in memory. What is of more importance is that in temporal and spatial apprehension it is just the absence of observable connecting links between a and b which joins these two together so closely and in so pre-eminent a degree, that we give the name of association to their conjunction alone, although there must be some conjunction between a and c, b and d, as well. I shall return to this point immediately; but, before going further, I will merely point out how superfluous it is to distinguish from the *indirect* reproduction of one idea b by another a—the case so far considered—the *direct* recalling of the same a by a. We should know nothing whatever of this fact, the reproduction of a former a by the present a, if the two were simply present, with no distinction between them, at the same time. To know the present a as repetition of the former a, we must be able to distinguish the two; and we do this because not only does the repeated a bring with it the former one which is its precise counterpart, but this former one also brings with it the ideas $c\,d$ which are associated with it but not with the present a, and thereby testifies that it has been an object of our perception on some former occasion but under different circumstances.

266. Respecting the great ease with which a successive series of ideas is reproduced in the order of their succession, a fact which it would be superfluous to illustrate, an attractive theory has been developed by Herbart. Let us suppose that the external impressions $A\,B\,C\ldots$ follow one another in time, and that the first of them awakens the idea A; on its appearance in consciousness, which is never empty, this idea A will at once sustain a check from the contents already present in consciousness; and, owing to this check, its strength will have been reduced to a at the moment when the new idea B is aroused. The only association formed therefore will be between a and B—the association $a\,B$—and there will be no association $A\,B$ in consciousness at all. The combination $a\,B$, again, sustains the same check, and will be weakened to the degree $a\,b$ at the moment when C makes its impression C: the association that arises will be $a\,b\,C$, and no other will arise. Again, when D acts, it finds $a\,b\,C$ checked into $a\,\beta\,c$: it is this therefore, and only this, that connects itself with D. If now the series of external impressions, or that of their ideas, is repeated, A will not call up all the rest forthwith, nor will it call them up with the same degree of liveliness, for it never was in actual fact connected with them: not until it itself has sunk to the strength a, will it reawaken B with which alone it was associated; not until $a B$ in its turn has sunk to $a b$, will it reproduce C; and in this way the series is repeated in memory in its original order.

The advantages of this view are not indissolubly connected with the conception, which we were unable to accept, of a variable strength of our ideas. Associations are not formed between those impressions alone which we hold apart as separate *ideas*, each having its distinct content; but every idea connects itself also with the momentary tone G which characterizes our universal vital *feeling*, or the general feeling of our whole state, at the instant when the idea appears; and, as many experiences testify, the recurrence of the general feeling G reproduces with no less liveliness the ideas which were formerly connected with it. But, again, the arrival of a new idea A changes this feeling G into g_1: then the second idea B connects itself with this association $A g_1$, and in its turn changes g_1 into g_2: with this new association, and with it alone, is connected C; and in this way the succession of these $g_1 g_2 g_3$ becomes the clue by help of which the reproduction of the ideas, in their turn, arranges itself; G must be changed again into g before B can be again produced by the association $g B$.

In the next chapter I shall mention other considerations which recommend this point of view to us; I content myself here with the remark that it promises to be of use when we come to consider the reproduction of the component parts of a spatial image by one another. If we assume that the perception of the spatial image $A B C D$ is brought about by the eye traversing this whole successively and repeatedly in various directions $A B C D$, $A C D B$, $A D C B$, ..., the question will still remain, how does it come about that a later consciousness understands the various series, arising from these voluntarily chosen directions, to be merely various subjective apprehensions of the single objective order $A B C D$? If this understanding is to be attained, it will be necessary that, at every step we choose to take within $A \ldots D$, the position of each element relatively to its neighbour should be indicated by a definite general feeling g arising in the course of this movement; and this feeling must be of such a kind that the various g's, which arise in the different directions of the movement from part to part, when compared and adjusted, give as their result these fixed actual positions of the single ideas in the total order $A B C D$. How we are to conceive this process more in detail, I shall show later on.

I close here these brief remarks on the forces which are active in the course of our ideas. I have not noticed the more general share taken in it by the body. Highly significant as that share is, I should seek it in a different direction from the present one. There are no physical analogies either for associations or for reproductions; and although it is asserted that they too are merely products of co-operating nervous currents, those who make this assertion have not yet been able to show, even in a general way, what we should require to have shown,—how these processes can be mechanically construed at all. But this again is a point to which we shall have to return at a later time.

CHAPTER III.

On the Mental Act of 'Relation[1].'

267. If we glance at a number of coins laid side by side in no particular order, each of them produces its image in the eye, and each image produces the corresponding idea. And yet it often happens that, when we look away, we cannot tell how many coins we have seen. That, nevertheless, we have seen each and all of them, and, therefore, that their images have been conscious ideas, we know from the fact that *sometimes* we succeed in counting them over in memory, without needing to have the external impression repeated. This and countless similar experiences convince us that we have some ground for distinguishing between feeling and the perception of what is felt[2]; but at the same time they show that we must not press this distinction further than the statement that the consciousness of the relations existing between various single sensations (among which relations we reckon here the sum formed by the sensations when united) is not given simply by the *existence* of these relations considered as a fact. So far we have considered only single ideas, and the ways in which they either exist simultaneously in consciousness and act on each other, or else successively replace one another; but there exists in us not only this variety of ideas, and this change of ideas, but also an idea of this variety and of this change. Nor is it merely in thought that we have to distinguish that apprehension of existing relations which arises from an act of reference and comparison from the mere sensation of the individual members of the relation; experience shows us that the two are separable in reality, and justifies us in subordinating the conscious sensation and repre-

---

[1] ['*Von dem besiehenden Vorstellen.*' Cp. sect. 80, end. There is no English verb for 'to put in relation;' to 'refer' has been used where a verb seemed indispensable.]

[2] [In this sentence *Empfindung*, elsewhere translated 'sensation' to distinguish it from *Gefühl*, which is translated 'feeling,' 'emotion' (see § 266), is rendered 'feeling,' because we have no verb in English corresponding to the substantive 'sensation.']

sentation of individual contents to the *referring or relating act of representation*, and in considering the latter to be a higher activity,—higher in that definite sense of the word according to which the higher necessarily presupposes the lower but does not in its own nature necessarily proceed from the lower. Just as the external sense-stimuli serve to excite the soul to produce simple sensations, so the relations which have arisen between the many ideas, whether simultaneous or successive, thus produced, serve the soul as a new internal stimulus stirring it to exercise this new reacting activity.

268. The possibility of all reference and comparison rests on the continuance in an unchanged form both of the members which are to be referred to one another, and of the difference between them. When once two impressions *a* and *b* have arisen, as the ideas 'red' and 'blue,' they do not mix with one another, disappear, and so form the third idea *c*, the idea 'violet.' If they did so, we should have a change of simple ideas without the possibility of a comparison between them. This comparison is itself possible only if one and the same activity at once holds *a* and *b* together and holds them apart, but yet, in passing from *a* to *b* or from *b* to *a*, is conscious of the change caused in its state by these transitions : and it is in this way that the new third idea γ arises, the idea of a definite degree of *qualitative* likeness and unlikeness in *a* and *b*.

Again : if we see at the same time a stronger light *a* and a weaker light *b* of the same colour, what happens is not that there arises, in place of both, the idea *c* of a light whose strength is the sum of the intensities of the two. If that idea did arise, it would mean that the material to which the comparison has to be directed had disappeared. The comparison is made only because one and the same activity, passing between *a* and *b*, is conscious of the alteration in its state sustained in the passage ; and it is in this way that the idea γ arises, the idea of a definite *quantitative* difference.

Lastly : given the impressions *a* and *a*, that which arises from them is not a third impression $= 2a$; but the activity, passing as before between the still separated impressions, is conscious of having sustained no alteration in the passage : and in this way would arise the new idea γ, the idea of *identity*.

We are justified in regarding all these different instances of γ as ideas of a higher or second order. They are not to be put on a line with the ideas from the comparison of which they arose. The simple idea of red or blue, as it hovers before us, does not suggest to us any activity of our own which has contributed to its existence ;

but, in return for this loss, it gives us a directly perceptible content. The ideas γ, on the contrary, have no content at all of their own which can be perceived by itself. They are therefore never *represented* in the strict sense of the word, as the simple idea is; never represented, that is, so that they stand before us now as resting perceptible images. They can be represented only through the simultaneous reproduction of some examples or other of a and b, and through the repetition of the mental movement from which they arose.

269. I may look for the objection that this description of the way in which the relating activity proceeds is strange and incapable of being clearly construed. I admit the objection, but I see no reproach in it. It is possible that better expressions may be found, to signify what I mean: my immediate object is to indicate what happens at least with such clearness that every one may verify its reality in his own internal observation. It is quite true that, to those who start from the circle of ideas common in physical mechanics, there must be something strange in the conception of an activity, or (it is the same thing) of an active being, which not only experiences two states a and b at the same time without fusing them into a resultant, but which passes from one to the other and so acquires the idea of a third state γ, produced by this very transition. Still this process is a fact; and the reproach of failure in the attempt to imagine how it arises after the analogies of physical mechanics, falls only upon the mistaken desire of construing the perfectly unique sphere of mental life after a pattern foreign to it. That desire I hold to be the most mischievous of the prejudices which threaten the progress of psychology; and at this point, which seems to me one of the greatest importance, I once more expressly separate myself from views which are meeting now with wide-spread assent: first (a), from the attempts to construe the life of the soul materialistically, psycho-physically, or physiologically, without regard to its specific peculiarities; and, secondly (β), from a view which must always be mentioned with respect, that view of the psychical mechanism, by which Herbart rendered, up to a certain critical point, great services to science.

As to the first point (a), these attempts either persistently pass over the problem whence that unity of consciousness comes, which is testified to by the most trivial exercise of the activity of representation in comparison; or they deceive us by the apparent ease with which single formulas, believed to have been discovered for single psychical events, gives rise in their combination to new formulas, in which even the

desired unity is supposed to be attained. But this whole super-
structure of oscillations upon oscillations, of embracing waves upon
partial waves, this discovery of unities in the shape of points of intersection for different curves,—all this leads to pleasing wood-cuts, but
not to an understanding of the processes they illustrate. Mathematical formulas in themselves determine nothing but quantitative
relations, between the related points which have been brought into
those formulas by means of universal designations. Such formulas,
therefore, subsume the definite real elements or processes, to which they
are applied, under a universal rule; and no doubt these elements or
processes may really fall under the rule in respect of those properties
in virtue of which they were subsumed under it. But the universal
rule in its formal expression no longer reminds us of the special
nature of the object to which it is applied; and thus, partly owing to
the different values given to the quantities contained in it, partly
through its combination with other formulas, a number of consequences can be drawn from it, respecting which it remains entirely
doubtful whether they mean anything whatever when they are applied
to the definite object in question; or, if they do mean anything, what
the actual processes and agencies are which in the real thing lead to
an occurrence corresponding to the result of the calculation. The
first of these two cases I will not discuss further, though examples of
it might be adduced. If we have begun by calling the conditions
under which an effect appears, a threshold, we must, of course, have
something that either passes over it or fails to reach it; and then
these portraits of the deductions drawn from a metaphor easily pass
for self-evident facts. If in a calculation, in which x signifies the
liveliness of a sensation, we come to a negative x, we consider ourselves justified in speaking of negative sensations too. There are
various ways of making mythology: at present the mathematical turn
of imagination seems to take the lead. Respecting the second case
I shall meet with a readier assent. Formulas do not produce events;
they copy them after real causes have created them, and they copy
only individual aspects of them. No coincidence of formulas, therefore, can ever prove that the events which meet or fuse in them, also
fuse as a matter of course in the real thing without the help of any
particular cause to bring about this union. If this cause, without
which the event is metaphysically unintelligible, could be included in
the calculation, and that in such a way that every peculiarity of its
procedure found a precise mathematical expression,—then, and only
then, would these quantities be rightly denominated, and only then

could the calculus securely predict from their universal relations the further consequences which may be drawn.

270. In opposition to Herbart, again (β), I must repeat the doubts I expressed long ago in my *Streitschriften* (I, Leipzig, 1857). When Herbart calls that which goes on in the simple real being when it is together with others, its self-preservation, he raises hopes that in his general view the specific conception of activity will get its rights; a conception which we shall always believe to signify something special and something really to be found in the world, although we find it quite impossible to define what we mean by it, when we oppose it to a mere occurrence, in any way approaching to a mechanical construction. Did we deceive ourselves in this view of Herbart's intention? Ought we to have taken self-preservation for an active form of speech describing a mere occurrence, which, without anything being done by anybody, simply ends, as a matter of fact, with the result that something continues in preservation, the non-preservation of which we should rather have looked for as the probable end of the occurrence?

The further course of the Herbartian psychology would confirm this interpretation. For, according to this psychology, if the soul was ever active at all, it never was active but once. It asserted itself against the stimuli which came from without, by producing the simple sensations: but from that point it became passive, and allows its internal states to dominate its whole life without interference. Everything further that happens in it, the formation of its conceptions, the development of its various faculties, the settlement of the principles on which it acts, are all mechanical results which, when once these primary self-preservations have been aroused, follow from their reactions; and the soul, the arena on which all this takes place, never shows itself volcanic and irritable enough to interfere by new reactions with the play of its states and to give them such new directions as do not follow analytically from them according to the universal laws of their reciprocal actions.

But the limitation of the soul's activity to these scanty beginnings was neither theoretically necessary at starting, nor is it recommended by its results. It was due to Herbart's quarrel with an earlier psychology, with the assumption of a number of original faculties which, doubtless to the detriment of science, were then considered to contain everything necessary to the production of results, whose causes are in reality formed only by degrees and ought to have been made the object of explanation. Here lie Herbart's unquestionable merits, and

I need not repeat that I fully recognise them. But they lie side by side with that which I regret to have to call his error. The mere plurality of these faculties, even the view of them as mere adjacent facts the real connection of which remained unintelligible, could not, taken alone, justify Herbart in going so far the other way as to base the development of the mind upon a single kind of process and the consequences flowing from it. For he himself both knew and said that the simple sensations from which he started are just as independent of one another as were the faculties he rejected; that we cannot conceive any reason why a soul that feels ether-waves as colours must, in consistency, perceive air-waves as sounds; that therefore the soul has just the same number of primal faculties irreducible to one another as of single sensations different from one another. He did not on that account surrender the unity of the soul, or doubt that in it this multiplicity is bound together by some connexion, albeit that connexion entirely escapes us. Now if this one nature of the soul can produce simultaneously, or, so to speak, on the same level of its action, such manifold expressions of its essence, why should it not in the same way produce manifold expressions successively at different periods of its development? Why should not its own internal states, through their increasing multiplicity, win from it new reactions, for which in their simpler forms they gave no occasion? There is certainly nothing impossible in the idea of a constantly renewed reaction, in which that whole essence of the soul that is always present casts new germs of development into the machinery of its internal states; and a view that rejects this source of aid could have proved it to be superfluous only by its own complete success. That I do not find this view everywhere thus successful, I shall have to mention again; here I will refer to three points.

First, the deduction of the perception of Space. I have already spoken of its impossibility and will not refer to it again at length. We must content ourselves with regarding this perception as a new and peculiar form of apprehension, which, proceeding from the essence of the soul, attaches, as a reaction of the kind just described, to a definite manifold of impressions, but does not of itself issue from that manifold. The second point is attention: I shall have to mention it directly in the course of the present discussion. Thirdly, in the case of any act of reference or comparison, Herbart's psychology seems to me to take no account of the eye which perceives the relations obtaining between the single ideas; the consciousness of the investigator which has performed this task of perception everywhere takes the place of the consciousness investigated, which is required to perform it. It is of no avail to answer that it is implied in the very notion of the soul as something that represents, that it perceives everything that exists and occurs in it, and therefore that it perceives the relations in which its single ideas stand to one another: the need of a deduction of the perception of space is by itself sufficient to disarm this rejoinder. For Herbart agrees that the impressions which muster in the simple essence of the soul, are together in the soul in a non-spatial way. A consciousness which as a matter of course perceived their reciprocal relations, could only apprehend them as they are, as non-spatial. But this is not what happens: consciousness changes them and reproduces in perceptions of something side by side in space what in themselves are only together with one another in a non-spatial way. Here then the perception is at the same time a new creation of the form in which it takes place: but even in those cases where there is nothing novel in the reaction to surprise us, the perception of relations is no mere mirroring of their existence, but at the least the new creation of the very idea of them.

271. Expressed in Herbart's terminology, my view would take the following form. The soul is stimulated by the external sense-stimuli s_1, as stimuli of the first order: and in consequence it forms the simple[1] sensations which we know, and to which perhaps the simplest feelings of sensuous pleasure and pain ought to be added as creations which arise with equal readiness. But the various relations (whether of simultaneous multiplicity or of temporal succession) which exist between the sensations or the images they have left in the memory, do not simply exist, they form for the soul new stimuli s_2, stimuli of a second order, and the soul responds to them by new reactions. These reactions differ according to the difference of their stimuli, and cannot be explained from these secondary stimuli themselves, but only from the still unexhausted nature of the soul, which they stir to an expression of itself for which there was previously no motive. Among these reactions we count the perception of Space, which holds a certain simultaneous manifold together; the time-ideas of a change, which are not given by the mere fact of temporal change; lastly, not only these ideas[2] of the kind γ, which measure theoretically the existing relations between different contents, but also among other things, the feelings of pleasure and pain which are connected with these relations. Obviously, on this view, any condition of feeling or any series of referring activities, directed in the way of comparison or judgment to

---

[1] ['Einfachen sinnlichen Empfindungen,' v. note on p. 464.]   [2] [v. §. 268.]

different contents of given ideas, may become in its turn a new stimulus to the soul, an object of a still higher reflexion; but it would be mere trifling to reckon up reactions of a third and fourth order, unless a detailed psychology, for which this is no place, had succeeded in pointing out distinctly in internal observation the processes which would justify us in assuming this ascending scale of orders.   And for the purposes of metaphysic such a course would bring us no further than we are brought already by the recognition, once for all, that the soul is in no case a mere arena for the contentions of its internal states, but the living soil, which, in each instantaneous creation that it brings into being, has produced at the same time new conditions for the generation of still higher forms.

272. There is only one point, therefore, with regard to which I will continue these remarks.   Those ideas[1] γ, the origin of which I touched on above, were, so far as they were then considered, in themselves no more than definite single ideas of a quantitative or qualitative difference, or of a single case of identity.   It is only when we suppose this same referring activity of knowledge to be applied to many repeated cases of a similar kind, that we understand how the *general* ideas of quantity and quality arise in the same way.   As to the origin of universal conceptions generally, we are sometimes told that they arise from our uniting many single examples : those parts of the examples which are like one another are accumulated, those which are opposed cancel one another, those which are dissimilar dim one another.   But this mechanical mode of origination presupposes that the individual ideas, in balancing one another so as to produce the universal, have disappeared and been lost; and the contrary is the fact.   They continue to exist; and it is not out of them that the universal is produced, but side by side with them : it could not be felt at all *as* universal, as something that is true of them among others, if they had vanished and simply left it behind as their production.   The structure of the different kinds of universal conception is very complex, and it is the business of Logic to analyse it.   Psychology can do no more than base their origin on a more or less intricate exercise of the referring activity through which we apprehend the different relations of the constituents which have to be united in them.   The idea produced by this group of activities is not of the same kind as those *ideas*, which, as the direct result of external impressions, represent a perceptible fixed content; it is a *conception*, and the apparently simple name which language gives to it is never more than the expression of

[1] [v. p. 471.]

a rule which we require ourselves to follow in connecting with each other points of relation which are themselves conceived as universal. We can fulfil this requirement[1] only if we allow our imagination to represent some individual example or other, which answers to this rule, while at the same time we join to our perception of this individual the consideration that many other examples, and not this one only or exclusively, can with equal justice be used as the perceptible symbol of that which cannot in itself be perceived.

273. The fact of *attention* still remains to be mentioned. It was depicted by psychologies of an earlier date as a moveable light which the mind directs on to the impressions it receives, either with the view of bringing them for the first time to consciousness, or else in order to draw the impressions already present in consciousness from their obscurity. The first of these alternatives is impossible ; for the supposed light could not search in consciousness for something which is not there : the second at least leaves the obscurity in which the ideas are supposed (without any reason being given for it) to find themselves, very obscure. The necessary complement of this view would lie in the perception that the direction of this moveable light cannot be accidental, but must depend on fixed conditions, and that therefore it must naturally be the ideas themselves that attract attention to themselves. But I think the view I am speaking of was right in regarding attention as an activity exercised by the soul and having the ideas for its objects, and not as a property of which the ideas are the subjects. The latter notion was the one preferred by Herbart. According to him, when we say that we have directed our attention to the idea *b*, what has really happened is merely that *b*, through an increase in its own strength, has raised itself in consciousness above the rest of the ideas. But, even were the conception of a variable strength free from difficulty in its application to ideas, the task which we expect attention to perform would still remain inexplicable. What we seek to attain by means of it is not an equally increasing intensity of the represented content, just as it is, but a growth in its *clearness*; and this rests in all cases on the perception of the relations which obtain between its individual constituents. Even when attention is directed to a perfectly simple impression, the sole use of exerting it lies in the discovery of relations ; it could achieve nothing, and a mere gazing at the object, even if it were heightened to infinity, would be utterly fruitless, if there were nothing in the object or around it to compare and bring into relation. If we wish to tune a string exactly, we compare its sound

---

[1] [On the nature of Universal Ideas, cp. Logic, sect. 339.]

with the sound of another which serves as a pattern, and try to make
sure whether the two agree or differ; or else we take the sound of the
string by itself and compare it at different moments of its duration, so
as to see that it remains the same and does not waver between different pitches. We shall assuredly find no case in which attention
consists in anything but this referring activity; and, on the other hand,
there are moments when we cannot collect ourselves, when we are
wholly occupied by a strong impression, which yet does not become
distinct because the excessive force of the stimulation hinders the exercise of this constructive act of comparison. So closely is the
distinctness of a content connected with this activity that, even after
the eye has repeatedly traversed the outlines of a sensuous image, we
use a new expedient to secure the image in our memory : we translate
its impression into a description, in which, through the aid of the
developed forms of language, the internal relations of the image are
subsumed under the conceptions of position, direction, connexion,
and movement (all of them conceptions of relation), and which prescribes a rule enabling us to re-create the content of the impression
through successive acts of representation or thought.

274. The interest which the idea *a* possesses at a given moment,
has two factors,—the stable value of the idea for emotion, and the
variable significance which this value possesses for our total state at
the particular time. And this interest is the condition which on the
one hand awakes attention and enchains it, and on the other hand
diverts and distracts it. The latter case occurs when the associated
ideas *b, c,* which *a* reproduces, exceed *a* in momentary interest; then
it is that the course of our thoughts moves in those strange leaps,
which we know so well, which we understand in their general conditions, but the direction of which we can seldom follow in any particular instance. It is however in this fact, that the idea *a* is, to a
greater or less extent, able through its associations to attract and bring
back our attention to itself, that the greater or smaller force consists
which it exerts on the course of our ideas; and further, it is in this
that there lies the measure of strength which we are accustomed to
ascribe to the idea as an inherent quality. If *a* has merely served as
a point of transition to the more rapid awakening of other ideas *b, c,*
neither of which reconduct us to *a*, we regard *a* as an idea that was
weak, or that only raised itself slightly above the threshold of consciousness; and alas! by this figure of speech we too often suppose
ourselves to have described the real condition on which the slight
influence of the idea depends. But it is not at the moment when *a* is

passing through consciousness that we rate it as clear or obscure, strong or weak; it is only at a later time, and when other occasions reproduce it and convince us that it must have been in consciousness at a former moment, that it appears to us as an idea that was weak ; and it appears so, because we do not remember any referring act of attention which at that time, by analysing its content, made it strong, or which, by pursuing its relations to other ideas, assigned it a determinate position in the connexion of our inner life. Lastly it is obvious, according to our general view, not only that every activity of attention that has been put forth may become an object to a higher consciousness, but also that there need not be any such reflexion on what has been done. The oftener we have made like relations between a number of points of reference the object of acts of comparison and reference, the more is there connected with the new example, in the manner of a fixed association, the idea of the universal relation under which its relations are to be subsumed. When impressions first occur, we are often unable to connect and judge them without consciously considering how we are to use our ideas in order to do so ; but, when the impressions are repeated, these acts of connexion and judgment frequently take place without any such considerations being necessary : and so we are easily deceived into thinking that in these cases there was really no operation to be performed, and that the mere existence of the relations between the single impressions makes the perception of those relations a matter of course.

My object in devoting this chapter entirely to the referring activity was to emphasize its decisive importance. I may remind the reader that it is really this activity whose delicacy is directly measured by the psycho-physical experiments respecting our capacity for distinguishing impressions, and that all assertions as to the strength of sensations, are, in so far as they rest on these experiments, theoretical deductions drawn from this immediate result of observation.

CHAPTER IV.

The Formation of our Ideas of Space.

275. The concluding remarks of the last chapter may serve to introduce the discussion which is to follow on the psychological genesis of our ideas of space and the localisation of the impressions of sense. In this discussion I must use the freedom claimed by every one who holds, as I do, that our perception of space is merely subjective. In consistency no doubt we should have to consider our own body, as well as the organs of sense by means of which it takes possession of the external world, to be nothing but appearances in ourselves; to be, that is, the ordered expression of a different non-spatial order, obtaining between those super-sensuous real elements, which the all-embracing meaning of the world has made into a system of direct immediate links of connection between our soul and the other constituents of the world. It is not impossible to make this point of view clear to oneself in a general way, and to see that the questions, now to be dealt with, respecting our sensuous commerce with the outer world, might be expressed in the language of that view; but to carry it out in detail would lead to a prolixity as intolerable as it would be needless. Needless, for this reason, that, if the perception of space is once for all fixed by the nature of our mind as our mode of apprehension, this perception has a rightful existence for us, and we can hardly propose to look down upon that which has the power of shedding clearness and vividness upon relations which can be perceived by us only in this way and not in the form they actually possess. It is enough to have assured ourselves at a single point in metaphysic that spatiality is only our form of apprehension, perhaps also a form belonging to every being that has a mind. After it has been shown in a general way how the true intelligible relations of things admit of an ordered manifestation within this form, we may again merely in a general way, subsume this special instance of those relations, the structure of our own body, under that general demonstration; but in the further course of our enquiry we

shall everywhere substitute for the conception of the system of intelligible links of connexion between ourselves and the world that spatial
image of our body which, unlike the conception, can be perceived.
Accordingly, we presuppose here the ordinary view; for us, as for it,
the world is extended around us in space; we and the things in it
have determinate places in it; the actions or effects of those things on
us are propagated in determinate directions up to the surface of our
body, and, passing somehow to the soul, produce in its perception a
spatial image; the component parts of which have the same reciprocal
positions—either exactly the same or within definite limits the same—
as the external things by which they, as sensations, were produced.

276. Owing to the directness of the impression we receive from
the external world, it seems as though the spatial perception of that
world came to us without any trouble on our part, as though we need
only open our eyes to take possession of the whole glory of the world
as it is.   Yet, as we know, and as many experiences at once remind
us, it is not by merely existing that things are objects of our perception, but solely through their effects upon us.   Their spatial
relations, no less than others, come to our knowledge not by the
mere fact of their existence, but only through a co-ordination of their
effects upon us, a co-ordination which corresponds to the relative
position of the points from which those effects proceeded.   And,
conversely, the possibility of correctly concluding from the impression
these effects produce on us to the spatial relations of their causes,
depends on the extent to which those effects preserve their original
co-ordination in being propagated up to the point at which they
impress us.

But here begin the misunderstandings which obscure the way
before us.   Our bodily organs offer an extended surface, on the
various points of which these impressions may be grouped in positions
similar to those held by the points in the outer world, from which
they came.   It is therefore possible for an image to be produced
which has the same aspect as the object whose image it is; and this
possibility has often seemed enough to make all further questions
superfluous.   But in fact it has only doubled the problem.   If it was
not clear how we perceived the object itself, it is no more clear how
we perceive its image; and the fact that one resembles the other
makes matters no plainer.   So long as this image consists simply in
a number of excitations of nervous points arranged in a figure corresponding to the figure of an external object, it is no more than a
copy, brought nearer to us or diminished, of that which *may* be the

object of a future perception, but it does not give us any better rationale of the process through which that thing *becomes* the object of perception.

The question how this fact of nerve-excitation becomes an object of knowledge for the soul at once gives rise to divergent views. We may imagine the soul to be immediately present in the eye : there, as though it were a touching hand, with its thousand nerve-points it apprehends the individual coloured points exactly in the position they actually have in the eye ; and to many this view seems to make everything clear. They forget that it would be just as difficult to show how the feelings of touch which the hand receives justify us in referring the various points apprehended to definite positions in space: before they could do so we should have to presuppose that each position of the hand in space was already an object of that perception which was precisely what we were trying to explain ; then, no doubt, it would be certain that every point of colour lies at that spot in space where the hand apprehends it. Others appeal to the physiological fact that stimulations of the nerves are conducted to the brain by isolated fibres, which may be supposed to lie (where they end in the central portions) in the same order in which they begin in the organ of sense. Thus, it is said, each impression will be conducted by itself and free from intermixture with others, and all the impressions will retain, in being conducted, the same geometrical relations of position which they possessed in that organ. All that this idea accomplishes, again, is to bring the copy, which has taken the place of the distant external thing, rather nearer to the spot where we suppose the mysterious transition of the physical excitation into a knowledge of that excitation to take place.

But how does this come about ? How does the soul come to know that at this moment there is a stimulation of three central nervous points, which lie in a straight line or at the corners of a triangle ? It is not enough that that which happens in these points should have differences in its quality, and produce on the soul an effect corresponding to those differences : but it would also be necessary that the spatial relation of the stimulating points should not only exist, but should also produce an effect on the soul and so be observed by it. Perhaps at this point we might conceive of the soul itself, or of its consciousness, as an extended space, into which the excitations of the nerves might be continued in their original order and direction : and then the whole solution of the riddle would consist in a mere transition. But, even if we supposed the many impressions to have thus really appeared in the soul in exactly that shape in which they came from the external objects, still this fact would not be the perception of this fact. Even if we regarded each of these excitations not simply as the condition of a future sensation, but as a present state of the soul, a conscious sensation, yet, in spite of this, the perception of the relations between them would remain to be accomplished by a referring consciousness, which in the unity of its activity excludes the spatial distinctions holding between its objects. When we perceive the points *a*, *b*, *c*, in this order side by side, our consciousness sets *a* to the left and *c* to the right of *b* : but the *idea* of *a*, through which we thus represent *a*, does not lie to the left, nor the idea of *c* to the right, of the idea of *b* ; the idea itself has not these predicates, it only gives them to the points of which it is the idea. And, conversely, if we still suppose consciousness to be a space, and further that the idea of *a* lies in it to the left of the idea of *b*, this fact would still not be the same thing with the knowledge of it ; the question would always repeat itself, How does the extended soul succeed in distinguishing these two points of its own essence, which at a given moment are the places where that essence is stimulated ; and by what means does it obtain a view of the spatial line or distance which separates the two from one another ?   The connecting, referring, and comparing consciousness, which could perform this task, could never be anything but an activity which is unextended, intensive and a unity—even if the substantive being to which we ascribed this activity were extended.   In the end the impressions would have to pass into this non-spatial consciousness ; and therefore we gain nothing for the explanation of the perception of space by interposing this supposition,—a supposition which in any case is impossible for us to accept.

277. Let us return then to the other idea, that of a super-sensuous being, characterised only by the nature of its activity.   Now it is doubtless incorrect to think of the soul under the image of a point, for, if a thing is non-spatial, its negation of extension ought not to be expressed in terms of space ; still the comparison may be admitted here where we only wish to draw conclusions from that negation. This premised, it is obvious that all those geometrical relations which exist among the sense-stimuli and among the nervous excitations they occasion, must completely disappear in the moment when they pass over into the soul : for in its point of unity there is no room for their expansion.   Up to this point the single impressions may be conducted by isolated nerve-fibres which preserve the special nature

of each impression; even in the central portions of the nervous
system similar separations may still exist, although we do not know
that they do so ; but in the end, at the transition to consciousness, all
walls of partition must disappear.   In the unity of consciousness
these spatial divisions no more exist than the rays of light which fall
from various points on a converging lens continue to exist side by side
in the focal points at which they intersect.   In the case of the rays
indeed the motion with which they came together makes it possible
for them to diverge again, beyond the focal points, in a similar
geometrical relation ; in the present case, on the other hand, the
required continuation of the process consists not in a re-expansion
of the impressions into a real space, but in the production of an idea
—the idea of a space and of the position of the impressions in that
space.   This perception cannot be delivered to us ready-made.   The
single impressions exist together in the soul in a completely non-
spatial way and are distinguished simply by their qualitative content,
just as the simultaneous notes of a chord are heard apart from one
another, and yet not side by side with one another, in space.   From
this non-spatial material the soul has to re-create entirely afresh the
spatial image that has disappeared ; and in order to do this it must be
able to assign to each single impression the position it is to take up
in this image relatively to the rest and side by side with them.   Presupposing then, what we do not think need be further explained, that
for unknown reasons the soul can and must apprehend in spatial
forms what comes to it as a number of non-spatial impressions, some
clue will be needed, by the help of which it may find for each impression the place it must take, in order that the image that is to arise in
idea may be like the spatial figure that has disappeared.

278. We may illustrate this requirement in a very simple way.
Let us suppose that a collection has to be arranged in some new
place in exactly the same order that it has at present.   There is
no need to keep this order intact during the transport ; we do whatever is most convenient for the purposes of transport, and when it is
finished we arrange the several pieces of the collection by following
the numbers pasted on them.   Just such a token of its former spatial
position must be possessed by each impression, and retained throughout the time when that impression, together with all the rest, was
present in a non-spatial way in the unity of the soul.   Where then
does this token come from ?   It cannot be the point in external space
from which the sense-stimulus starts, that gives to it this witness of its
origin.   A blue ray of light may come from above or from below, from the

right or from the left, but it tells us nothing of all this ; it itself is the same in all cases. It is not until these similar stimuli come in contact with our bodies that they are distinguished, and then they are distinguished according to the different points at which they meet the extended surface of our organs of sense. This accordingly may be the spot at which the token I am describing has its origin, a token which is given along with the stimulus in consequence of the effects produced by it at this spot, and which in the case of each single stimulus is distinct and different from that given along with any other stimulus.

And now that fact regains its importance, which we could not admit as a short-hand solution of these problems ; the isolation of the conducting nerve-fibres. I cannot help remarking in passing that physiology is mistaken when it finds the exclusive object of the structure of the nervous system in the unmixed conduction of individual excitations. In the optic nerve we find this structure devoted to that purpose ; but the olfactory nerve, which possesses it no less, shows very little capacity for arousing such a multiplicity of separate sensations as would correspond with the number of its individual fibres. Nor is it only in the nerves that we meet with these elongated unramified fibres ; we find them in the muscles also, and yet the isolated excitation of a single fibre certainly cannot be the object here, where the simultaneous and like stimulation of many fibres is required for the attainment of any useful result. Thus we must suppose, I think, that the wide diffusion of this structure of the fibres has a more general explanation. Perhaps their forms were the only ones possible to the forces which shape an organic form, and a foundation for greater effects may have been producible only through adding together such elementary organs. Perhaps again the physical processes, on which the activities of life rest, are necessarily connected, within narrow limits, with the fineness of the fibre, and could not take place in masses of a thickness discernible to the naked eye. But however this may be,—however this structure came into being,—when once it is present, it can without doubt be used for the purpose of separating the impressions of sense. Each single fibre, at the spot where it receives the stimulus, can attach to it the extra-impression described, and can transmit it to consciousness, stamped with this character, and preserved by the isolation of the fibre from mixture with other physical excitations.

279. A further assumption is necessary before we can make use of this process to explain the localization of impressions. We must suppose that similar stimuli give rise in each nerve-fibre to a special

extra-impression, an extra-impression which is different in the case of every single fibre, and which connects itself, in the manner of an association, with that main impression which depends on the *quality* of the stimulus,—connects itself, therefore, in such a way that neither of the two impressions, the main one and the extra one, interferes with the peculiar nature and tone of the other. It must be confessed that we have no anatomical knowledge of a diversity in the single nerve-fibres so manifold as this assumption requires. But this diversity may consist not only in properties which escape all the expedients our external observation can employ, but in the very spatial position of the fibre; we might suppose, that is, firstly, that in a number of fibres lying side by side interactions take place which produce different states of susceptibility in the fibres lying at different spots in this system; and, secondly, it is no less possible that the excitations of each fibre may acquire a particular tone from the effect produced on it by occurrences in the surrounding tissues. But this question of detail, again, we must leave undiscussed; what is certain is that no other view of the matter can dispense with an assumption similar to that for which we have suggested an explanation. In order to know whether a push we felt when our eyes were shut came against our hand or our foot, it is necessary that, the two pushes being in other respects of equal measurement, the total impression should be different in the two cases. In such a case it is of no use to appeal to associations, and to say that on a former occasion the impression of the push was connected with a simultaneous visual perception of the place that received it, and that now when the push is repeated it reproduces this perception. For in the course of life we unfortunately so often receive pushes on all parts of the body, that the impression in question will have associated itself almost indiscriminately with the images of all of them: it will be impossible, therefore, in the case of a repetition to decide to which of these parts the impression is to be referred, unless in this new case the impression itself once more tells us to which of them we are to refer it: it is necessary, in other words, that the impression now recurring should be provided with a clear token of its present origin. Let $A\,B\,C$, then, stand for three diverse stimuli, $p\,q\,r$ for three different spots in an organ of sense, $\pi\,\kappa\,\rho$ for three specific extra-impressions, which those spots connect with the main sensations occasioned by $A\,B\,C$: then the difference between these connected *local signs* $\pi\,\kappa\,\rho$ will be the clue by means of which the sensations falling upon $p\,q\,r$ can be localised in separate places in our perception of space. The associations $A\pi\,A\kappa\,A\rho$ will signify

three *similar* impressions which have fallen on the *different* spots $p\,q\,r$
of the organ of sense, and which are prevented by this very difference
in their local signs from being fused into one sensation, a fusion
which could not have been prevented if the three A's had been perfectly
identical [1]; since, where no distinctions exist, no activity of consciousness can make them.   The associations $A\pi\, B\kappa\, C\rho$, on the other hand,
will signify three *dissimilar* impressions which affect those three
different spots in the organ at the same time ; these impressions, owing
to their qualitative difference, need nothing further to prevent their
fusion into one sensation, and all that $\pi\,\kappa\,\rho$ give them is their spatial
arrangement.   Lastly, $A\kappa\, B\kappa\, C\kappa$ would be the same three *dissimilar*
stimuli, acting on one and the *same* spot q in the organ of sense, and
therefore, as we seem obliged to suppose, appearing *successively* at
the same point in our perception of space.

280. I have no desire to conceal the difficulties which arise when
these considerations are pursued further.   As long as we abstain
from considering the differences between the organs of sense, and
only try to fix in a general way the requirements we have to satisfy,
it is possible to form many different views respecting the nature
and genesis of local signs.   The simplest would be one I have mentioned in passing, that the extra-process destined to accompany
the main impression takes place directly, and at first as a physical
excitation, in the same nerve-fibre which is affected by an external
stimulus.   In that case it would depend on the form of excitation
which was found to be the general mode of activity in the nerves,
whether it could permit the simultaneous conduction, without intermixture, of two different processes: and even if we found that this
was not possible in the case of two excitations of the same kind, still
it might be so when one of the two processes was the extra-excitation
which accompanies the main movement issuing from the stimulus,
but is not of the same kind with it.   However, the whole supposition
of a double conduction fails to attain its object.   For it involves the
tacit presupposition that two processes, which, without being otherwise connected, proceed along the same nerve-fibre, thereby acquire a permanent association : and this presupposition rests in the
end on a mode of thought which we were unable to accept, on
the notion, I mean, that the mere fact of the excitations existing side
by side in space is sufficient to give rise to the idea of their intrinsic
connexion.   On a former occasion I compared consciousness, by
way of figure, to a single vessel; and the various excitations, which

[1] [I. e. if $A\pi$, $A\kappa$, $A\rho$ had been simply A, A, A or $A\pi$, $A\pi$, $A\pi$.]

were conducted to this vessel and flowed into it through different pipes, were supposed at last to meet in it and mix indiscriminately together. We need not keep to this figure; but however we like to picture the transition into consciousness, the mere fact that A and π B and κ, C and ρ *were* together in space, will give consciousness no token that it is to connect them exactly in this way instead of joining A to κ, B to ρ, and C to π.

Accordingly, the other supposition seems to me more natural than this of a double conduction,—the supposition that the main impression and the extra-excitation really give rise in each fibre to one total state, which is conducted as a total state and occasions nothing but one total sensation. If this sensation remained by itself alone, we should feel no occasion to distinguish different elements in it, any more than violet, if we knew nothing but it, would suggest to us to separate red and blue from one another in it. But many different stimuli are in process of time connected with one and the same extra-process; and it may be that the comparison of these cases would arouse an activity of separation which would analyse the total impressions into their component parts, but which would at the same time learn to refer the local sign, thus separated, in each case to that qualitative impression from which it was parted in thought and in thought alone. It is possible to find instances in which this actually occurs. The effect produced by a tone must be apprehended as an excitation which is at first one and total, in the sense above described: but not only does the comparison of many successive tones enable us to distinguish the quality from the height of each, but further, when we hear several tones at once, we are able to attach each quality to the height of the note from which it was thus separated. The artificiality of this point of view may make us distrust it, but we shall not find it easy to escape this artificiality by taking another road.

Let us put aside altogether, what is quite unessential, the image of the soul as a point at which all the impressions conducted to it discharge themselves. Let us suppose, what we shall afterwards find confirmed, that the soul perceives the physical nervous processes directly at the spot where they reach the final form in which they are destined to be objects of its perception. Still there remains the question we would so gladly avoid: supposing the soul has apprehended many impressions in this way, either at the same time or one after the other, it can analyse each of them into the components described above; what determines it then not to allow these components to fall asunder but to hold them together in a way corresponding to the connexion from which it has previously disjoined them ? We see, then, that the artificiality lies in the fact itself, or in that view of the fact which alone remains open to us now,—in the enigmatical nature of associations generally. I reminded the reader in a former passage that in using this name we are merely designating a fact we are obliged to assume, without being able to give any account of the means by which it is brought about. And now it seems to me that the source of the doubts that beset us here is that we cannot persuade ourselves to renounce our search for a mechanism which would bring about this connexion of states after the analogy of physical processes. Such a construction we shall certainly never find. On the other hand, if we are tempted to regard associations as a peculiarity of the psychical activity, to which there is no analogy elsewhere, we are held back by the undoubted fact that we do not yet possess any intelligible general point of view which would exhibit the *ratio legis* in every case, and which would explain not only the connexion at this point but its absence at that point : instead of this we are forced in each particular case to make assumptions which appear artificial because they are always constructed *ad hoc*. I believe then that the hypothesis I have been speaking of here, that of the origin of the local sign in the stimulated nerve itself, might be maintained ; but, later on, I shall substitute for it another hypothesis, according to which the extra-production of the local sign is less direct : and my reason is not that the second hypothesis is free from the difficulties of the first, but that it offers other advantages. At present we will proceed for a moment with our general remarks.

281. If the local signs $\pi \kappa \rho$ merely differ generally in quality, it is true that they would suffice to prevent three perfectly similar stimuli from coalescing, and to make them appear as three instances of the same felt content. But the only result would be an impulse to hold the sensations apart in a general way ; there would be nothing to lead us on to give to the sensations thus produced a definite localisation in space. It is this that is left unnoticed by those who regard the isolated conduction of three impressions by three fibres as a sufficient reason, taken by itself, for their being perceived as spatially separate. Even if (in the absence of the extra local signs) this isolation were a sufficient condition of the three impressions being distinguished as three, yet the question whether they were to be represented at the corners of a triangle or in a straight line, could only be decided by a soul which already possessed that capacity of

localisation which we are trying to understand. In this case the soul would stand, as it were with a second and inner vision, before the open key-board of the central nerve-terminations, would see them lying ready side by side, and doubtless would very easily refer the arriving excitations to the places occupied by those keys on which they produce some observable motion. If this is impossible, as it is, just as little would it be possible for the local signs given along with sensations to produce a real localisation of the sensations, if these local signs simply differed without being also *comparable*. If they are to lead to this localisation they must necessarily be members of series or of a system of series, in each of which there must be some general characteristic in common, but within its limits a difference, measurable in some way, of every individual from every other. If

$$\kappa = \pi + \Delta, \rho = \pi + 2\,\Delta, \text{ or } \kappa = \rho - \Delta,$$

then, but only then, can these signs be the reason why a perception, which can and must apprehend these arithmetical differences in some spatial way or other, should place $B\kappa$ nowhere but in the middle between $A\pi$ and $C\rho$. And if more than one series of this kind is involved, so that the general character of the local signs in the one is qualitatively distinguished from that of the other, still even in the transition from series to series this alteration of quality must somehow proceed by measurable differences; otherwise we should not know how great, in terms of space, is the declination of some of the impressions from the straight line which is the shape others are to take in the perception.

282. This postulate is closely connected with the settlement of another question. Wherever the local signs may arise, there is no doubt that, to start with, they are physical excitations which arise, on occasion of the stimulus, in the stimulated spot, this spot having an individuality or special nature of its own. We have gone on to assume as self-evident, that they then produce sensations, states of consciousness, just as the main impressions do which they accompany; and that, from a comparison of the associations which have thus arisen, a referring activity decides what relative position each impression is to take among the rest. Now is this necessary? Or does it suffice to regard $\pi \kappa \rho$ simply as physical processes which do not themselves appear in consciousness, and merely determine the direction in which consciousness guides each impression to its place in the perceived space? Now, supposing we adopt the second of these alternatives, the difficulty remains that it will be just as necessary for the unconscious faculty of localisation as it was for the conscious,

that the local signs should stand in the reciprocal relations we have indicated; otherwise this faculty will have nothing to determine it to the definite directions spoken of. This will at any rate be necessary if we are to hold in this case to that general rationality of the phenomena which alone gives any interest to attempts to explain them : for of course it is possible to take a purely fatalistic view, and to say, It simply is the fact that, if the spot p is stimulated the ensuing sensation must take the place x, and if the spot q is stimulated the ensuing sensation must take the place y; and there is no *rule* or reason why the existence of one of these relations should involve that of the other. On this assumption any further hypotheses as to the nature of the local signs would be superfluous : but then on this assumption *all* investigation would be superfluous, for there would be nothing to investigate. If however that general rationality of phenomena is admitted, then I find no sufficient clearness in the theory that our determination of place in perception is conducted unconsciously. For this reason: according to the theory there is something which determines the position to be given to each single impression in the space perceived. This something, this ground of determination, must remain conjoined with this single impression and with it alone (for it holds good of it and of no other impression). It cannot be merely a prior process determining the future localisation ; it must be a permanent definite mark attached to that idea whose localisation it is to further. And, since the idea now appears in consciousness, it is difficult to imagine how the grounds of determination can leave such an after-effect attached to the idea, as would operate in consciousness and yet not *appear* in that same consciousness.

Here again there lie more general difficulties which interrupt our course. It is, once more, because we are accustomed to observe the external world, that we naturally separate any occurrence produced by causes into a preceding impression on the one side, and a subsequent reaction on the other. In a chain of processes, in which each link is the sufficient reason only of the next, we may make this distinction between the first link a and the last link z; but it is useless to interpose between the next neighbours a and b another impression, which it is supposed that a must have already made before it can call forth b as a reaction. We are separating what is really a unity, the occurrence which is at once reception of an impression and reaction against it ; and it is this false separation which in the present case makes it seem natural that the external stimulus should first produce in the soul an impression which is not yet consciousness, and that the conscious sensation should afterwards follow on this impression as a reaction. But it is easy to see that this interposition can be carried on *ad infinitum*. On such a view, the activity of sensation, in its turn, could not react in consequence of the unconscious impression till it had been stimulated by it—if, that is, the impression had produced in it a second unconscious state: and it would be only to this second stimulation that the activity of sensation would respond with its conscious manifestation. On this point I accept Herbart's opinion: a conscious idea is *directly* an act of self-preservation against a disturbance. This disturbance does not first appear apart, and then call forth the idea as a reaction. The disturbance only threatens, its threat is only effective, it itself only exists in so far as it asserts itself in the idea itself which, but for it, would not have existed. But I will not pursue these doubts. They cannot be definitely set at rest. We have assuredly no right to interpose some mere lifeless impression between two adjacent links of a causal connexion: but still it remains undecided whether, as a matter of fact, the physical excitation in the nerve, and the psychical process of sensation, do form such adjacent links of a chain. It is not necessary that the sufficient ground for the arousal of a sensation or idea should consist in the connecting link of an unconscious state of the soul; but it is possible that it may consist in this. Accordingly I do not put forward my view as anything more than the hypothesis that I prefer. It may be stated thus: if the physical processes $\pi \kappa \rho$ are the local signs directly used by a referring activity when it determines the position of the sensations in the perceived space, they are so used not as physical processes, nor through the instrumentality of unconscious impressions aroused by them in the soul, but in the shape of conscious sensations resulting from them. I shall return to the objections which stand in the way of this supposition, and consider them in detail.

283. These then are the general postulates to which the local signs have to conform. And it is these postulates alone that I regard as a necessary metaphysical foundation for our spatial perceptions. Shortly expressed, they come to the one requirement, that all the spatial relations of the stimuli acting on us should be replaced by a system of graduated qualitative tokens. In adding some instances in fuller detail I am quite aware of the many abiding difficulties which could only be removed by an accurate consideration of all the experience that is available to us, or that may become so. Nothing but experience can disclose to us the means by which the local signs we require

are really produced; and I do not think this production takes place
in the same way in the case of the two senses which have to be
considered.

In the first case, that of sight, the first of the suppositions mentioned appeared to me improbable, I mean the supposition that the
local signs arise directly in the spot stimulated. Even supposing
that the same kind of light L, falling on various points of the retina,
produced sensations of colour somewhat differing from each other, C
in the point p, and c in q, still there will always be another kind of
light l, which occasions in q that same sensation C which L excites
in p. Accordingly it cannot be this difference of quality in the
impression that gives the reason for referring that impression to
a definite spot p or q. On the other hand, there seemed to me to be
a real importance in the fact that, from the yellow spot on the retina
—for our purposes let us say, from the central point E of the
retina—where the sensitiveness is greatest, there is a gradual diminution of irritability in all directions, until at the edges of the hemi-
spherical distribution of the nerves this irritability entirely disappears.
This fact, again, taken by itself is not sufficient for our purpose: for
a weak stimulation of a spot lying near the point E would necessarily
have the same effect as a stronger stimulation of a spot at a greater
distance from E. But if a stimulus in the way of light falls on one
of these side-spots p, it also makes the eye turn to such an extent and
in such a direction that the ray meets, instead of p, the point of clearest
vision E. This direction of the glance, as it is commonly called, is
accompanied by no idea of the end it actually serves, or of the means
by which it is brought about. It must therefore be regarded, at any
rate originally, not as an intentional act, but as an automatic movement, a physical effect due to the stimulus and unknown to the soul.
Accordingly the following hypothesis seemed to be admissible : in the
central organs the single fibres of the optic nerve are mechanically
connected with the motor nerves of the muscles of the eye in such
a way that the stimulation of each of the former is followed by
a definite excitation of the latter, from which it results that the eye is
turned in a particular way. How this mechanical connexion of the
sensory and motor nerves is effected, is a question which does not
touch our present object; and the settlement of it may be left to
Physiology, which has to raise the same question in regard to many
other reflex motions.

284. The motions just described would satisfy the requirements to
be fulfilled by the local signs. If p is the point stimulated, $p E$ would

be the arc which has to be traversed in order that the point of clearest
vision E may be stimulated instead of p; if q is stimulated, the corresponding curve is $q E$; these motions will be different in every case,
but the difference between them will be merely one of magnitude and
direction.   But then, on my hypothesis, it was not these motions
themselves, but the sensations excited by them, which were to be
directly used as the signs $\pi \kappa \rho$ of the spots $p q r$.   Now a movement,
in occurring, occasions a sensation or feeling of our present state,
which is different from the feeling of the non-occurrence of the movement: and we even when at rest distinguish the momentary position
of our limbs, produced by former movements, from that position
which is not now present: these are facts which need no proof, however simple or however complicated may be the conditions which give
rise to these feelings. But a further assumption is necessarily involved.
We must suppose that the perceptible differences of the feelings in
question correspond in their turn to the slightest differences of those
movements which the eye needs in order to turn its glance from one
point of the field of vision to its next neighbours: and this hypothesis
may arouse graver doubts.   These doubts, however, really apply, I
think, only to a point which is of no decisive importance here.   No
doubt, as a matter of fact, we notice those minimal movements, which
the glance has to make in passing from one point of the field of
vision to the next point, and from that again to the next; but to our
immediate *feeling* they seem merely a greater or smaller alteration of
our state, a greater or smaller degree of a change which does not alter
its character. We cannot here, any more than in the case of our other
sensations, reduce the magnitude of these steps to comparable arithmetical values, so as to judge that one of them is double or half as
great as another.   The reason why this becomes possible is that the
movements described bring a number of distinguishable points one
after another to the spot of clearest vision, and the images of these
points, instead of at once disappearing again, remain for sensation
side by side with one another: and it is only the number of these distinguishable points which enables us to interpret the differences in
our feelings of movement as expressive of equal or unequal spaces
traversed, or of definite differences between these spaces.   Thus if the
eye were shut or did not see, it would doubtless be aware, from the
immediate feeling of movement, that the curve $p E$ is smaller than the
curve $q E$ (which it would describe if it continued the same movement), and that $q E$ is smaller than $r E$; but these feelings would not
enable it to determine the co-ordinates of that point x in the field of

vision which would meet its glance if it were opened or began to see. It is only the series of images which pass before the seeing eye while it moves, and which remain side by side for some time so that they can be compared, that enable us to give an accurate quantitative interpretation to the different sections of a series of feelings of movement. If we follow with our eyes from beginning to end a line of one colour drawn before us, doubtless we are conscious of a continuous and homogeneous movement of the glance; but suppose there is a stroke drawn across the line near the beginning, marking off a small part of it, we cannot guess how many more fractions of the same size the rest of the line will contain: it is only by marking them off that we can tell their number and be sure that they are equal. How is it again that we learn this last fact, the equality of the distinguished parts? Is it by keeping the head fixed and turning the eye in such a way that these parts of the line, from a to s, are brought one after the other into the direction of clearest vision? And do we then judge that the movements bc, cd, de, up to ys, in each of which the eye starts from a different position, and which really would not be equally great, *are* equally great, and therefore that the parts ab, bc ... ys are also equal? We cannot ascertain their equality in this way. Any attempt to do so accurately is really made thus: in looking at the starting-point a, b, c of each line ab, bc, cd we place the eye so that the direction of its glance forms in every case the same angle with the direction of the piece to be judged, e.g. a right angle: the movements which the eye has then to make in order to go from a to b, b to c, y to s, are not only equal in magnitude (supposing the lines to be equal), but they are identical, since the position from which they start is in each case the same, and the position in which they end is in each case the same. If, on the other hand, the lines are unequal, one of the movements is readily felt in a general and inexact way to be smaller or greater than another, since the position of the eye, at any rate at starting, is the same in each case.

Thus, as with all sensations, our original capacity of estimating impressions quantitatively would (apart from the results of practice) rest on the possibility of generally recognising what is exactly like *as* like, and what is different *as* different. And I do not think that for our purposes any more delicate sensibility is required. I do not mean that the two local signs $\pi = pE$ and $\rho = rE^{1}$ would enable the soul forthwith to set the two sensations A and B connected with them at

¹ [The letters on the right hand stand now not for the movements themselves but for the feelings answering to them.]

definite points in a circular field of vision: it suffices that these signs
secure to the impressions their positions in relation to one another;
that, for example, they make it necessary to set *B between A* and *C*
and nowhere else. With these explanations as to details, I think we
may hold to the theory that the feelings of movement $\pi \kappa \rho$ are the
direct local signs of the sensations. But each of these feelings themselves is at bottom a series of momentary feelings of position answering
to the various places traversed by the eye in its movement. In order
to keep the signs as simple as possible I merely mention this here, and
shall use π to indicate the whole series of the successive sensations
$\pi_0, \pi_1, \pi_2 \ldots$, which follow each other as the eye turns along the
curve $p E$.

285. The further application of these ideas will be as follows. If
we assume that the first impression of light felt in our lives affected the
lateral spot p, it will follow that there succeeded an actual movement
$p E$, and that, during this movement, there took place the series π of
successive feelings of the position of the eye. If the *same* impression
is repeated, the same movement will ensue; and the fact that an
identical stimulation has occurred in the past will make no difference
to the present one. But the case will be otherwise if at the moment
of their second stimulation *another* stimulus affects the spot q, and
solicits, with a force equal to that of the first stimulus, a movement of
the eye directly opposite to that which is required by p. The result •
here will be that the eye remains at rest: but at the same time the
two impulses to movement, which in their effects cancel one another,
will not on that account be a mere zero; as excitations of the nerves
they will remain, just as the force of gravity in two masses remains,
although those masses counterbalance one another in the scales and
therefore do not move them. The operation of that force consists in
the bending of the beam and in the pressure exerted on the point of
suspension. And I see no reason why, in the case before us, the two
excitations, which are prevented from producing an effect in the way
of movement, should not still be represented in the soul by two definite
feelings, so that the equipoise of opposed forces would be something
different from the repose due to the mere absence of excitation.

No doubt, if this is so, we must once more reform our idea of π or
κ. So far we have regarded them as feelings which arise from the
movement set up; thus they will not occur unless the movements do.
But I do not doubt that the stimulation of the spot p, apart from the
actual movement connected with it, can arouse a feeling by its mere
existence and occurrence, and that by means of this feeling the

presence of a thwarted impulse may be indicated to consciousness and so distinguished from the mere absence of the impulse. This feeling we should now regard as the first link π_0 in that series $\pi = p\,E$, which is produced during the movement $p\,E$, and of which each link r_m will now stand for a momentary feeling of position and also for the momentary remnant of a thwarted impulse to movement. Now, taken by itself, π_0 will be simply a feeling, a way in which we are affected, and it will not of itself point to its causes or its possible effects. But then in that first experience the whole further series π connected itself with the first link; this series is associated with π_0 and, on the repetition of π_0, it also will be reproduced. Accordingly, though there is no movement of the eye, there arises the recollection of something, greater or smaller, which must be accomplished if the stimuli at p and q, which arouse only a weak sensation, are to arouse sensations of the highest degree of strength and clearness. This is what happens at first; but if the soul has learnt that the movements of the eye, reported by its feelings, *are* movements,—are, that is, alterations of the relation in which the organ of sensation stands to a number of what may be treated as fixed simultaneous objects; and if finally the soul both can and must apprehend the differences between such relations in a spatial form,—in this case the idea of that something to be accomplished will be transformed into the idea of a greater or smaller spatial distance between the impressions falling on p and q and that middle point of the perceived space which corresponds to the point E in the eye. If, lastly, we add that to each of the many stimuli which at one and the same time excite the spots $p\ q\ r\ \ldots$ of the retina, there is now conjoined the corresponding series $\pi\ \kappa\ \rho$ of reproduced feelings, the result will be that owing to movements once performed and now remembered, the eye, even when at rest, will be able to assign to each impression its position among the rest.

286. I should be very prejudiced if I felt no alarm at the artificiality of these ideas. But my intention was not to recommend the hypothesis at all costs, but honestly to recount all the presuppositions it involves; and, further, I do not know that it is possible to reach the end we aim at in any simpler way, or that the artificiality lies anywhere but in the facts themselves. The fact itself is strange enough—and it cannot be got rid of—that we can see an unnumbered mass of different-coloured points at once, and can distinguish them. It must be possible, therefore, that what we require should be effected: it must be possible for a large number of impressions

to be in consciousness without mingling together; there must be
in each of them something, some 'reason,' which makes it appear
now at one point in space, and now at another point; and these
various 'reasons' again, which are present simultaneously, must
operate without intermixture, each of them in exclusive relation to the
definite impression it belongs to. In other words, the same complicated relations which we assume between the feelings of movement,
must exist between any other possible elements which we might
substitute for those feelings. The only question, therefore, is whether
internal experience witnesses to the truth of our hypothesis, or
whether any other source of knowledge opposes to it objections
which are insuperable.

As to the first point, of course, I cannot tell whether others find in
themselves what I find in myself. If I ask what meaning an im-
personal knowledge (if the phrase may be used) would attach to the
words 'two elements p and q are at a distance from one another,'
I can imagine an answer by means of the idea of a universal space
in which I myself have no fixed position. But for my sensuous
perception of the seen points p and q, the only possible meaning of
the statement that these points are at a distance from each other
is that a certain definite amount of movement is necessary if I am to
direct my glance from the one to the other; the different positions
of the single points are felt by me simply and solely as so many
solicitations to movement. But then I can base nothing on this
experience. My individual disposition cannot be communicated.
I cannot therefore contradict those who tell me that they observe
nothing of these feelings of movements, however much I may be
convinced that they deceive themselves and, though they really have
the feelings, do not recognise them for what they are. I must content
myself therefore with pointing out to them that, in my view, the spatial
perception of the world is not something suddenly given us by nature
as soon as we open our eyes, but is the result of successive experience
and habituation; only this habituation goes on at a time in our lives
of which we have no distinct recollection. The skill of the piano-
player, once acquired, seems to us a natural gift that costs no trouble;
he glances at the notes, and complicated movements of the hand
immediately follow: in this case we know what a laborious process
he has gone through, and with what difficulty practice has set up
these associations of ideas with one another and with the movements
we see,—mere links of connection which no longer show themselves
in the consciousness of the practised artist. Exactly the same thing

may happen in the case under discussion; and there need be no distinct recollection in consciousness of the actual movements through which we once learnt to localise our sensations. But, it will be answered, this may be a probable account of the slow development of a child, and as a matter of fact we see that its eyes turn towards any light that is brighter than usual: to an animal on the other hand the spatial knowledge of the world comes with so little trouble that we cannot in its case believe in such a prolonged process of learning. To this I reply that in reality we do not at all know what it is that an animal sees directly it is born, nor what sort of perception of space it has. In order merely to account in general for the early use it makes of its limbs we have to assume a number of mechanical reflex movements. It is therefore conceivable that the unhesitating way in which it makes for an object lying in the direction of its glance may really rest merely on a reflex movement set up by the stimulus; and the fact that many of its other earliest movements are unsuccessful would then go to show that it, like man, only gradually acquires an ordered knowledge of that remaining part of the spatial world which lies outside the direction of its glance. Again, the small amount of experience we possess respecting the rise of an optical idea of space in persons born blind and afterwards operated on, will not suffice to decide the question. In all cases the patient has already learnt, through touch and movement, to find his way in the spatial world. Doubtless the ideas of space thus developed may be very unlike the space that manifests itself to a man who can see: for a touch can apprehend only a few points at once, and can only approach distant objects by means of considerable movements; and therefore the space of the blind man may be not so much what we mean by space, as an artificial system of conceptions of movement, time and effort: and, as a matter of fact, the few reports we possess tell us of the astonishment with which the blind man, after a successful operation, learns what the appearance of space or the spatial world is. Still, in spite of such differences, we cannot tell to what extent this previous practice may assist the formation of the visual perception which ensues: in any case it cannot be analogous to the first formation of *all* ideas of space; and finally, there is even a difficulty in discovering what it really is that is seen at first, since the patient who is just beginning to see, cannot express his first experience in the language of sight.

287. There are many questions which psychological optics would have to settle respecting the further development of the spatial ideas:

but it is not the business of metaphysic to discuss them. I will only briefly remark that there is no foundation for any of those views which ascribe to the soul an original tendency to project its impressions outwards, and that in one particular way and in no other; all this has to be learnt through the combination of experiences. How it is actually learnt piece by piece we cannot discover; how it *may* be learnt, it is easy to understand in a general way; but there are particular points in the process which cannot at present be understood at all. What we have accounted for so far is nothing more than the arrangement of the points in the field of vision, the internal drawing of the total image; but this image itself as a whole has as yet no place and no position, for the perception of the total space, in which its place and position are to be, is still entirely wanting. The movements of the eye as it opens, shuts, and turns, make the seen image appear, disappear, and change. We therefore naturally associate this image with the eye in such a way that we conceive it as lying in any case in front of us—to use the later language of the developed perception of space: what is behind us—an expression which at this stage has really no meaning—does not exist at all, and has no more to do with space than the general feeling we have in the hand or foot has to do with clearness or dimness. And so it would remain, if we could not move our bodies and could only turn our eyes to a very slight extent. But as soon as we have learnt to turn on our axis and to refer the consequent feelings of movement to their true cause, the movement, we discover that our first field of vision $a\,b\,c$, instead of suddenly disappearing altogether, passes successively into $b\,c\,d$, $c\,d\ldots x\,y\,z$, $y\,z\,a$, $z\,a\,b$, and $a\,b\,c$. The unbroken series of images which returns into itself awakes in us the idea of a complete circular space with no gaps in it; and this idea, by the help of similar movements of the eye in other directions, soon passes into the ordinary perception of the spherical space that surrounds us on all sides.

At the same time, this idea could neither arise nor attain any clearness unless the idea of the third spatial dimension, that of depth, were being simultaneously formed. In its own nature the soul has certainly no impulse to project its visual impressions outwards; it does not yet know this 'outside'; and in any case it could not project anything merely *generally* outwards, it could only project impressions *into a definite distance;* and that definite distance it has as yet no means of determining. Just as little is it possible, as has been supposed, for the soul to represent its impressions as lying directly on the eye; for this again means simply the negation of distance, and distance must be known if it is to be negated. The simple fact is really that the impressions are there, and are seen, but they have no assignable position in the third spatial direction, for this is still unknown. That there is such a third direction, we learn only from experience; and we learn it most easily from our finding ourselves moving *through* the images we see, and from the fact that, in consequence of this movement, the single images undergo various displacements, some of them being hidden, and others which were hidden coming into view. And this greatly increases the difficulty of applying the general idea we have thus acquired, in estimating the degree of distance in any particular case; a problem which we leave to physiology and the special psychology of sense-perception.

Lastly, I will touch very briefly on one vexed question; why do we see objects upright, although the image of them on the retina is upside down? We must remember that we do not observe the image on our eye with a second eye, which further could compare its own position with the position of the object. There is nothing before us but the image itself; all the geometrical relations of the picture on the retina utterly disappear as it passes into consciousness; and, in the same way, the fact that as a whole it has a certain position in the eye does not in the least prejudge the question how it is to appear later in a spatial perception gained through some further means. We are absolutely dependent on this other perception. If there were only a *seen* space, we could give no answer at all to the question what is above and what is beneath in that space. These expressions have a meaning only if we presuppose another idea of space, an idea for which these two directions are not merely generally opposed to one another, but are uninterchangeably different. When we have this idea, and not till then, we can say that that in the visual world is 'above,' the image of which we find or have to seek in the fixed direction towards the 'above' of the other space. It is our muscular feeling or general sense which (even when unaided by the sense of sight) instructs us respecting the position of our body, that gives us the other perception of space. For, the body being in its usual upright position, the downward direction means the direction of weight, and when we oppose our forces to it the result is a number of feelings of effort; and by these feelings the downward direction in this other, non-visual, perception of space is uniformly and uninterchangeably distinguished from the upward. Consequently, if *a* and *b* are places in the field of vision, *b* appears to us as beneath *a*, when the sight or touch of *b* is attained through a movement which, in the language of the muscular sense, is a downward movement; or when (our body being upright) the image of *b* always enters the field of vision along with the images of the lower parts of our body, and never along with those of the upper. This last requirement is satisfied by what is commonly called the reversed position of the image on the retina, since the imaging surface of the eye lies behind the centre of rotation; and it would equally be satisfied by an upright position of the image, if that image arose in front of the centre of motion and on the anterior convex surface of the eye. Thus there is a contradiction between the reports of the eyes and of the muscular sense when we use an inverting telescope which gives an upright position to the image on the retina. In such a case, even if we have no other visual image to compare with the telescopic one, we at once notice an opposition to the reports of the muscular sense : we feel that in order to reach the tops of the trees we see, we should have to move our hand in a direction which, for that sense, is downward.

288. I have still to mention that localisation of impressions which we obtain through the sense of touch. Here again the basis of our view is given by *E. H. Weber's* attempts to fix experimentally the conditions under which we can distinguish two impressions on the skin, which are qualitatively alike but locally different. The skin is lightly touched with the two blunted points of a pair of compasses: and the experiments showed that the extent to which the two points have to be separated in order to be distinguished as two, is very different at different parts of the body. For the finger-ends, the edges of the lips, the tip of the tongue, a distance of half a line suffices : while at many parts of the arm, leg, and back, one of twenty lines is necessary. An explanation seemed to be offered at once by the structure of the ·
nerve-fibres. The sensory nerve-fibre, though isolated and unramified during its conduction, separates at its peripheral end into a number of short branches, and so distributes itself over a small space of the skin for the purpose of receiving stimuli from without. It was thought, then, that all the excitations which affect one of these nerve-ends simultaneously would, through the unity of the fibre which has to conduct them further, be destined to form one resultant, and to be incapable of being distinguished from one another. If, again, these excitations occurred one after the other, they might be distinguished in their qualitative character, but would give no ground for local distinctions. On the other hand it was supposed, if two impressions fell on two different nerve-spaces, this alone would not make it possible to

distinguish them as two: this possibility arising only if, between the two stimulated spaces, there lay one or more of such spaces which remained unstimulated. This last supposition is in any case inadmissible; for at every moment there are a great many unstimulated nerve-fibres; if any particular ones among them are to be used for the purpose of distinguishing two impressions *a* and *b*, there must be something in them which shows that they lie between the two stimulated nerve-spaces; and this presupposes the possibility of accomplishing what has to be explained, the localisation of the sensations.

In other respects too the point of view described fails to give a sufficient basis for this localisation. Nor was this exactly its purpose: it was intended only to explain why two impressions can sometimes be distinguished and sometimes not. But even in this point I found myself unable to accept it. Two points of the compasses which when they touch the skin simultaneously give only one impression, often leave two distinguishable impressions when they are laid against the skin in turn; and their two impressions appear as locally distinct, though no accurate estimate of the distance can be given; moreover, within one radius of sensation the onward movement of a point can be distinguished from its continued pressure on the same spot. Lastly, the conduction of the excitations by the same or by different nerve-fibres did not seem to me to decide anything; the partitions of the fibres are not continued into consciousness, and there all the impressions must in the end come together, qualitatively distinguishable, if they were different, and indistinguishable if they were not. But neither for the like impressions nor for the unlike did the theory assign any ground of local separation, still less any clue by means of which each of them might have its own place given to it.

289. Thus I found myself obliged in this case; no less than in that of the impressions of sight, to look for local signs, abiding certificates of local origin; these local signs would be attached, in the form of qualitatively distinguishable extra-impressions π, κ, ρ, to all excitations A, B, C, according to the particular spots p, q, r of the skin which they affect. Let us suppose that a stimulus, strictly limited in its local extent—say the prick of a needle—affects the spot p. Owing to the connexion between different parts of the skin it is impossible that the operation of this stimulus should be confined to a point destitute of any extension: whatever alteration it produces directly at the point of contact will produce in the neighbourhood of that point a number of little stretchings, pressings, and displacements. Now, though there is a general uniformity in the structure of the skin, it

is by no means exactly alike at all parts of the body. The epidermis is thicker at one place, finer at another ; when the skin is attached to the points of bones it is stretched, at other places the extent of its possible displacement is greater. It differs again not less widely according to the nature of its substratum : it is not the same when spread over a cushion of fat as when it is stretched over bones, flesh, or cavities. Lastly, at different places in the body these various situations may pass into one another either suddenly or slowly. We may therefore perhaps assume that at any point p in the body the wave π of little extra-agitations, called forth by the stimulation of that point, will differ from any other wave κ which accompanies the stimulation of a spot q. But these extra-excitations would avail us nothing if they simply occurred without becoming objects of our perception ; and this last requisite will depend on the distribution of the nerve-fibres. Let us suppose a case. Within the field of distribution of one and the same fibre, let $p\ q\ r$ be the single ends of that fibre : then the local sign π of the spot p will consist in the sensations of those extra-impressions which the direct stimulation of p calls up in its neighbourhood, and the conduction of which to consciousness is secured by the nerve-terminations q and r that receive them. Now if the structure of the skin within this field of distribution were perfectly uniform, the nerve-fibre which unites $p\ q\ r$ would reach precisely the same final state whichever of these terminations were the place directly stimulated : the impressions could not be distinguished, whether they were simultaneous or successive. But if the structure of the skin varies within this field, the stimulation of p will produce different extra-excitations in q and r from those which the same stimulation of q will produce in p and r. Accordingly, if one and the same impression A affects different places in succession, the uniting fibre will bring this impression to consciousness in company with *different* local signs $\pi\ \kappa$ ρ ; and we shall have a motive for the separation of three sensations, although as yet no motive for a definite localisation of them. If the impressions are simultaneous, the uniting fibre may either conduct them side by side without intermixture, or it may be only capable of conducting a single resultant of their influences : which of these alternatives is correct is a question we cannot discuss.

Let us now return to the other idea. Let $p\ q\ r$ stand for three different nerve-fibres ; but let the stimulus A act on a spot of the tissue where there is no nerve-termination : then the effect produced must distribute itself until it finds a nerve-termination on which it can discharge itself. Now if in the whole field of $p\ q\ r$ the structure of

the skin was uniform, I should say that it matters nothing whether it
is one or two of these fibres that receive the like impressions, which
would be accompanied by like local signs; for in no case could the
impressions be distinguished, and the only use of the multiplicity of
the fibres would be the general one of securing the entrance of the
stimuli into the nervous system; for there can be no doubt that the
excitation of the tissue could not propagate itself to any very considerable distance. On the other hand, if the texture and state of the
skin within this whole field varies rapidly, the different local signs
which arise at point after point would be useless unless there are a
great number of closely congregated nerve-terminations, each of which
can receive the wave of excitation of a small circuit, before that wave
has lost its characteristic peculiarity by meeting with others which
began at different places and spread over the same field. It seems to
me that these suppositions answer to the results of observation. On
the back and trunk there are long stretches where the structure of the
skin is uniform, and here impressions can only be distinguished when
they are separated by wide distances. In the case of the arm and leg,
the power of distinction is duller when the stimuli follow another in
the direction of the longitudinal axis of those limbs—the direction of
the underlying muscles; it is sharper when the stimuli are arranged
round the limb, in which case the skin is supported alternately and in
different ways by the swell of the muscles and the spaces that intervene between them.

290. The name local signs, in its proper sense, cannot be given
to these extra-excitations themselves, but only to the sensations they
occasion. Now it strikes us at once that there is one of our postulates
which those sensations altogether fail to satisfy. It is true that they
differ in quality, while at the same time they admit of resemblances;
for example, if we touch any part of the skin that is stretched above a
bone, whether it be the forehead, the knee-cap, or the heel, feelings
are distinctly aroused which have a common tone. But these feelings
are not quantitatively rateable members of a series or system of series.
They cannot therefore serve directly to fix the locality of their causes;
and, besides, what we require in this case is not the localisation of the
sensations within an absolute space, but within that variable surface of
the body, to the various points of which they are to be referred. We
must have learnt the shape of this surface beforehand, and have discovered through observation to what point p in it that impression A
belongs, which is characterised by the local sign π: until this is done
we cannot refer a second stimulus $B\pi$ to the same point in the surface

of the body. This can be done easily enough if we can use our eyes; but how is it to be accomplished by the blind man, who, beyond these feelings, has nothing to help him except movement? Without doubt the help that movement gives him is of decisive importance; but how it is possible to use this help is not so easy to understand as is often supposed. While the movement is going on, we have of course a certain definite feeling which accompanies it; but then this feeling is in itself nothing but a manner in which we are affected; it itself does not tell us—we have to guess—that it is caused by a movement of the limbs. This discovery, again, is easy when we can use our eyes, and so notice that our hand is changing its place while we are experiencing the muscular feeling; but the blind man has to make out in some other way that the alteration of his general feeling is not a mere change of his internal state, but depends on the variable relation into which he or his bodily organs enter towards a series of permanent external objects.

Now it seems to me that the condition which makes such a knowledge as this attainable, consists in this,—that the skin, like the eye, has a number of sensitive and moveable points. If an organ of touch in the shape of an antenna possessed in its tip the sole point at which the skin of the whole body was sensitive; and if its capacities were strictly limited at every moment to the power of bringing one single object-point A to perception, the result would be that, when a movement of this organ led from A to B, the perception of A would altogether disappear and the wholly new perception of B would take its place. No doubt while this was going on a muscular feeling x would have been experienced; but how could it occur to us to interpret that feeling as the effect of a spatial movement? However often we passed from A to B and from B to A, and experienced the feelings $\pm x$, we should never discover what those feelings really signified; this transition would remain a perfectly mysterious process, of which all we knew would be that it transformed our idea A into B. On the other hand, if the hand, like the eye, can feel the three impressions $A\ B\ C$ at once; if this image of pressure changes during the movement by regular stages into $B\ C\ D$, $C\ D\ E$; and if by a movement in the opposite direction we can again reach the parts that have disappeared, or grasp them with one hand while the other moves away from them, these facts must certainly tend to suggest the idea that the muscular feelings which accompany the succession of sensations arise from a variable relation of ourselves towards independent objects—that is, from movement. As soon as this is discovered, it is possible—in a way which I need not further describe—for the limitless variety of combination between the sensations of that part of the body which touches, and the not less sensitive part which is touched, to conduct us to a knowledge of the surface of our body, and to the localisation and arrangement of our single sensations in that surface.

CHAPTER V.

The Physical basis of Mental Activity.

In passing on to consider the forms in which soul and body act on one another, I must observe that there are a number of special questions for the answers to which there is not as yet any sufficient foundation; and of these I do not consider it my duty to treat. All that can be considered proper to this metaphysical discussion are the fundamental conceptions used by various theories in interpreting the facts. We may leave out of sight an infinity of so-called experiences, all of which are not by any means equally well attested, and which alter every day with the progress of observation. They will gradually define the object of some future theory, but, so far at least, they do not contribute to the criticism of these metaphysical foundations.

291. It has been said that the soul is the same thing ideally that the body is really; or that the two are the different sides of a single whole. Such wide expressions will not give us what we want. When once we have distinguished body and soul as two parties between which manifold interactions take place, we need ideas more definite and more capable of being pictured, in order to conceive the processes through which these reciprocal influences make themselves felt. And among the questions which require a clear and unambiguous answer is that concerning the spatial relations of the soul—the question, to adopt the current phraseology, of the *seat of the soul.* There was a time when some philosophers looked down with pity on the maladroitness supposed to be involved in the very asking of this question. Nevertheless, unprejudiced persons will always raise it afresh; and therefore it must be answered and not ignored. I might attempt to answer it at once, by connecting it with the preceding discussions; but I prefer to leave them out of sight, and to repeat the considerations by which on other occasions I have attempted to indicate my view. Let us take, then, the various ideas which are really intelligible to us respecting the spatial relations of anything

capable of action, and which we are in the habit of applying to them, and ask which of them answers to the special case of the human soul.

292. To be in a place means simply and solely to exert action from that place and to experience the actions or effects that reach that place: if we put these two powers out of sight, it is impossible to attach any meaning to the assertion that a thing is at this place p and is not at that other place q, where, as at p, it neither exerts nor experiences any action. Now it is possible to conceive an existence standing in a direct, and at the same time an identical, relation of interaction with all the other elements of the world. There is one case in which this is a current idea; it expresses what we mean by the omnipresence of God. No element of the world needs to travel a long road, or to call in the help of other things in order to bring its own state to the presence and knowledge of God; nor have the divine influences to make a journey in order to reach distant things: the interaction here is perfectly direct. But then it is also one and the same in all cases, and has not different degrees; at any rate there is no measure of distance, according to which the interaction is necessarily stronger or weaker; the only reason why its work may be greater in one case than in another is that the meaning of things, or of what goes on in things, gives a reason for an interaction of greater weight in one instance and of less weight in another. In this alone consists our conception of omnipresence: the infinite spatial extension which forms the theatre of that omnipresence we are far from ascribing to God as an attribute of His nature; and on the other hand we see no contradiction between the plurality of the points at which His activity manifests itself, and the perfect unity of His nature.

Now the attempt has often been made to ascribe this omnipresence to the soul, within the limits, that is, of the body in which it resides: and the cause of this mistaken idea is most commonly to be found in the aesthetic impression which makes it seem as though the whole of the body were penetrated by a psychical life, and every part of it were the immediate seat of sensation and a direct organ of the will. But there are some simple physiological facts which show us that this beautiful semblance of omnipresence is the result of a number of intermediating agencies; that the soul knows nothing of the stimuli that reach the body, and loses its power of setting up movements the moment the continuity of the conducting nerve is broken; that therefore the space within which body and soul act directly on one another is limited, and must be found somewhere, though we cannot yet define its limits, within the central portions to which all impressions are conducted, and from which all impulses to voluntary movements start. We may refuse to believe this; we may answer that a natural feeling tells us all that the soul feels directly in the touching hand, and that this natural feeling cannot be created by such intervening agencies. But the objection will not help us. There are certain peculiar double feelings of contact which arise when we touch an object with an instrument held in the hand; but we do not consider ourselves justified in concluding from this that the soul can occasionally prolong its activity to the end of a stick or a probe. And yet we fancy that we have a direct feeling at that point of their contact with a foreign body.

293. The natural sciences have familiarised us with the idea of another interaction, which is direct, but also graduated. This is our notion of the attractive and repellent fundamental forces of masses. These forces need no intermediation; they send their action to infinite distances, whether the space traversed by that action is full or empty; but the intensity of the action diminishes with the increase of the distance. If we applied this notion to the present case, we should conceive of the seat of the soul as a point, or at least as a limited district of the brain, on which the interactions of the soul with the surrounding parts would be at the maximum of intensity, while the further they left it behind the more they would diminish in strength, although actually extending to infinity. But a sober observation finds no witness to this outward activity. The slightest intervening space that separates things from our senses makes them simply non-existent to us, except where there are verifiable processes through which we act on things indirectly, and they on us, and which therefore help us over this spatial interval. Any amount of freedom being permitted in suppositions of this kind, the assumption might be suggested that the force of the soul diminishes in the ratio of a very high power of the distance; in this case it might exert no observable influence upon the lengths of nerve which extend even a slight distance from its mysterious seat. All that is certain is that, however close to the root of the nerve a breach of its continuity may be, the outgoing force of the soul is never able to produce on the other side of this breach the effects which it commonly produces in the nerve. But, be this as it may, to assume that there is a fixed limit—whether the surface of the body, or the smaller zone within which the roots of the nerves lie—at which the outgoing force ceases to operate, is simply equivalent to a surrender of this whole point of view. There is nothing in

one spherical surface of empty space that can make it, rather than any other such surface, the limit at which an activity ceases to diffuse itself. If there is any such limit, the reason of its existence must lie in the fact that the force does not stream outward aimlessly through empty space, but that there are other real conditions on which its activity and the absence of its activity depend.

294. But it is not worth while to pursue any further this idea of limited action at a distance. There is a more decided view, which has always been preferred to it, and to which many natural processes bear witness. According to this view, action never takes place except in contact, and therefore we must assume one single seat of the soul, fixed or variable, in the form of a point; and apart from other reasons a local habitation of this kind appeared most suited to that which is immaterial and a unity. Yet this idea was at once found to involve a crowd of difficulties. Let us first suppose the seat of the soul to be, not changeable but fixed. In this case we must assume either that all the nerve-fibres join at this point of intersection, or else that there is a formless space—whether parenchyma or cavity—into which all nervous excitations discharge themselves, and are able to reach the soul which resides at some point of this space. But as to the point of intersection, anatomy, instead of discovering it, has simply made its existence incredible; and as little is it possible to discover a formless space, having edges where all the nerve-fibres terminate, and offering a field within which the excitations of these fibres can spread until they reach the soul. It might possibly be the case that the soul needs no such primary assembly of all the primitive fibres, but stands in direct interaction with a few of them, which would be, as it were, the delegates of the rest: but, so far, we know of no anatomical fact which makes this probable. Secondly, then, we may suppose that the place where the soul resides is not fixed but moveable. This idea leads us back to the notion of limited action at a distance. At any given moment the soul would have to be at the particular spot, where an excitation is arriving—an excitation which cannot become a sensation unless the soul is there; and if it is to be at this spot, it must have been already acted upon *from* this spot and so induced to move to it. Finally, if the soul is to impart an impulse to the root of a motor nerve, it must move to the spot from which it can exert this impulse: but as the motor nerve is not yet active it cannot solicit the soul to move to this spot, and therefore the soul must itself choose its line of movement and follow it: and this implies a knowledge of locality which no one will admit.

But is all this really necessary? Is it really necessary to assume any one of these alternatives?—either that the activity of the soul penetrates indiscriminately the whole body or that it penetrates, again indiscriminately but with decreasing intensity, space simply as space; or finally that the soul is confined to one point and acts only in contact? The root of all these difficulties seems to be a confusion in our idea of the nature of an acting force and of the relation of this force to space. And there is no lack of other examples which will enable us to arrive at a more correct conception.

295. Any force arises between two elements out of a relation of their qualitative natures; a relation which makes an interaction necessary for them, but only for them and their like. It is altogether a mistake to regard a force as a hunger for action, spreading itself throughout a space and seizing indiscriminately on everything it finds in that space. We should do better to think of the magnetic force, which within the provinces over which it extends operates on no bodies but those which can be magnetised, and remains indifferent to those with which, though they lie within the same space, it has no elective affinity. Or we may think of the chemical reagents which, when poured into a fluid, pass without acting by the substances which are indifferent to them, while they supplement those with which their chemical nature makes it necessary for them to join. These examples prove nothing, and the idea they are meant to illustrate is intelligible without them, but they enable us to picture it. It is not their spatial position that compels the elements to act on one another or makes such interaction impossible; but it is their own natures and the relations between them that make some elements indifferent to each other and impel others to a vigorous copartnership. If we apply this general idea to the present case, our first assertion must be this: wherever the soul may have its local habitation (for it may be still held that we must assume that it has such a habitation), the extension of its activity will not be determined by its position there: this position will not confine the soul to an interaction with those nerve-elements which surround and touch that habitation: nor will its activity start from this centre, and, like a physical force acting *in distans*, extend with a decreasing intensity to all the elements which are grouped at an increasing distance around that centre. On the contrary, wherever there are elements with which the nature of the soul enables and compels it to interact, there it will be present and active; wherever there is no such summons to action, there it will not be or will appear not to be.

Now doubtless it is pleasant to the imagination to represent the elements that stand in this sympathetic relation to the soul as in spatial proximity to one another, and, where this is possible, to picture a small extended province of the brain, best of all, a single point, where they are all assembled. But there is no necessity in real earnest for this hypothesis. We have reached the conviction that spatial positions and spatial distances are not in themselves conditions of the exercise or non-exercise of forces, and that they form such conditions only because they themselves are the manifestation of forces[1] which are already active and determine the continuance and progress of the action. We have seen that to be in a place means nothing but to exert action and to be affected by action in that place, and that the sufficient grounds of this action and affection lie nowhere but in the intelligible relations of existences in themselves non-spatial. With this conviction of this insight we can now take up again, in a better defined shape, the idea of that omnipresence of the soul in the body which, as we explained in dealing with it, we could not help rejecting. The soul stands in that direct interaction which has no gradation, not with the whole of the world nor yet with the whole of the body, but with a limited number of elements; those elements, namely, which are assigned in the order of things as the most direct links of communication in the commerce of the soul with the rest of the world. On the other hand there is nothing against the supposition that these elements, on account of other objects which they have to serve, are distributed in space; and that there are a number of separate points in the brain which form so many seats of the soul. Each of these would be of equal value with the rest; at each of them the soul would be present, with equal completeness, but not therefore without any distinction; rather we might suppose that at each of them the soul exercises one of those diverse activities which ought never to have been compressed into the formless idea of merely a single outgoing force. In using the current conception of omnipresence we refused to attribute to God, as a predicate of His nature, the infinite cubic extension which His activity fills; and we could see no danger to the unity of His nature in the infinite number of distinct points which form the theatre of that activity: and there is just as little conflict between the unity of the soul and the multiplicity of its spatial habitations. Each of them is simply an expression, in the language of our spatial perception, for one of the manifold relations in which the soul as taking part in the intelligible connexion of things is at one and the

[1] [Cp. §§ 116 and 203.]

same time involved. Our imagination naturally and unavoidably symbolises this unity, no less than the variety, in a spatial way. We shall therefore be inclined to oppose to these many places a single one which is really and truly the seat of the soul. Perhaps it will be the fixed geometrical central point of all the rest ; perhaps it will be a variable central point, and then we must conceive it to be determined not geometrically but dynamically as the joint result of the spatial co-ordinates of the distinct places on the one hand and the intensities of the psychical activities going on in them at the given time on the other. Such ideas do no harm and they act as supports to our per-ception : but they have no objective meaning ; for the point arrived at by such a calculation as the above, would not express a real fixed position of the soul in that point at the given moment, nor would it give us grounds for determining anything whatever as to the behaviour of the soul in the next succeeding moment.

296. But our view has to meet an objection coming from another side, and will therefore have to undergo another and a final revision. Observation discloses no such differences among the elements of the brain as would give some few points in it the exclusive privilege of forming the seat of the soul. And yet we have to suppose the existence of such a special qualification. For if we were to widen our idea into the supposition that the soul can stand in the direct relation of inter-action, above described, with *all* the constituent parts of the brain, the laboriously intricate structure we find in it would become wholly unintelligible. But is it necessary, is it even possible, to suppose that a real existence A stands once for all in the relation of interaction with other real existences B and C, simply because B is B and C is C, while it stands in no such relation with D and E, just because they are D and E? In the first place, what is it that makes B to be B and C to be C but this: that under different conditions (these conditions forming a series) B experiences the states $\beta_1 \beta_2 \beta_3 \ldots$, and not $\gamma_1 \gamma_2 \gamma_3 \ldots$, whereas under the same conditions C experiences the latter states and never the former? And, in the second place, we have to suppose that at one time an interaction takes place between A and B, and at another time does not take place ; and yet what would this inter-action mean, if A and B were simply A and B, and if A did not undergo certain variable states a_1 or a_2, which formed signals to B to realise forthwith β_1 or β_2, and no other of the states possible to it? Without doubt, then, our conception was still incomplete, when we sought to place the soul S in a direct and ungraduated connexion of interaction with different nerve-elements $B\ C\ D$, considered simply as such. Things cannot stimulate one another in respect of their un-changing natures; they can only be stimulated in respect of what goes on in them, and that reciprocally. Accordingly it is the events $\beta\gamma\delta$ which occur in BCD that, in virtue of their occurrence, make these points and no other points the seats or localities of a direct interaction with the soul.

Starting from this point of view, then, we should be led in con-sistency to the following metaphysical conception of the significance of the central organs. The interlacing of the nerve-fibres serves two ends. First, it has to act upon the excitations which arrive from without through the organs of sense, so to connect, separate, and arrange them, that as the result there arise those final states $\beta\gamma\delta$, which now for the first time, and in their present shape, are in a condition to be brought to the knowledge of the soul, or by which alone it is capable of being stimulated. The second function is the converse of this. The excitations which come from the inner nature of the soul, have to be transformed into physical occurrences in such an order and arrangement, that their centrifugal action on the moveable members of the body will allow of an influence, answering to a conceived end, on the shape of the external world. At the point where these duties are fulfilled, lies a seat of the active soul, the locality of one of the different functions, in the connected whole of which its life consists. In an earlier passage I spoke of this point of view as one of the hypotheses which might be framed in accordance with the facts to be explained : it will now be seen that it is only the continuation of our ontological views. We have left far behind us the theory which conceived the world as based on a number of elements, beings, or atoms, which simply 'are' and form a primary fact, and between which we then suppose actions to take place, the nature and occurrence of these actions being thus of necessity grounded in something external to the fixed existence of the primal elements. We found that there is nothing in the fullest sense actual but the one reality which is in eternal motion, and in the development of which any member of the whole is connected with any other only in accordance with the meaning of the whole, and stands in no such connexion where the meaning of the whole does not warrant it. It is only this connexion of events that gives to single stable conjunctions of these manifold occurrences the appearance in our eyes of beings with an independent existence; in reality these conjunctions are only the meeting points, or crossing points, of in-going and out-going actions, which the significance of the course of events keeps in being, and they form actual beings or existences only when, like the soul, they do not simply appear to others as such centres, but really make themselves such centres by opposing themselves, in consciousness and action, to the external world.

297. From the preceding account of the functions of the central nervous organs we might conclude that their only business is to bring about the commerce of the soul with the external world ; the internal activity of the mind would seem not to need their co-operation. Taken as a whole, I do not disclaim this inference, though it must be limited in essential respects ; rather I regret that no further explanation is possible regarding those other operations, in which it is agreed on all hands that the help of the body is needed. There are a very large number of cases in which unfortunately we are not simply unable to point out the means which would render the required service, but we do not even know exactly what services are required. And I mean this admission to apply not only to my own view, but to many others which would be very unwilling to make a like confession. We studied the retina of the eye, and the nerve-terminations found in it : dioptrics revealed to us the passage of the rays of light, and their point of meeting on the nerve-terminations : What more did we want ? Were we not in complete possession of all the conditions (so far as they can be fulfilled in the eye) implied in the occurrence of visual perception ? And yet further investigation has discovered new layers of a strange structure in the retina, of the use of which we know nothing, and which yet can scarcely be useless. It is certain then that we made a mistake in supposing our knowlege to be complete, when we cannot tell the function of what is afterwards discovered : and yet even now we cannot guess what part it was we overlooked in the work the eye has to perform. Now in the case of the brain we are equally at a loss : it is not merely that in the greater part of its structure we find everywhere arrangements of the most remarkable kind, and yet cannot tell their purpose : but even where experience has disclosed to us with sufficient certainty the existence of relations between psychical functions and particular parts of the brain, we cannot get further than this very general result : no one can specify the exact physical function their elements have to perform in order that this or that definite expression of psychical activity may be possible. Thus we talk in a highly perfunctory way of organs of this or that mental faculty, without knowing very well what there is to prevent the soul from manifesting itself without this organ, what intelligible properties there are which enable this organ to supply

the conditions lacking to the soul, and lastly in what way the soul is enabled to make use of this organ as its instrument. This last idea indeed, the idea of an instrument, is the most unsuitable of all that could possibly be applied to the case. We may call the limbs of the body instruments: for though we do not know *how* they follow out our ideas, we are at any rate able consciously to connect the movements, which we do not understand in detail, so that they form the means of carrying out an intention. But when we are told that man cannot think with a frozen brain, it is only the obliging preposition 'with' that gives these words the appearance of meaning something; for it seems to indicate that we are able to understand how gloriously thought goes to work with an *un*frozen brain as its instrument. If for the preposition we substitute the conditional sentence which forms its real meaning,—'*if* the brain is frozen, man cannot think,'—the words remind us only of what is perfectly familiar, the many conditions on which life in general and therefore every mental activity depends, but they tell us absolutely nothing of the *nature* of the service which these conditions render to the realisation of these activities. Nothing can help us over this state of ignorance, but the multiplication of exact observations: all that remains for us to do here is to touch on the few general ideas which we should wish not to be neglected when the new knowledge we hope for comes to be interpreted.

298. The older psychology used to speak of a *sensorium commune*: but it was not able to point it out, and the motive for assuming its existence was probably only an indefinite desire for a place where all sensations could be collected into a common consciousness. It may be that in this matter we are in the position described in the last section: perhaps there really is some function we have overlooked, which is necessary to this end, and has to be performed by the physical organs. But all that is certain is that we do not know of any such function. So long, therefore, as we cannot point to definite processes of modification, to which all impressions must submit before they can become objects of consciousness, we have no ground at all for supposing such a place of assembly for these impressions.

Modern physiology has sometimes spoken of a *motorium commune*, and supposed it to be found in the cerebellum. But the movements of the body show the utmost degree of variety; and their classification under the head of movements connects them no more closely with one another than with other functions of the mind to which they are conjoined in the economy of our life. We may suppose that the manifold excitations of the muscles, which each species of animal needs for its characteristic kind of locomotion, and for the preservation of its equilibrium in different positions of the body, are really dependent on a central organ, which compels them to occur in company, and grouped in a way that answers this special purpose. But I know no reason why we should make the same centre a condition of all the other movements, which are excited for other purposes and by other occasions in the various limbs of the body. Thus the idea of this general motory organ, again, seems to me to owe its origin to a logical division of the psychical activities, and not to a consideration of the connexion in which these activities have to stand in supporting each other for the purposes of life. It is much more likely that sensory and motor nerves are combined with one another in various ways, so as to form central points for whole complexes of exertions dependent on one another. Even the motorium to which we ascribed the preservation of the equilibrium, would be unable to perform its task unless it received at every instant an impression of the threatening position which it has to counteract by a compensating movement. And even if it is possible for this movement to be carried out in a perfectly mechanical way, and without the participation of the soul, it is, in the ordinary course of events, at the same time an object of our perception. It seems to me probable, therefore, that this organ, too, consists in a systematic connexion of sensory and motor fibres; although the former do not always communicate their excitations to consciousness, but sometimes simply produce a movement by transferring their excitation to motor fibres. Now among the organs which I should suppose to be formed in this way, I should place first an organ of the perception of space : and I am completely satisfied, although utterly unable to prove it, that in all the higher kinds of animals this organ, dedicated in each case to a function which appears everywhere the same, forms a considerable part of the brain. If the hypotheses I have ventured respecting the local signs of the sensations of sight be correct, the function of this organ would be to connect the optical impressions with the motor impulses of the eye. But how this function can be performed, and in what form the efficient connexion of the sensory and motor nerves is established,—these are questions on which I will offer no conjecture.

299. In the second division of the functions of the central organs— those functions which consist in the physical working out of the internal impulses of the soul—there is one process with respect to which the observations of the most recent times seem to have led to a

secure result. It has been proved with sufficient certainty that an organ of language is to be found at a particular spot in the large hemispheres of the human brain. In order to understand the office of this organ, let us glance at the different modes in which our movements in general arise. I put aside the purposeless twitchings which occur in particular muscles, owing to internal irritations for the most part unknown to us : but even with respect to the movements which we produce at will in accordance with our intentions, we must confess that we do not understand how they take place. We do not know by nature either the structure of the limbs which gives the movement its form, or the position of the muscles and nerves which carry it into execution. Even if we did, there would remain a further question as to which we are still in darkness, and which science also is not at present able to answer : what is it exactly that we have to do, if we are to give to the nerve that first impulse which produces in all this preparatory mechanism the desired state of activity? It takes the newly-born animal but a short time to acquire that control over its limbs which characterises the genus to which it belongs ; and this fact compels us to assume, not merely a succession of chance experiences which gradually teach the animal that its limbs can be used, but also internal impulses which call these experiences into being. On the one hand, the external stimuli, by transferring their excitations to motor nerves, will at once call forth connected groups of movements combined in conformity with their common end ; on the other hand, the central apparatus, on which this combination depends, may be stimulated to activity from within by variable states of the body. The sensory excitation then will produce in consciousness a sensation of the stimulus, and at the same time the movement that occurs will produce in consciousness the sensation of its occurrence, and the perception of its result ; and in this way the soul, playing at present the part of a mere spectator, will have acquired the different elements of an association which it can reproduce at a later time with a view to its own ends. The soul cannot always produce of itself the efficient primary state that would recreate the movement: sometimes this movement demands, for its repetition, the complete reproduction of the corporeal stimulus from which it sprang originally as a true reflex movement. For example, up to a certain point one can imitate coughing and sneezing at will, but one cannot bring about an actual sneezing or vomiting without a fresh operation of their physical excitants. Even the movements which depend on states of emotion are only to a slight extent conjoined to the renewal of the mere ideas of a pain or pleasure; they depend on the renewal of the pain and pleasure themselves. I refer to the familiar facts of bodily expression and gesture—an endowment due to nature, and not to our invention—involuntary manifestations of its internal psychical states, which the soul simply witnesses without willing them, and, for the most part, without being able to hinder them.

300. But what is the starting-point which the soul must produce in order that the motor mechanism may execute exactly that movement which at the given instant answers to the psychical intention? I speak simply of a starting-point, because we certainly cannot suppose that the soul exerts an independent and conscious control over the details of the process, and metes out to the particular nerve-fibres, which must be called into action in the given case, those precise quantities of excitation which will secure the direction and strength of the desired movement. In place of thus generating homogeneous impulses, and merely giving them different directions in different instances, it has to produce for different movements A and B qualitatively different internal states a and β; and these, instead of being guided by it, seek and find their way for themselves, simply because they are themselves and no other states. Let a and b be two different motor central points, of which a connects into a whole the single excitations necessary to A, and b those necessary to B: then a will find its efficient response only in a, β only in b, while to other nerves they will remain indifferent. If, again, both movements A and B depend on the same central point, only that they depend on different degrees of its excitation, then the strength of a and β will determine also the strength of this excitation. If, lastly, one movement requires the simultaneous activity of both organs, then the internal state γ, which is to set up that activity, must contain the two components a and β, and these two components will determine the share taken by a and b in the joint-result they have to produce. This view of the origin of movement corresponds but little to ordinary notions; it leads us back to the often-repeated idea, that the ultimate ground or reason of every action or effect lies in the fact that the two elements which stand in this relation of interaction exist for one another directly—that they stand, if the word may be used, in a direct sympathetic *rapport*, which makes each receptive to the moods of the other. There may be many intermediating processes producing the conditions on which this *rapport* depends, or removing the hindrances to it, but they are all mere preparations for the action; the action itself, which comes when they are finished, cannot be explained in its

turn by a similar machinery, between every pair of whose parts this immediate sympathy would again be necessary. Our theory presents difficulties to the imagination only if we take in literal earnest the expression in which the internal state *a* or *β* is described as finding its way to *a* or *b*. The internal state has not really any way to traverse; for the soul in which it arises is not placed at some distant spot in space, from whence it has to send out its influence in search of the organs that are to serve it. The soul, without its unity being on that account endangered, is itself everywhere present where, in the connexion of all things, its own states have attached to them the consequent states of other elements.

301. When the soul then reproduces within itself these starting-points, they proceed, without any further interference or knowledge on its part, and in obedience to a mechanism which was not invented by us and remains concealed from us, to produce as a final result the actual movement. We now naturally ask the further question in what precisely do these starting-points consist? A very close approach has already been made to our view when it is asserted that, if the movement is to become actual, we must will, not the movement itself, but the end of it, and that then the movement will take place of itself. But the question is, What is this willing of the end? The imitative movements with which the devout spectator accompanies the actions of the fencer or skittle-player, or by which an unskilful narrator tries to portray the objects he speaks of, might convince us that, in the absence of hindrances, the mere idea of a movement passes of itself into the actual movement. And if we take this point of view, we may really leave the influence of the will out of account. For whatever else it may consist in, and whatever positive contribution, over and above the mere absence of resistance, it may make to our movements, still its function in reference to a given movement *a*, distinguished from another *b*, will consist essentially in this,—that it favours the definite ground or reason *a* or *β*, which leads to the one or the other of these movements; and the nature of this starting-point or ground is precisely the question we were concerned with. On the other hand, I certainly do not think we need look for this starting-point in the idea, at any rate not in the visual ideas of the movement; although innumerable little acts of our daily life are directly conjoined, without any consideration or resolution of the will, to the ideas arising in us of a possible and desirable movement; and though they even seem to be conjoined, without the intermediation of an idea at all, to the mere perception of the object with which the act may deal. Taken

by itself the visual idea would signify nothing more than the somewhat abstract fact that a moveable limb is at this moment at the spot p in space, and at the next moment at the spot q; but it would contain none of the concrete interest for us which is given to this fact by the circumstance that *we* are the cause of the visual idea and that *our* limbs are the object, whose spatial positions are in question. Thus the starting-point or state, which the soul has to reproduce in itself in order that, conversely, the actual movement may be conjoined to that state is not, I conceive, the idea of the movement, but rather the feeling which we experience during the execution of the movement and in consequence of its execution. It is common in physiology now to speak of feelings of innervation, but I should not choose that name to describe what I mean. The case is not, I think, that there is an act, consisting in an influencing of the nerve, and directed now here and now there, but in other respects always of the same nature; and that this act is on the one hand what we feel, and on the other hand what according to the direction given to it produces this movement *a* or that movement *b*. The case is rather that this feeling itself, its mere unhindered existence, constitutes that internal condition of the soul which effects an innervation proceeding from it and affecting in all cases a particular complex of nerves. There are some very simple facts of experience which seem to me to confirm this view. A beginner finds it difficult to hit a certain musical note or a given uttered sound, and then there is this special difficulty that the necessary movements are not completely visible; but we also find·that any other movement which is at all complicated, continues, even though it be fully measured by the eye, to be difficult to us until we have once succeeded in it. Then we know how we must *feel* if we wish to repeat it, and that feeling π,—or, to state the matter as we did in the case of the local signs, that first link π_0 in the series of momentary muscular feelings which followed one another during the actual movement,—has to be reproduced if the movement is to be repeated; and we consider the movement to be successful, and to answer our intention, if the repeated series π is identical with the series we remember.

802. If, taking these results as our presuppositions, we now return to the organ of language, our account will be as follows: the idea of that which we wish to designate awakes the idea of the sound of its name, and this idea awakes the idea of the muscular feeling π which is necessary to the utterance of the name; and to this last idea is conjoined the movements of the organs of speech. But here we

come to a standstill; we cannot determine what contributions the organ has to make to this end. Since the feeling π arose from the physical excitations experienced by the muscles when first the movement was executed, it seems a tenable hypothesis that the reawakening of this feeling in the soul must produce (to begin with) a general state of physical excitation in the organ, and that this state then, in conformity with the structure and internal states of the organ, divides into the various components which give their particular impulses to the executing nerves and muscles. The morbid phenomena produced by an injury to the organ, as well as many simple phenomena of daily life—those of passion, intoxication, and others—show that this chain of processes may be interrupted at various points; there may be a correct image of the object, though the idea of sound united with it is false; or the latter may be still distinct to us, but we are annoyed to find that the spoken word does not correspond with it. But these disturbances again give us no exact information respecting the function of the organ in its healthy state. It is easy to talk of telegraphic conductions and perverted connexions of them, but this is nothing but a way of picturing the observed facts; and images are useless unless one can confront every single line of them with the real process which corresponds to them point for point. The other movements of the body are subject to similar disturbances; but these I must leave to the pathological works in which interesting descriptions of them may be found. Whatever anatomical basis is given to that feeling which instructs us respecting the position, the movement, and the amount of exertion of our limbs, the fact remains that, wherever this *feeling* is diminished or disappears, we find it difficult or impossible to execute movements, the *idea* of which is none the less present to consciousness, as the idea of a task to be accomplished.

303. Phrenology has attempted to connect with corporeal bases the activities commonly ascribed to the higher faculties of the mind. We cannot say that the observations on which this attempt rests have no significance; but phrenology should have confined its efforts to talents whose nature is unambiguous, such as can scarcely conceal themselves where they really exist, and never can be simulated where they do not. It was of little use to speak offhand of peculiarities of disposition and character, respecting which our knowledge of mankind is easily deceived, and which, where they are actually present, may owe their existence to the co-operation of very various influences of life and education. If this limitation were observed, an accurate

comparison might then give us, not indeed an explanatory theory, but trustworthy information establishing a connexion between particular facts of bodily and of mental development. These facts would then have to be interpreted; and we cannot tell what the result of a conscientious attempt to interpret them would be. But at any rate it is quite impossible to put any faith in the cherished notion that every one of the capacities and inclinations enumerated in the phrenological plans has a local subdivision of the brain assigned to it as its particular organ : for each of these peculiarities, considered psychologically, is the final outcome of the co-operation of a number of more general psychical functions, and any one of them is distinguished from any other by the different proportions in which the manifestations of these more general activities co-operate. It is only in the case of these general activities that phrenology can hope to discover a dependence on the structure of the brain or skull ; and even this hope depends on the very doubtful assumption that fundamental faculties, whose business is a constant and close interaction, would find their needs answered by a localisation of their organs at different spatial positions.

But I pass from these questions, for no one can decide them ; I may hold it to be in general a natural assumption that, supposing a material mass to be necessary to the manifestation of a mental function, that manifestation will be more intense according to the size of the mass ; but for the higher mental life I believe much more importance is to be attached to the quantity, multiplicity, and intensity of the stimuli afforded by the body to the excitation of an activity, which in its innermost nature or work seems neither to need nor to be accessible to any further physical help. But the contributions which the bodily organisation thus makes to the vivacity and colouring of the psychical life, need not consist exclusively in structural relations of the brain. They may come from all parts of the body; from those delicate mechanical and chemical differences of texture which are not less real because we imperfectly describe them as contrasts between tense and lax fibres ; from the architecture of the whole which allows to one organ a more extensive and to another a less extensive development. For all these peculiarities of the solid parts give a special stamp to the play of the functions and the mixture of the fluids, and in this way they are continually bringing to consciousness a large quantity of small stimuli, the total effect of which is that dominating tone or general feeling, under whose influence the labour of the mental forces is always carried on. A part of these bodily influences we know by the name of the temperaments, which need not be described here, and the definite assignment of which to physical bases has never yet been achieved. As peculiar forms taken by our internal states, in accordance with which the excitability of our ideas, emotions, and efforts, is greater or smaller, one-sided or many-sided, passing or continuous, and their changes are slower or more rapid, the temperaments condition in the most extensive way the whole course of mental development. And although the body does not by the physical forces of its masses directly create the faculties of the soul, it forms in this indirect manner one of the powers which control their exercise.

304. We in no way share the view which conceives the activities of the soul materialistically as an effect of its bodily organs, and, as a matter of fact, every attempt hitherto made to connect its higher functions with given substrates has proved fruitless : yet there are many facts which require us to consider the general dependence of *consciousness* on states of the body. The name consciousness cannot now be withdrawn from use ; but it has this inconvenience, that it seems to represent as an independent existence something which is really only possible in inseparable union with those variable states which we conceive as occurrences happening to it. We all know that consciousness [1], or being conscious, means only being conscious in oneself of something ; the idea of consciousness is incomplete if we omit from it either the subject, or the something which this subject knows or is conscious of. But in handling special questions we often forget this, and lapse into various fancies ; sometimes we imagine a bodily organ, which prepares consciousness in general for the use of a soul which is to employ it, in application to a content that may come into it ; sometimes we dream of a special faculty of the soul itself which produces the same curious result ; or at any rate we figure consciousness itself as the natural and constant state of the mind— a state which is not, properly speaking, unreal and inoperative even when it is completely prevented from appearing. In opposition to these ideas we are ready to admit that it is only in the moment of a sensation that consciousness exists as that activity of the soul which directs itself to the content felt ; and that it forms a continuous state only in so far as the multiplicity of simultaneous or successive exertions of this activity does itself, as before described [2], form the object or exciting

---

[1] [The German word *das Bewusstsein*, which we translate 'consciousness,' means literally 'conscious-being,' or 'the being conscious.']   [2] [Sect. 271.]

cause of a new act of representation—an act by which we form an idea of this multiplicity. Accordingly we should agree that a soul which never experienced a first stimulus from without, would never, as we say, awake to consciousness : but the question remains whether, when once the play of this internal activity has been started, it can carry on an independent existence, or whether it remains as dependent on bodily causes for its continuance as it was for its excitation.

Now the states of unconsciousness offered to observation by natural sleep, swooning, diseases, and injuries of the central organs, have made the conclusion seem probable to many minds that nothing but the constant continuance of physical processes contains productive conditions of consciousness. By this we need not understand that the activity, in which consciousness at every moment of its actual presence consists, is the private and peculiar product of a bodily organ ; the functions of this organ may be no more than stimuli which, but for the particular nature of the soul, would be unable to win from it an activity which is possible to it alone : yet, even so, this activity will still be the production of the organ, so long as its exercise has for its indispensable cause the excitation of that organ. Now on a previous occasion[1] I thought it necessary to remind my readers that the cessation of an activity previously in a state of exercise can, generally speaking, be explained in either of two ways ; it may be that the productive conditions of its appearance are absent, or, again, that there is a hindering force which opposes its exercise. None of the phenomena mentioned above seemed to me to preclude the second of these ideas. When a sudden fright interrupts consciousness, the physical impression made on the senses by the fact that causes terror may be perfectly harmless, and the reason of our disquietude lies in the interpretation which our judgment puts on the perception : in this case we can see no reason why this psychical movement should not be the direct cause which makes the soul incapable of a continuance of its consciousness, no reason for the supposition that the bodily fainting, which can have its cause only in itself, must intervene and produce, as a secondary effect, the loss of mental activity. When disease slowly clouds over the consciousness, this final result is commonly preceded by a series of feelings of discomfort in which we can see the beginning of the check that is going on, just as in health trifling depressions of mind make a continuance of mental activity distressing though not impossible. But it is not, we may generally say, necessary that the influences which check consciousness should

---

[1] [See Lotze's *Medicinische Psychologie* (Leipsic, 1852), § 388 ff.]

at the beginning of their hindering action be themselves an object of our consciousness. We must remember that of that which is going on in our nerves and of the mode of their influence on the soul we experience nothing: it is only the final result of these processes, the sensation, or the feeling of pleasure and pain, that appears in consciousness; and, when it does appear, it tells us nothing of the mode in which it was brought about. In the same way then, when bodily excitations, instead of producing consciousness, check it, it is possible for their action to remain unnoticed until unconsciousness suddenly supervenes. Injuries of the brain, lastly, can hardly be defined with any probability as the clean disappearance of an organ and the excitation dependent on it; they will probably always include positive changes in the organs that remain, and in the activity of those organs, and from these organs they will develope forces that check consciousness.

These were the general remarks on which I formerly relied; but at bottom they only had a significance in opposition to the view which took consciousness to be the direct product of the work of a bodily organ, and they have not much to say against the other view which conceives activities, in their own nature mental, to be evoked anew in every moment by the constant excitation of the nerves, and to be capable of continuance in this way alone. Many facts, which have been more accurately observed in late years, favour this idea. We know that animals can be sent to sleep, if a compulsion, lasting some little time but causing no pain, deprives them of all movement, and if at the same time all external sense-stimuli are shut out, and so any new sensation prevented: it follows that the internal changes conditioned by the transformation of substances by tissue-change, and by nutrition, are not sufficient to preserve in them the waking state which preceded the experiment. It is not quite safe to argue from brutes to men; but in any case it is certain enough that men too fall asleep from ennui, and quite lately a remarkable case of prolonged anaesthesia (Dr. *Strümpell*, Deutsches Arch. f. Klin. Med. XXII) has proved that in the case of men also the same experimental conditions that were applied to animals can rapidly produce sleep. Nevertheless it remains doubtful whether all these facts tell us anything new, or whether they only present, no doubt in highly remarkable circumstances, what we knew before. With regard to the animals successfully experimented on, we do not know whether there is any impulse in them tending to extend the course of their ideas in any considerable degree beyond the contents of their sensuous perception;

in the case of ennui, we know that for the moment this impulse is absent, while the sensations of the special senses are not absent, and it is only the lack of interest in them that removes the stimulus to follow up what is perceived with an attention that would find relations in it. Thus we seem to have found nothing but what needs no explanation : where the external and internal impulses which stir the soul to activity are absent, this activity is absent, and the lack of it may form the point of departure for that further depression of nervous irritability by which at last sleep is distinguished from waking.

305. Before I attempt to give some final view on this subject, I have still to mention that alternation of consciousness and unconsciousness which is presented to us in the forgetting of ideas and their recollection. Everyone knows the views which regard memory and recollection as possible only by means of a corporeal basis; according to this view some physical trace of every perception must have remained in the brain, a trace which, it would be admitted, would gradually entirely disappear if no occasion for its renewal occurred. It would be unjust to require a closer description of these abiding impressions ; but a consideration of the precise requirements they must fulfil does not, as it seems to me, reveal the advantages which this hypothesis is thought to possess when compared with a theory which regards these processes as merely psychical. I raise no objection to the idea that the simultaneous stimuli traversing the brain in extraordinary numbers, leave behind them an equal number of traces which do not intermix : that for a moment, at least, these traces can remain unintermixed is proved by the fact that they help us to form an equally large number of separate perceptions ; but this very fact at the same time proves that the unity of the psychical subject holding these perceptions together in its consciousness, is, no less than the brain, capable of a simultaneous multiplicity of states which remain apart from one another. This, however, was the very point respecting which these theories at starting expressed mistrust : a material system, consisting of a large number of parts, seemed to them better adapted to the purpose of receiving and preserving a number of impressions than the indivisible unity of an immaterial substance. But the theory does not get rid of the necessity of ascribing these capacities to such a substance, as well as to the brain ; unless indeed we are prepared to return to the old mistake of confusing a multiplicity of impressions distributed in the brain with the perception of this multiplicity. As we proceed, the duties demanded alike of brain and soul are multiplied at the same rate for both. If we approach an object, there is only one

point of it—that which our glance continuously fixes—that throws its image constantly on one and the same element of the retina ; all the other points, as the apparent size of the object increases with our approach to it, make their impression from moment to moment on fresh spots in the nerve.   Thus, if this one object is to be perceived, countless images must be represented within a short time, and that in such a manner that every part a of the object leaves traces in countless elements $pqr\ldots$ of the brain, while each of these elements again receives such traces impressed upon it by all the parts $abc\ldots$.   An intermixture of these latter images would be of no service to the act of representing the object ; each single material atom will in its turn have to preserve countless impressions without intermixture—the very same task which this theory refused to entrust to the unity of the soul—and on both sides the functions to be performed multiply immeasurably when, instead of one object, there are many to be perceived.

But the important point was not this preservation itself, but the service it can render to memory when only a part $a\,b$ of a composite image is given by a new perception, and the parts $c\,d\,e$ which belong to it have to be supplied.   If we suppose that the new impression $a\,b$ now affects the same nerve-elements p and q which it affected before, it is conceivable that the trace of it still remaining may be somehow called to life again in those elements ; but how does it come about that p and q renew in other nerve-elements, r and s, the traces of the impressions c and d which formerly affected *them*—these impressions c and d being precisely those which united with a and b will form the image that has to be recollected ?   It may be answered that the psychological view of the matter equally demands that a peculiar connexion should be established between those impressions which occur simultaneously, or, if successively, with no intervening link : that the very same solidarity obtains between the abiding remnants of the nerve-excitations ; that, if time be conceived as a line of abscissa's those of equal abscissa form such an associated group.   And this stratified deposition of the impressions, supposing it admitted, might indeed explain why their reproduction would take the direction from $a\,b$ only to $c\,d$, and not to any $p\,q$ belonging to another stratum ; but the mechanical possibility of the process itself which takes this direction would remain in obscurity.   For we cannot misuse the metaphor to such an extent as to regard the simultaneous states of all the nerve elements as a connected stratum, the continuity of which produces the result that a vibration of one point sets all the rest vibrating in those ·forms in which they formerly vibrated in this stratum, and not in those forms in which they vibrated in other strata. It could be nothing but the nature of the impressions *a* and *b* that in its turn revives the others *c* and *d* which are connected with them : and since there is no reason why *a* by itself or *b* by itself should reproduce *c* or *d* any more than many other impressions, it can be nothing but the concurrent existence of *a* and *b* that limits the selection to those impressions that really belong to them. This implies not only that the single nerve-elements in which *a* and *b* are revivified, interact on each other, so that the fact of the concurrent existence of those two impressions is transformed into an efficient resultant, by which the reawakening of *c* and *d* can be brought about; but, over and above this, those nerve-elements which are now to contribute *c* and *d*, can only add this definite contribution to the whole, if the fact of the previous simultaneity of their impressions *c* and *d* with *a* and *b* has left behind in them, too, a permanent disposition to answer this and no other solicitation with this and no other response.

I will not pursue the investigation further. Its final outcome seems to me clear: the hypothesis must transfer to every single nerve-atom precisely the same capacity of an ordered association and reproduction of all successive states which the psychological view claims for the soul. How these two occurrences (this association and reproduction) come about we have confessed that we do not know; but it is utterly vain to hope that a physical construction can enable us to understand them without presupposing that the same enigmatical process is repeated in every element of matter.

806. These considerations would all be useless, if interruption of memory occasioned by bodily suffering admitted of no explanation whatever in consonance with our views. Unfortunately I cannot maintain that what I have been saying makes such a satisfactory explanation possible; but this does not seem to me to diminish the impossibility of those other views which localise particular groups of ideas or particular remembrances off-hand at definite places in the central organs. All that we can, properly speaking, be said to observe is not an absence of memory, but merely the incapacity to reproduce ideas, which, according to the ordinary view, may nevertheless still be present as unconscious ideas, only that the associations are wanting, by help of which they might be restored to consciousness. This account, apart from a further definition, would do no more than explain the total forgetting of ideas of which there is nothing what-ever to remind us; whereas in the cases of morbid interruption of memory, the sensuous perceptions frequently go on unhindered, and

bring with them a quantity of impressions, associated in manifold ways with the forgotten ideas : and yet the restoration of these ideas to memory does not take place.

There is only one supposition that I can suggest, and I am not sure myself whether it does not push to exaggeration a conception which in itself is valid. Ideas are connected not only with one another, but also in the closest way with the general feeling g of our total state at the moment of their origin. If g changes into γ, and it is impossible to us to experience g again, the way is barred which might lead our memory back to the ideas connected with g : in whatever numbers single ideas among these may be reproduced by new perceptions, still the common bond is absent, which connected them together as our states, and thus made those contents of theirs, which in themselves were reciprocally indifferent, capable of reciprocal re-excitation. It is in this way that I should attempt to interpret the facts that, when we have recovered from severe illness, we do not remember what we experienced while it lasted, or while, before its outbreak, our general feeling was already changed; that, when we are free from the paroxysm of fever, we do not remember sets of ideas which accompany it, and that in particular cases these sets of ideas are carried on when the next paroxysm occurs, owing to the return of the morbid general feeling: that unusual depression sometimes brings long-forgotten things to remembrance, while in other cases of the same kind things familiar to us affect us so little that they seem like something new, unknown, and unconnected with the whole of our life. It is far harder to apply this explanation to those defects of memory that occur with regard to a certain definite subject-matter of our ideas; e. g. the forgetting of proper names, of a series of scientific conceptions, of a foreign language. But here again what other course is open to us than to refer these cases, so far as they are confirmed by observations, to similar causes? It is impossible to conceive of the activities which are here impeded as assigned to different organs; they could only be assigned to different ways of working on the part of the organs : we should have to come back to a general depression of the organs, preventing them from executing a group of functions, which, though they belong to one another, do not disclose even such a similarity of physical work as would correspond to their intellectual connexion, and would make it a matter of course that they should all be interrupted together. In that case there would be no greater impossibility in the further supposition, that this physical depression has for its consequence a mental general feeling, different from and superseding that which ordinarily accompanies these mental operations. For that which moves and forms connexions in us is not abstract truths: the course of our thoughts is always a course of *our* states, and every particular form of our intellectual activity gives us the feeling of a peculiar mental posture, which reacts again on the bodily general feeling. If a change originally set up in this latter feeling makes its mental echo impossible, the mental activities will be checked in their turn by the conflict of the tone of feeling which they find in existence with that which should normally accompany them.

307. Efforts to assign to the soul a sphere in which its activity should be independent of the body, commonly proceed from the desire to secure its substantiality, and thereby its endless continuance; though in reality the certainty with which we can infer the latter from the former is strictly proportionate to the energy with which at starting we have chosen to identify the two. No such motives have guided our present investigation: indeed what use would there be in securing to the soul all the rights of substance, if the exercise of these rights is not equally unrestrained? But no theories can change the facts. Whether we see in the central organs the creative causes of mental activity, or only, on occasion, the causes which impede it, in either case the facts remain, that a state of perpetual wakefulness is impossible to us; that the exhaustion of the body brings with it the total cessation of mental life; that, conversely, this life, in some way, whatever that way may be, consumes the forces of the body; that diseases and injuries of the brain either cripple particular faculties, or sink us in a complete mental night. When, then, we joined in the efforts alluded to, it was not with the hope of finding in the intrinsic substantiality of the soul any warrant for an independence of which so little does as a fact exist; but in the certainty that, even if exact observation should prove the activity of the soul to be still more closely bound up, than it is now proved to be, with the body and its agitations, still this dependence could in no way alter the essence of our conviction; and that essential conviction is that a world of atoms, and movements of atoms, can never develope from itself a trace of mental life; that it forms, on the contrary, nothing more than a system of occasions, which win from another and a unique basis the manifestation of an activity possible to that basis alone.

But even this expression of our view must after all be once more modified. We found it impossible to conceive the world as built up out of a disconnected multiplicity of real elements of matter: just as little, on the other side, have we considered the individual souls on which this system of occasions acts, to be indestructible existences ; both they and these occasions meant to us simply actions of the one genuine being or existence, only that they are gifted with the strange capacity, which no knowledge can further explain, of feeling and knowing themselves as active centres of a life which goes out from them. Only because they do this, only in so far as they do this, did we give them the name of existences or substances. Still we have so named them ; and now the question arises whether it would not—but for the exigencies of imagination—be better to avoid even that name and the inferences into which it will never cease to seduce men. Beginning by speaking of the souls as existences, we go on to speak of their states, and we even venture to talk of such states as betray nothing whatever of the essential nature of that to which we ascribe them. Thus we have not scrupled, any more than any psychology has so far scrupled, to use the supposition of unconscious ideas, or of unconscious states, which ideas have left behind, and which become ideas again. · Is it really necessary that they should so be left behind, and can we gather any intelligible notion from these words unless we take refuge, as men always naturally and inevitably have done, in the crassest metaphors of impressions that have altered a spatial shape, or of movements that are not conceivable except in space? There was nothing to compel us to these suppositions but the observed fact that previous ideas return into consciousness : but is there no other way in which that which once was can be the determining ground of that which will be, except by continuing to be instead of passing away? And if the soul in a perfectly dreamless sleep thinks, feels, and wills nothing, *is* the soul then at all, and what is it? How often has the answer been given, that *if* this could ever happen, the soul *would* have no being! Why have we not had the courage to say that, *as often as* this happens, the soul *is* not? Doubtless, if the soul were alone in the world, it would be impossible to understand an alternation of its existence and non-existence : but why should not its life be a melody with pauses, while the primal eternal source still acts, of which the existence and activity of the soul is a single deed, and from which that existence and activity arose? From it again the soul would once more arise, and its new existence would be the consistent continuation of the old, so soon as those pauses are gone by, during which the conditions of its reappearance were being produced by other deeds of the same primal being.

Conclusion.

I have ventured on these final hints because I wished to give a last and a full statement of that requirement which I believe we must lay on ourselves,—the total renunciation of our desire to answer metaphysical questions by the way of mathematico-mechanical construction. There can be no need for me to express yet again the complete respect I feel for the physical sciences, for their developed method and their intellectual force; the efforts of Metaphysic cannot in any way compare with their brilliant results. But it has sometimes befallen the investigation of Nature itself, that, at points which for long it thought itself warranted in using as the simplest foundations of its theories, it has discovered a whole world, new and never surmised, of internal formation and movement; and in this world it has at the same time discovered the explanation of occurrences, which had previously been connected, in a bare and external way, with these seemingly simple points of departure. It is a like discovery that Metaphysic has always sought, only the distance which separated its goal from anything that can become the object of direct observation was still greater. It sought the reasons or causes on which the fact depends, that we are able to pursue with confidence throughout the whole realm they govern the fundamental conceptions of the natural sciences, and which at the same time would determine the limits of this realm. It is a true saying that God has ordered all things by measure and number, but what he ordered was not measures and numbers themselves, but that which deserved or required to possess them. It was not a meaningless and inessential reality, whose only purpose would have been to support mathematical relations, and to supply some sort of denomination[1] for abstract numbers: but the meaning of the world is what comes first; it is not simply something which subjected itself to the order established; rather from it alone comes the need of that order and the form in which it is realised. All those laws which can be designated by the common name of mathematical mechanics, whatever that name includes of eternal and self-evident truths, and of laws which as a matter of fact are everywhere valid,—all these exist, not on their own authority, nor as a baseless destiny to which reality is compelled to bow. They are (to use such language as men can) only the first consequences which, in the pursuit of its end, the living and active mean-

[1] [Cp. § 214. end.]

ing of the world has laid at the foundation of all particular realities as a command embracing them all. We do not know this meaning in all its fulness, and therefore we cannot deduce from it what we can only attempt, in one universal conviction, to retrace to it. But even the effort to do this forces upon us a chain of ideas so far-reaching that I gladly confess the imperfections which, without doubt, can be laid to the charge of this attempt of mine. When, now several decades since, I ventured on a still more imperfect attempt, I closed it with the dictum that the true beginning of Metaphysic lies in Ethics. I admit that the expression is not exact; but I still feel certain of being on the right track, when I seek in that which *should* be the ground of that which *is*. What seems unacceptable in this view it will perhaps be possible to justify in another connexion: now, after I have already perhaps too long claimed the attention of my reader, I close my essay without any feeling of infallibility, with the wish that I may not everywhere have been in error, and, for the rest, with the Oriental proverb—God knows better.

INDEX.

THE END.

Printed in the United States
212481BV00003B/1/A

9 780559 650925